The Shell Builders

The SHELL BUILDERS

TABBY ARCHITECTURE *of*
BEAUFORT, SOUTH CAROLINA,
and the SEA ISLANDS

COLIN BROOKER

Foreword by Lawrence S. Rowland

THE UNIVERSITY OF
SOUTH CAROLINA PRESS

© 2020 University of South Carolina

Published by the University of South Carolina Press
Columbia, South Carolina 29208

www.uscpress.com

Manufactured in the United States of America

28 27 26 25 24 23 22 21 20 19   10 9 8 7 6 5 4 3 2 1

Library of Congress Cataloging-in-Publication Data
can be found at http://catalog.loc.gov/.

ISBN: 978-1-64336-071-3 (hardcover)
ISBN: 978-1-64336-072-0 (ebook)

Dedicated to the memory of my beloved niece
Dr. Susan Guignard Guion
1966–2011

CONTENTS

FOREWORD

The Sea Islands, which form the Atlantic shore from Charleston, South Carolina, to Jacksonville, Florida, are places of exceptional natural beauty and bounty, as well as great mystery. Edgar Allan Poe was captivated by the mysterious Sea Islands when he visited Charleston in 1843 and wrote "The Gold Bug." When Francis Griswold (1902–2001), author of the best-selling historical novels *Tides of Malvern* (1930) and *A Sea Island Lady* (1939), was asked why he chose to write about such as an impoverished backwater of early twentieth-century America, he said, "Little justice it seems to me has been done to this section of the South . . . full of beauty and pain . . . full of profound implications for the heart of man." New York author and member of the famed "Algonquin Circle" Samuel Hopkins (1871–1958) was ensnared by the mystery of the sea islands in 1935 and made Beaufort his winter home for the next twenty-three years. For Adams, part of the allure of Beaufort was the "lovely belvedeered mansions, the picturesqueness of the old Arsenal, the older slave quarters, the ancient 'tabby' buildings constructed from a unique amalgam of oyster shell mortar."

For centuries, visitors to the beautiful Sea Islands have puzzled over the tabby ruins of long-forgotten civilizations. They have invented stories about who built these ancient structures and how they were constructed. Almost intuitively these stories harkened back to the Spanish explorers whose settlement of the Sea Islands predated Jamestown, Virginia, and Plymouth Rock, Massachusetts, by a century and whose exploits have been largely lost to history. Old-time residents kept alive this Spanish memory in sixteenth-century place names (St. Helena, St. Catherine, St. John) and in the names of the lush, subtropical vegetation. The forbidding yucca plants have been known since colonial times as "Spanish bayonets." The ubiquitous and iconic gray moss that festoons the ancient oaks is universally known as "Spanish moss."

This indeed turned out to be the origin of the "unique amalgam" of oyster shell, lime, and water that formed the only lasting structures on the sandy, windswept Sea Islands.

The book that follows solves one of the enigmas of the Sea Islands. The author, a British-trained architect, engineer, and historian, bought and restored a three-story tabby mansion on Carteret Street in Beaufort, South Carolina, in 1980. The two-hundred-year-old Barnwell Gough House was the childhood home of Beaufort's famous son Senator Robert Barnwell Rhett (1800–1876), the "Father of Secession." Through practical and painful experience, Colin Brooker made himself an expert on the chemistry, engineering, and practical limitations of tabby buildings. Since then, he has spent more than thirty years studying the "unique amalgam." He has studied its archeology. He has consulted on its preservation and restoration from South Carolina to Florida. In the process he has made himself one of the world's foremost experts on tabby construction. In recent years,

Colin Brooker has extended his expertise to the nearby Bahama Islands, where he has studied and consulted on historic construction with available local building materials. The transfer of methods, knowledge, and people from the Sea Islands to the Bahamas was common in the eighteenth century. So, evidently, was the transfer of technology and engineering.

On the sandy Sea Islands, masonry building material was nonexistent. No rocks were available, and bricks were too expensive to build and transport. But scattered throughout the bountiful Sea Islands are huge mounds of dried oyster shells, many cast into convenient piles on the lee shores of rivers, creeks, and sounds, and others left in "rings" by millennia of Native American occupation. This resource was exploited by Spanish settlers at Santa Elena (1566–87) as early as the 1570s. "Tapiyah" construction migrated from Muslim North Africa to Moorish Spain in the Middle Ages and found its way to America with the first explorers. A house built for Christopher Columbus in La Isabella, Santo Domingo, in 1493 used the stamped-earth method of construction called "tapia."

Colin Brooker traces this technological migration to South Carolina, Georgia, and Florida. He applies knowledge learned from dozens of his own archeological examinations of tabby ruins and describes with an engineer's detail the materials and methods used by settlers, planters, slaves, and military engineers over nearly five centuries of European and African settlement.

The Shell Builders: Tabby Architecture of Beaufort, South Carolina, and the Sea Islands is the most thorough and scholarly treatment of this subject yet published. It is the definitive answer to one of the enduring mysteries of the beautiful Sea Islands.

Lawrence S. Rowland

ACKNOWLEDGMENTS

My interest in tabby building was aroused in 1981 when Cynthia Cole Jenkins, then writing her survey of local historic resources for the Lowcountry Council of Governments, showed me photographs and a sketch plan dated 1966 by John D. Miller (Charleston Museum) of the Edwards House, Spring Island, then almost unknown despite its formal distinction and unrivaled landscape setting. In Beaufort my purchase of the old Barnwell Gough House (built ca. 1785), one of the town's largest tabby dwellings, had already prompted intensive search for historic records describing tabby fabrication, this task becoming urgent as my wife and I discovered the near-derelict house was severely compromised, if not endangered, after decades of neglect. It also became apparent that local contractors had only vague notions about how tabby might be repaired, their misplaced beliefs about the efficiency of impaired historic fabrics almost causing the building's ruin.

Besides finding William Rettew, PE, a resourceful structural engineer whose skill prevented disaster on more than one occasion, we were directed to Frederica, Georgia, where the National Parks Service has conserved tabby dating back to the 1750s. The site's superintendent and staff generously gave of their experience, which guided our own stabilization efforts. Gradually, knowledge gained was expanded by other documentation, survey and tabby conservation projects for private, corporate, and institutional entities. Professional interests allowed collection of information about the diffusion, adaptation, and evolution of building materials related to tabby from places once linked through those notorious triangulations engendered by the North Atlantic slave trade in North and West Africa, the Caribbean, Latin America, Europe, and the southeastern United States.

From the beginning owners, custodians, researchers, historians, officials, and other interested individuals provided every assistance. Colleagues with whom I have collaborated include Dr. L. (Larry) Lepionka, a pioneer investigator of Beaufort County's tabby building; Dr. Michael Trinkley (Chicora Foundation, Columbia, S.C.); Dr. Eric Poplin (Brockington and Associates, Atlanta, Ga., and Charleston, S.C.); David B. Schneider (former director, Historic Beaufort Foundation); Dr. James Miller (former Florida state archaeologist); Dr. Stanley South (University of South Carolina); Craig M. Bennett (Bennett Preservation Engineering, Charleston, S.C.); Sean Taylor (archaeologist, South Carolina Department of Natural Resources); and Ian Hill (Beaufort County preservation officer). Other professionals who aided my work in more ways than can be mentioned include the late Dr. Thomas H. Eubanks (Louisiana state archaeologist); Daniel J. Bell (S.C. State Parks Service); Sarah Fick; Bruce G. Harvey; Dr. Janet Gritzner; Meg Gaillard (South Carolina Department of Natural Resources); the late Dr. Lewis H. Larson Jr. (former Georgia state archaeologist); Dr. Lawrence S. Rowland (University of South

Carolina); Dr. Daniel L. Schafer (University of North Florida); Dr. Mary Socci; Dr. Steven Wise (director, Parris Island Museum); Maxine Lutz (executive director, Historic Beaufort Foundation); and the late Mills Lane, whose volumes on southern architecture first published by the Beehive Press, Savannah, remain an inspiration.

At different times, my wife, Jane Bruce Guion Brooker, whose lowcountry connections go back three or more centuries; Ramona Grunden, whose indigenous American lineage is much longer; Jean Leidersdorf; Ashley Heffner; Evan Thompson; and John Huntley helped with field surveys and measured drawing besides clearing vegetation on densely overgrown sites where structural remains were almost concealed from view. Beaufort County Planning Department personnel, including Jason Flake and Terri Norris, undertook GPS plotting and mapping. With its voracious insect life, venomous snakes, ubiquitous poison ivy, and saturated atmosphere in summer, the lowcountry is not always a benign place for survey work notwithstanding its great natural beauty. I'm grateful to these people and others pressed into service at short notice for assisting with stoic forbearance.

County, state, and federal authorities smoothed the way and opened doors that might otherwise have remained closed. I am especially indebted to former Beaufort mayor Henry C. Chambers and to U.S. National Parks Service personnel at the Castillo San Marcos, St. Augustine, Florida; Kingsley Plantation, near Jacksonville, Florida; and San Juan, Puerto Rico. The late Senator Strom Thurmond facilitated access to documents apparently lost but soon found thanks to his persistence at the U.S. National Archives, College Park, Maryland. Mary Ann Eaddy (Georgia Department of Natural Resources) kindly extended an invitation to speak at the 1998 Jekyll Island Symposium on the Conservation and Preservation of Tabby. Special thanks are owed Gary Kubic, Beaufort County administrator, for supply of scanned Historic American Buildings Survey photographs. The late Talbird Reeve Sams (1924–2013) permitted use of family papers, photographs, and drawings relating to Dataw Island while Alicia "Lish" Thompson generously provided copies of family papers, maps, and paintings relating to the Townsends of Edisto Island. I must also mention the support, encouragement, and continued interest provided by Jackson Brown and other members of Dataw Historic Foundation.

The expertise, dedication, and friendship of my colleague Richard Wightman in conserving tabby ruins is acknowledged with admiration and sincere thanks, his skill having ensured that numerous lowcountry structures will survive for decades to come. To list the numerous private owners with whom I have worked or who have allowed me to examine buildings in their care would be an invasion of their privacy. It goes without saying that without generous help by these individuals, this book would not have been written.

Investigation of southern Spain's Islamic *tâbiyah,* partially funded by a National Endowment for the Humanities postdoctoral fellowship through the American Schools of Oriental Research, Amman, was first undertaken with Dr. E. Axel Knauf (Heidelberg University), who, like myself, was displaced from archaeological and historic research in Jordan by violent events preceding the First Iraqi War, drove many miles along back and mountain roads of Andalucía using (for reasons I no longer remember) only Roman-era maps as a guide. An earlier kindness of the late Professor Titus Burkhardt is not forgotten.

On a freezing winter day replete with snow flurries, he took time from his conservation work in Fez (Morocco) to introduce the city's historic monuments. During my first visit to Marrakech, long before mass tourism impacted the city's social order, the then governor graciously provided local transportation and guides. Learning of restoration plans likely to impact many significant medieval structures, I made three additional journeys to Morocco between 2015 and 2018 with the object of recording structural features before they were concealed or permanently altered. Despite difficulties with my Bedouin-taught Arabic or schoolboy French, the good humor of Moroccan construction workers, supervisors, custodians, hoteliers, and drivers greatly assisted what became an increasingly urgent task. Fortunately, perhaps, plans for travel across the Sahara prompted by a fanciful suspicion that somewhere further south—in Mali perhaps or Niger—I would discover *tâbiyah*'s ultimate origins were frustrated by political turmoil.

Earlier Dr. Y. Zahran (UNESCO) sponsored brief visits to West African sites associated with the Atlantic slave trade, including Goree Island and Basse Casamance, a rice-growing region of Senegal bordering Guinea Bissau, which retains cultural links with South Carolina's coastal wetlands initiated two centuries ago. My brother, R. Christopher Brooker, and his wife, Margaret, helped explore cob and rammed-earth building traditions of the English West Country. Godela von Xylander's hospitality allowed extended stays in central London, thereby facilitating search for relevant documentary materials held by British National Collections. Across the Ecuadorean highlands, Jonathon Loftin, a Fulbright scholar, searched out individuals who still make tapia, besides navigating our way through Imbabura Province, where traditional rammed-earth and formed-earth construction is still commonly used.

Former prime minister the Right Honorable Perry Christie, PC, provided unique opportunities to visit historic sites scattered across the Bahamas. Dr. Keith L. Tinker, director, the Bahamas National Museum, aided with his habitual kindness, making funds available for journeys across the southern islands, where Kirkwood McKinney (chief councilor, Crooked Island/Acklins District) and his brother, Elton McKinney, found tracks across mangrove swamps, sailed small boats along coasts scarcely changed since first explored by Columbus, and cut paths through dense tropical thickets as I looked for abandoned Loyalist plantation buildings. Patsy Cartwright of the Long Island Museum (a branch of the Bahamas National Museum) and her husband drove me from one end of Long Island (formerly Fernandina) to the other, stopping at innumerable historic sites besides assisting with architectural recording. Keith Bishop (Islands by Design, Marsh Harbour, Abaco) sponsored fieldwork across the northern Bahamas, patiently answering questions regarding traditional building methods. Basil Mimms and the then district administrator first showed me the Hermitage, Little Exuma. On Grand Bahama, Jenius Cooper, an octogenarian when interviewed in 2008, described tabby structures he built at Old Freetown as a young man. The late June Maura, OBE; RVO (past president, Bahamas Historical Society), whose distinguished government service allowed access to records not usually available, spent countless hours researching sources and finding historic maps besides extending her friendship when I was new to the islands. Additionally Dr. Gail Saunders generously shared her unparalleled knowledge of Bahamian history.

Libraries and librarians and archives and archivists play an essential role in any historic research project. I am grateful for help from the U.S. National Archives, Washington, D.C., and College Park, Maryland; South Carolina Department of Archives and History, Columbia; Caroliniana Library, University of South Carolina, Columbia; South Carolina Historical Society, Charleston; Beaufort County Register of Mense Conveyance Office and its counterpart in Charleston County, South Carolina; Georgia Historical Society, Savannah; University of Pennsylvania Libraries, Philadelphia; U.K. National Archives, Kew; British Library, London; U.K. Hydrographic Office, Taunton, Somerset; the Gibraltar Museum; Department of Archives, Nassau, the Bahamas; and Library of the École Biblique et Archaéologique Française, Jerusalem, which supplied rare Spanish- and Arabic-language periodicals besides providing safe haven during turbulent times. Special thanks are due Dr. Nicholas Butler (Charleston County Public Library); Joanne Bloom (Fine Arts Library, Harvard College); and Grace Cordial and the staff of Beaufort County Library, who over an extended period gave much helpful advice besides access to important special collections of books, historic photographs, manuscripts, and maps.

Debts owed the Historic American Buildings Survey are innumerable. Paul Dolinsky (then director, Historic American Buildings Survey) encouraged, promoted, and ultimately made possible the survey of Beaufort County's tabby resources conducted in 2003. Virginia Price coordinated field and archival activities, transcribing by hand rare books from the University of Virginia Library judged too fragile to be copied otherwise. The extent of the late Jack Boucher's contribution will be evident from his photographs. For my own part, I highly valued his professional skill, willingness to tackle difficult subjects, and good humor, and the friendship that developed between us when traveling from site to site. Jack—as he insisted we call him—strongly encouraged publication of the Beaufort County Tabby Survey, using his photographs as illustrations. My one regret is that he did not see the outcome since sharing those records of historic buildings and landscapes made with so much care, diligence, and affection was his passion.

Introduction

With North African, Iberian, Sub-Saharan, Caribbean, and Spanish American antecedents, tabby construction found along the Eastern Seaboard of the United States from North Carolina to northeast Florida (with outliers in Alabama, West Florida, Mississippi, and coastal Texas) but most abundantly across the Sea Islands of South Carolina and Georgia was—before falling out of use during the 1860s—a quintessential product of the Atlantic World, carried from place to place by adventurers, merchants, military engineers, planters, and slaves. In the Carolinas few early settlers described fabrication of this material, it being too familiar perhaps for literary notice. Conversely visitors often thought tabby a novelty worth recording. The most complete late eighteenth-century account is given by a traveling French aristocrat, François Alexandre Frédéric, duc de La Rochefoucauld-Liancourt (1747–1827), whose interests were fueled by England's most influential agricultural reformers including Arthur Young, editor of the *Annals of Agriculture and Other Useful Arts* (published 1784–1815), which counted "improving" landowners on both sides of the Atlantic among its subscribers.

Visiting Beaufort, South Carolina, in 1796, the duke observed "le taby" (as he called it) "particulière" to the town. It comprised:

> a lime made from oyster shells mixed with water, a large proportion of whole oyster shells is mixed in. This mortar is poured into a wooden form the length and thickness of the wall to be constructed. These forms have no bottoms but their sides are joined at certain intervals at top and bottom by pieces of wood. The mortar is pounded in with force, and, when they are brim full, it is left to dry for two or three days.[1]

Forms were then struck, repositioned on top of the newly cast tabby, and refilled with mortar, the same process being repeated until the wall reached its desired height.

Oyster-shell lime has been reported as early as 1580 at the Spanish settlement of Santa Elena (now Parris Island), where it was used for flat roofs and daubing exterior walls of dwellings.[2] Formed tabby was not mentioned in South Carolina's official records before January 1727, when proposed for construction of a battery overlooking Charleston Harbor.[3] Near Beaufort the material made its first documented appearance in 1734 at Fort Frederick, Port Royal Island, where construction was overseen by two local planters, which suggests tabby was employed somewhat earlier hereabouts for domestic or plantation building.

Widely distributed along the southeastern Atlantic coast before the American Revolution, formed fabrics resembling those seen by La Rochefoucauld became commonplace soon thereafter across tidewater areas of what was then Beaufort District. Largely given over to cotton cultivation from 1800 to 1860, the district went through cycles of prosperity and depression as commodity prices fluctuated on international markets. Good years brought unheard-of prosperity for some planters—though by no means all—manifested by unprecedented growth of enslaved populations and expansive construction programs. Country residences, slave rows, and agricultural structures were built, rebuilt, enlarged, or otherwise improved. Profits also fueled erection of townhouses for the more affluent, who might use their own workers during slack times or hire gangs of specialist craftsmen contracted out by planter neighbors. Military construction absorbed large numbers of laborers too, but whether these individuals were recruited from local slave populations or from further afield I have not determined. It is certain that the number and variety of tabby building still represented by standing walls, ruins, or archaeological features is greater across present Beaufort County than anywhere else in the lowcountry, East Florida, or Gulf coast region. Although many of the largest local examples have disappeared above ground—notably successive forts guarding Port Royal Sound—the corpus of extant structures remains diverse and plentiful enough to merit detailed study. Thus an unprecedented trend toward multistory tabby construction is seen about Beaufort Town. On the islands linear or loosely organized houses illustrate an aesthetic developed to ameliorate hot, humid, sometimes deadly summer conditions. Rivalry can be detected among elite owners, who, flush with returns or the expectation of profits from long-staple cotton, built conspicuous county residences overlooking estuaries or navigable rivers. With fragmented floor plans linked by porches and exterior walls stuccoed, scored, burnished, or painted to look like stone, these buildings seen across open waterways, immersed among woods, or standing amid intensely cultivated fields gave the impression of mansions though enclosed spaces were actually modest and their skins built—not from marble—but from far humbler materials. Sham as they might be, it is their massing, scenic qualities and incomparable settings rather than any pretension toward "correct" or academic taste that makes these houses memorable even when ruined.

Tabby also became the vehicle for rural "improvement" of the kind espoused by late eighteenth- and early nineteenth-century European social reformers. Some plantation owners believed tabby-built slave houses, especially if laid out in neat rows or picturesque "villages," more comfortable and lasting than squalid timber-framed, log, or wattle-and-daub huts tolerated by less enlightened neighbors. Whether or not the enslaved shared similar sentiments is not recorded, but the fact remains that along the Atlantic Seaboard from South Carolina to Northeast Florida, tabby slave dwellings have a somewhat higher survival rate when compared with similar buildings constructed using alternative materials.

Through a convoluted mental process, Thomas Spalding (1724–1851) of Sapelo Island —an innovative planter whose principal writings were published in Charleston—drew analogies between American tabby and European *pisé de terre*. The latter material had excited the interest of rationalist architects near the turn of the eighteenth century, becoming fashionable for worker housing erected by progressive French, English, German, and even Russian landowners wishing to improve living conditions for their laborers. Ephemeral in Britain, enthusiasm for *pisé* lingered in the United States until the late 1830s, seeing several revivals thereafter, notably by the eccentric Orson S. Fowler, whose influence extended from New England south into the Carolinas and Bahamas during the 1850s.

Like pisé, tabby's popularity is explained by its relative cheapness, ease, and speed of construction compared with other incombustible materials. For most Sea Island planters, stone and brick were available only at prohibitive cost. Lacking quarries and easily exploited clay deposits, planters were obliged to import almost all masonry building products, the frustrations of transportation by sea and river compounding an already intractable situation. Although commercial brick making began southeast of Beaufort Town during the late 1830s or early 1840s, tabby making flourished until the Civil War disrupted local building traditions, dispersed indigenous craftsmen, and exiled almost the entire planter class.[4]

If tabby lingered after hostilities ceased, the introduction of Roman cements during the mid-nineteenth century and Portland cement in the 1880s rendered it obsolete. However, analogous fabrics made from Portland mixed with oyster-shell aggregates cast into timber forms did emerge. Hollybourne Cottage, built near the Jekyll Island Club in 1890 at a cost of over $19,000, is an early example from Georgia, inspired by the "old tabby" Horton House, situated nearby, albeit using modern materials that proved difficult to maintain over the long term.[5] Other large-scale tabby revivals included Thomas Carnegie's rebuilding of Dungeness on Cumberland Island, Georgia, in 1884, this project reusing foundations of an enormous four-story tabby house erected about 1803 by the widow of General Nathaniel Greene. Then there was the rambling mansion built about 1910 by Richard T. Wilson, a New York banker, at Palmetto Bluff (Beaufort County). Both houses were thought fireproof. Ironically both eventually burned and are now total ruins. As late as the 1920s similar materials were utilized for vaguely Mediterranean-style community buildings at Penn School, St. Helena Island, and several farmhouses and agricultural structures erected on the same island during the early twentieth century, displaying analogous construction.

Elsewhere antebellum tabby buildings had entered into cycles of neglect, decay, and destruction initiated by Beaufort's abandonment and subsequent occupation by Union forces in November 1861, an event seared in the collective memory of local residents. Lieutenant John A. Johnson of Beaufort's Volunteer Artillery (BVA), related a remarkable story. Landing in town exhausted, having spent the night rescuing wounded, dispirited, and disorientated men evacuated from Fort Beauregard on Bay Point Island (one of two Confederate forts guarding entry into the Broad River) after it had been bombarded and overrun by the Union Navy, Johnson "looked in vain for one living creature, human or other, throughout the length and breadth of Bay Street."[6] As in some fantastic dream he walked toward the Arsenal—headquarters of BVA—then moved on to his brother's unfinished Craven Street mansion, later called the Castle. Every person he knew had gone away, leaving most of their possessions behind.

Scenes of drunken revelry and wholesale looting followed, Union troops and now-masterless slaves ransacking the town before order was restored. Abandoned by their owners, outlying plantations were plundered by foraging soldiers seeking timber and other useful materials (brick and metals being prized) for their own encampments. Some buildings fell victim to random pillaging and fires set by half-starved "contrabands" seeking warmth and shelter if not revenge on former masters. Many cotton houses, gin sheds, and barns were deliberately burned with all their contents by planters themselves to prevent valuable agricultural commodities falling into enemy hands.

Beaufort was not destroyed, most larger buildings remaining intact though stripped of their goods and furnishings. Arriving by steamer in March 1862, Laura Towne observed that the view was "painful not only for the desertion and desolation; but more than that from the crowd of soldiery lounging, idling, growing desperate for amusement and occupation till they resort to brutality for excitement."[7] Another passenger, Edward S. Philbrick, said houses were "surrounded by heaps of broken furniture and broken wine and beer bottles." At Tidewater, a sumptuous house built in the 1830s by William Fripp, one of Beaufort's most prosperous and open-handed antebellum cotton planters, the new occupant remarked, "we kindle our fires with chips of polished mahogany."

Destruction left the architectural history of Beaufort and Beaufort County sadly impaired. Letters, diaries, plantation daybooks, building accounts, receipts, estate maps, deeds, plats, warrants, and other standbys of the historian are mostly missing, left behind or scattered during what a northern journalist, Noah Brooks, irreverently called the "great skedaddle."[8] Probate, mense, and court records disappeared en route to Columbia, where sent for safety, burned perhaps during freezing weather by desperately cold Confederate soldiers charged with carting cross-country loads of what must have seemed useless paper. Subsequently, private properties in Beaufort were foreclosed then auctioned off to satisfy unpaid direct taxes levied on Insurrectionary Districts by act of Congress on June 7, 1862. Union soldiers found they could buy confiscated holdings at a small fraction of their true value with minimal down payment, civilian speculators moving in when these buyers went into default or returned home. Later Beaufort's islands were divided into townships, each township subdivided into sections. Sections were further divided into lots and sold. Much of the best cotton land was acquired by federal agents and missionaries. Former plantation boundaries were mostly ignored by government surveyors, old property and ownership names disappearing from maps. Few antebellum proprietors ever returned to reclaim houses or lands, many petitions for compensation or restitution submitted after the war appearing unexamined when opened by myself at the U.S. National Archives in the 1970s.

Having survived hostilities an important group of tabby houses was destroyed in January 1907 as an accidental fire, fanned by stiff winds, burned out of control along parts of Beaufort's Bay and Carteret Streets, the town's chief commercial arteries then crowded with grog shops, retail stores, provisioning establishments, lodging houses, and other businesses, many licit, others less so. Fortunately these streets with their Federal-style and Greek Revival–style residences had been photographed by northern entrepreneurs who set up studios during the 1860s. Few documents can summon more immediate response than images made by Samuel A. Cooley, self-styled "Photographer Tenth Army Corps" who by January 1863 was employing three assistants from New York in his studio—located next to Beaufort's Arsenal—printing portraits, landscapes, *cartes de visite,* and stereoscopic views for military personnel, missionaries, and government agents who had descended "like the locusts of Egypt" on the town and nearby plantations.[9] Today photographs by Cooley and the less well-known firm of Hubbard and Mix provide almost inexhaustible sources of architectural information. Cooley employed a horse-drawn traveling studio when documenting outlying areas, but rural sites received less attention from itinerant photographers than more easily recognized urban structures occupied by federal authorities or converted into military hospitals. The dozen or so country places pictured give only tantalizing samples of the island's architectural heritage, which widespread poverty and extreme hardship largely destroyed, rendered uninhabitable, or altered beyond recognition before the Great Depression came to an end.

In the late 1970s unprecedented growth transformed Beaufort County. Natural, cultural, and historic resources were all impacted by residential and resort development, new bridges, new roads, and an expanding population. In February 1998 U.S. National Parks personnel; state preservation officers from South Carolina, Georgia, Florida and Louisiana; architects; engineers; archaeologists; and other specialists gathered at Jekyll Island for the Symposium on the Conservation and Preservation of Tabby organized by the Georgia Department of Natural Resources. After much discussion, many presentations, and comparison of notes, the assembly concluded that tabby represents an "irreplaceable cultural heritage" that has "long been neglected and is rapidly vanishing."[10]

Aware that the variety, quantity, and evolved character of Beaufort County's tabby resources were exceptional yet threatened, local government officials and preservation groups led by the Historic Beaufort Foundation (Brenda Norris, president; Jefferson G. Mansell, executive director) subsequently took up the challenge of sponsoring survey work, seeking assistance from the Historic American Buildings Survey (HABS) to document representative tabby structures selected by the author located in the City of Beaufort, in the Town of Port Royal, and on several islands—Daufuskie, Hilton Head, St. Helena, Dataw, Spring, and Callawassie. Agreements were reached and contracts made between the various parties in 2003. The late Jack E. Boucher (chief photographer, HABS) completed two weeks of field recording in April of that year, photographing fourteen separate sites with large-format plate cameras, this complementing earlier, more general HABS surveys made by Charles N. Bayliss. Subjected to rough roads and the hazards of transporting improbable amounts of heavy photographic equipment by small boat, setting up in buildings teetering on the edge of structural failure, and making the very best of structures exhibiting greater archaeological than architectural interest, Jack Boucher was accompanied by Ian Hill (Beaufort County preservation officer) and myself during his expeditions. Afterward I prepared captions for images produced and wrote histories of sites visited based on fresh documentary research or investigation previously conducted over some twenty-five years of local architectural and preservation practice.

The present contribution draws heavily from this material, now held by the Library of Congress, along with reports written, drawings made, articles published, and notes

A moment of discovery on Daufuskie Island, S.C. Dr. L. Lepionka (*left*) and the author
(*right*) find tabby foundations of Haig Point House (ca. 1833) reused as foundations
for the present late nineteenth-century lighthouse and keeper's dwelling.

collected during a number of prior projects. Previously unrecorded tabby structures coming to notice since completion of the HABS survey and further investigation of known
sites have allowed modification, correction, and amplification of previous perceptions
about building modes indigenous to the southeastern coastal plain. In this book information concerning the history, construction, use, and aesthetics of tabby building as
seen in Beaufort and on its once-dependent islands is synthesized. The goal as set by the
Jekyll Island Symposium is "to determine general and specific construction methods and
tabby formulations."[11] I have also attempted "to determine if, and to what extent, parallel,
independent, or transferred methods existed on the local, regional, national and even
international levels," the path toward this end having zigzagged from South Carolina to
the West Indies, Central America, North Africa, and Europe.

Yet questions of exactly how and from where tabby came into the lowcountry remain
open. Most commentators agree diffusion via Spanish colonial America is likely, the
pathway passing through St. Augustine, Florida. This is certainly not the whole story,
introduction of the material perhaps having occurred more than once and from other
places. Technically tabby is closely related to the tamped- and rammed-earth traditions of
the Old World in general and specifically the type of preindustrial form-cast *tâbiyah/tapia*
construction found across Northwest Africa and southern Iberia. Whether or not the

evolutionary line is direct, structures built of the latter material in Morocco and Spain offer parallels with American tabby building, an observation often repeated by Thomas Spalding in correspondence with his fellow planters.

Construction modes resembling tabby developed across the Maghreb and Andalucía in princely, military, and vernacular contexts over a time frame extending back over one thousand years. Commentary draws on personal observations collected during numerous study trips to these regions, supplemented by reviews of the relevant literature. Diffusion of similar building techniques into the New World is an untold process that, considering the area involved and dearth of published information, will take years to unravel. For the southeastern United States, I suspect two Mediterranean enclaves—Gibraltar, captured from Spain in 1704, and Minorca, first occupied by Britain in 1713—were gateways, medieval tapia having excited the interest of several observant British military officers stationed there before 1750. Spanish and Italian engineers played other parts, spreading knowledge of the material across the Caribbean region, into Central America and vast, sometimes trackless lands to the south.

The mechanics and operational procedures of tabby building slowly evolved and developed across the lowcountry from the early eighteenth century down until Reconstruction. Tabby fortifications played important roles in coastal defense both north and south of Charleston. Physical remains—if extant—are much damaged. However, contemporary documentation preserved in state and national collections provides information not otherwise available from the civilian sector about costs, specifications, the mechanics of finding suitable supplies, and labor required for large-scale tabby construction.

Descriptions of urban and rural structures follow, drawn primarily from fieldwork conducted either independently or in association with archaeological investigation by professional colleagues. Size is important when building in tabby. Structural dimensions are limited by the exigencies of formwork fabrication, material availability, and the intuitive difficulty of estimating whether cast walls—especially high ones—will or will not stand. Though they interrupt or encumber the narrative, measurements sufficient to calculate the footprint of most buildings mentioned are supplied in standard imperial form except where publication of features not seen or surveyed by myself are given in metric values. Dimensions also offer clues concerning living conditions, useful when studying slave dwellings. Viewed from modern perspectives these appear woefully crowded (about two hundred square feet per family among the "best" examples) to downright inhumane. Several archaeologists and anthropologists suggest contemporary occupants viewed their situation differently through inherited memories of spatial utilization shaped in West Africa, a region where names given by early modern mariners to those few strips of land they knew—the Windward, Rice, Ivory, and Gold Coasts for instance, or half-imagined kingdoms of Guinea, Dahomey, Benin, Kongo, and Angola—underscore its diversity. These places do not, I believe, allow sweeping generalizations about the cultural preferences of their past, or for that matter, present inhabitants. Discussion about such views, especially those with racial or ethnic overtones, is deferred therefore to specialists. Additionally, although tedious to read, information about the size of formwork and formwork supports—often valuable temporal indicators—is included as well.

Structures erected before and after the American Revolution are treated separately. For coastal plantations this arbitrary arrangement reflects fundamental shifts away from indigo, the chief prerevolutionary cash crop, to long-staple cotton at the end of the eighteenth century. Components of the antebellum landscape directly associated with cotton production are all but absent from Beaufort County. To supplement what little structural evidence is left, I have tapped journalistic literature for descriptive accounts by planters and those mostly northern correspondents who recorded processing operations observed during or just after the Civil War. Period photographs provide occasional portraits of enslaved or formerly enslaved workers along with views recording the infrastructure of plantation life: its fields, gardens, pens, paths, docks, and outbuildings. Mercifully the actual world of the "quarters" has passed. But images of slave dwellings survive, while meticulous maps made by professional surveyors help recover the location, layout, and history for a handful of plantation settlements out of the many hundreds lost from the Sea Islands.

Rather than discussing patronage, genealogies, gut-wrenching barbarities surrounding slavery, or hackneyed mythologies tacked onto local sites, I am primarily concerned with the history and practical realities of building. Questions of style are bypassed, tabby having been shaped more by environmental factors, expediency, and craft traditions than dictates of pattern books or professional arbiters of fashion.

Tabby structures in Beaufort County not documented by HABS or known from old photographs are introduced where these provide more comprehensive information than ruined or fragmented ones can offer. The text also strays away from county boundaries and crosses state lines. Historic links between Beaufort District and the so-called Golden Isles of Georgia from the time of James Oglethorpe and founding of Frederica on St. Simons Island in 1736 cannot be ignored. Nor can the fact that writings by Thomas Spalding aided the material's acceptance among residents of the Southeastern Seaboard over opening decades of the nineteenth century, reviving (or so he claimed) tabby construction on St. Simons and neighboring parts of coastal Georgia where it had fallen out of use.

Sites north of Beaufort County must also be taken into account when discussing production of Sea Island cotton. Reaching Edisto Island's southern extremity from Coffin Point on St. Helena Island requires a long, circuitous journey by road, even though it is clearly visible, being no more than a dozen miles away by sea. As might be expected, the history of plantation agriculture is mirrored on both sides of St. Helena Sound, now dividing Charleston County from Beaufort County. Plenty of tabby was made on Edisto before the island was occupied by Union forces during the Civil War (and possibly for some short time afterward), but, incompletely preserved or otherwise, it is overlooked and inadequately recorded. The same is true further up the coast on Wadamalaw, Johns, and James Islands. Consequently few examples from these latter places are mentioned.

Architecture can illustrate the standing, aspirations, pretensions, and even follies of its patrons. Social interactions and passions are less easily gauged from building programs executed in the enclosed plantation world, where so much was hidden from public view.

Catastrophic loss of records from antebellum Beaufort has left few contemporary voices describing daily life of either planters or their enslaved people, forcing an academic detachment in the narrative presented. Nevertheless the list of characters passing through this book is large. Besides landowners, it includes soldiers, sea captains, pirates, merchants and speculators, politicians, governors, an occasional clergyman, one or two Signers of the Declaration of Independence, several heiresses (about whom novels could be written), and women widowed with fortunes large enough to fuel ambitions of suitors and new husbands alike. In the background extends a huge diaspora spanning oceans and continents, which produced cadres of mostly anonymous enslaved workers on whose labor the whole edifice of plantation agriculture stood, a secondary diaspora after the American Revolution seeing new voyages to different islands by slaves and their Loyalist masters.

For an architectural historian, attribution of specific building methods or styles to particular nationalities is dangerous. Yet the possibility that exiled southern Loyalists and their slaves brought their own building traditions into the West Indies is, I think, entirely plausible considering the number of emigrants engendered by the American Revolution. For the Bahamas alone, tallies show about sixteen hundred whites and fifty-seven hundred blacks settled in the British colony, thereby tripling its population.[12] How many came from coastal Carolina or Georgia is undetermined, but there can be no question that most newcomers, white and black, were transported with various goods and chattels from St. Mary's, Georgia, or St. Augustine, after British East Florida reverted to Spanish rule in 1784.

At the Jekyll Island Symposium, the possibility that building techniques were carried by this exodus was raised by Walter S. Marder, who brought to our attention the Adderl(e)y House, Key Vaca, Florida, a "Bahamian styled single room dwelling believed to be of tabby," which Marder proposed could "possibly indicate that there was an awareness of tabby in the Bahamas."[13] I was reminded of his conjecture on Grand Bahama in 2008 after finding that materials locally called tabby were manufactured there well into the 1950s, several elderly informants still remembering practical aspects of the craft. Similar fabrics are scattered from one end of the Bahamian chain to the other, even across southern islands, which, except for transient woodcutters and shipwrecked mariners, were depopulated soon after Columbus's first voyage and remained so until the Loyalists came. Analogous certainly, yet in some ways unlike tabby found on the Sea Islands, parallel traditions from the very edge of the Atlantic World are examined in the epilogue bringing this book to its close.

When first created (July 29, 1769), Beaufort District comprised an area of about 1,920 square miles extending inland from the Atlantic Ocean between the Combahee and Savannah Rivers. It incorporated substantial parts of four Anglican parishes, St. Helena's, St. Luke's, Prince William's, and St. Peter's. This area was subdivided in 1785, in 1791, and again in 1878, when southerly portions of the original district became Beaufort County while areas to the northwest were absorbed into Hampton County. In 1912 Jasper County was created from parts of Beaufort and Hampton Counties. Another portion along Beaufort County's western boundary was absorbed by Jasper County in 1951.

Landscapes are dominated by water, numerous rivers and tributaries flowing from the west and northwest crossing the area. Coastal zones are heavily indented, sounds and smaller waterways allowing penetration of the Atlantic far inland. A string of sandy barrier islands, some densely wooded, others not, parallel or nearly parallel the ocean. Always shifting southward grain by grain, the largest islands, Hunting, Fripp, and Pritchard's, saw little exploitation except for hunting and ranching until recent times. Several marsh islands formed about old dune and beach ridges, worked and reworked by coastal processes since the Pleistocene era, also fringe the coast, these including Little Capers, St. Phillip's (Edding's) Island, and Bay Point Island, the last a summer resort for privileged families in the 1840s and 1850s who built themselves cottages on or near its wide beach. Indications for a quarantine station or pest house also exist, swallowed by the sea perhaps or by construction (1861) of Fort Beauregard near the island's southern tip. Then there are what might be termed river islands, entities removed from the ocean yet regulated and shaped by tides. Some are quite tiny (Palmetto Island), others sizeable. Dathaw (Dataw) Island, with its spread of ruins built by the Sams family overlooking Jenkins Creek (a tributary of the Morgan River), is notable as an important antebellum cotton producer and supplier of oranges carried by schooner as far as New York City. Callawassie has Indian mounds and an old tabby sugar mill, a confluence of channels formed by the Oakatie, Chechesee, and Broad Rivers linking it to Spring Island, the richest of the lesser islands in the 1850s and today perhaps the most varied and attractive of them all.

True Sea Islands represent Pleistocene sand bodies eroded by tidal or river action. North of St. Helena Sound, just outside the principal study area, Edisto Island (Charleston County) is one link in a chain of largish islands including Hilton Head and Daufuskie boasting long beaches facing the ocean. Like that of Bay Point Island, the topography of Hilton Head is enlivened by several high, relict dune fields. Relatively flat, St. Helena and Pinckney Islands are separated from barrier islands or marsh islands by lagoons and tidal creeks, Coosaw, Morgan and Parris Islands being shaped more by river systems than by the sea.

Marshes surround and protect Port Royal Island and neighboring Lady's Island from the worst effects of storm surges. Laid out on Port Royal in 1711, Beaufort is most directly reached by coastal vessels via the Beaufort River and Port Royal Sound. From the mid-sixteenth century onward many observers thought the sound wide enough and sheltered enough for a great port. No port of more than local note ever materialized except during the Civil War, when an armada of Union vessels anchored off Hilton Head and what is now the Town of Port Royal. This same fleet's earlier capture of Confederate forts guarding the sound's entrance brought what might be called the classic plantation era to its close.

St. Helena Sound, Port Royal Sound, and Calibogue Sound are dominant estuarine features. Cutting up and dividing land forms, these waterways evolved as river valleys draining inland areas when they stood much higher than now. Eventually they were drowned by rising ocean levels initiated during the Holocene period's glacial-warming episodes. Nurseries for fish, crustaceans, mollusks, and other marine invertebrates—supporting flocks of wildfowl and the predators that take them: alligator, eagle, mink, and otter—have been harvested by man since early times. Occurring along shorelines where the balance of intermingled freshwater and seawater is right, oysters (*Crassotrea virginica*) and southern quahogs (*Mercenaria campechiensis*) were dietary standbys for prehistoric peoples whose numerous middens and shell rings formed convenient quarries for eighteenth- and nineteenth-century tabby makers.

Beaufort's Sea Islands have seen significant exploitation. Their original cover of maritime forest dominated by live oak and pine was harvested before the Revolution to make way for indigo and cattle and to provide spars for the British Navy and timber for local ship builders. Later cotton required extensive clearance, as did plantation crops—corn, peas, okra, sweet potatoes, and greens—along with livestock necessary for sustenance of enslaved workers. During the early twentieth century, truck farming reclaimed many old fields abandoned following the breakdown of plantation agriculture during the Civil War. Today St. Helena, the most productive antebellum cotton producer, is still essentially rural and retains large farms growing tomatoes, harvested by an annual influx of migrant workers who bring mobile food stands serving tacos Vera Cruz–style and bodegas selling Latin foodstuffs to an area where Gullah traditions are alive but fading, where sweet-grass baskets are still made, greens grown, and sugarcane sold along main roads for one dollar per piece in the fall.

Lady's Island, formerly crisscrossed by properties owned by established planters and several Beaufort merchants (including the Verdiers and the McKays) seeking the gentility that only landed estates could bring, is largely covered by tract housing, scraps of regenerated woodland now reverting to pine barren. Hilton Head, where local cultivation of long-staple cotton began, retains rural sections, but these are squeezed by resorts and their inevitable golf courses, gated communities, and shopping malls proliferating along major highways even in the face of restrictive zoning. Port Royal Island's relationship between urban, semiurban, and rural activities is uneasy, a strong military presence plus population growth fueled by predominantly northern immigration tipping the balance toward expansion of commercial, light industrial, residential, and resort building rather than farming.

Over the course of the twentieth century, Beaufort steadily lost industrial facilities, wharves, warehouses, and maritime installations clustered along the Beaufort River. Passenger steamers linking Charleston and Savannah disappeared too. Today shrimp boats, occasional cruise vessels, pleasure craft, a small marina, and day docks are reminders of long-gone shipyards and coastal commerce. Designated in 1973, Beaufort's National Historic Landmark District resisted incompatible development for many years, but commercial pressures have become too strong for maintenance of traditional building densities and the original character of older neighborhoods. An inherent conservatism has stultified modern architectural design, which, with one or two exceptions, displays little of the innovation and occasional daring exhibited by the best of the area's legacy building, retreating instead into bland, ultimately sterile, historicism.

Inland, north and northeast of Whale Branch—where long-established ferries once linked Port Royal Island and Beaufort Town with what passed for highways connecting Charleston and Savannah—an older man-made landscape exists along waterways draining into the Combahee River and along the river itself. Rice plantations were established here by the mid-1730s, yet few prerevolutionary or antebellum architectural features still stand. Rather networks of enclosed rice fields, dykes, causeways, canals, trunks, and sluices installed at unimaginable human cost are the chief testament to gangs of enslaved people who gradually tamed this wetland wilderness and their masters, who, hungry for profit, risked fortunes on the task. Tabby was rarely favored perhaps because of abundant lumber and the effort required to bring large quantities of shell upriver from intertidal zones. What is more important, elite landowners, afraid of summer fevers (often malarial), lived elsewhere for most of the year. There is evidence of brick production, though it was limited, wealthier planters barging brick from Charleston when required.

Now part of Jasper County, lower St. Peter's Parish exhibited similar patterns from the late 1780s down until (occasionally beyond) the Civil War, its wetlands bordering the Savannah River north and northeast of Savannah, attracting established planters with patrician surnames—Elliott, Giroud, Huger, Heyward, Manigault—and motley collections of speculators, opportunists if not scoundrels intent on making new fortunes or sometimes losing old ones though timbering and the cultivation of rice. By 1860 eighteen large rice plantations (the majority in absentee ownership) located along the river's South Carolina side were populated by about twenty-five hundred enslaved people. Mills, storehouses, sheds, stables and wharfs, and settlements and dwellings accompanying this development have gone. Chimney bases, timber revetments, and pilings concealed by marsh grasses are among the few remaining built elements. Indications are that tabby was unusual, even rare hereabouts, playing minor roles in the architectural scheme of things, locally made brick constituting a convenient alternative.

Located just outside Savannah at the end of a half-mile-long oak allée, the Hermitage, founded in 1819 by Henry McAlpin (a former Charleston resident), made "Savannah Greys," beautiful, well-fired bricks of gentle color prized by architects and contractors building neoclassical structures on Savannah's streets, squares, and outlying plantations.[14] HABS photographs of the demolished regency-style main house probably designed by William Jay (onetime partner of McAlpin who promoted establishment of an iron foundry here) and its slave quarters give an idea of cultural losses sustained when shifting economic patterns left rice cultivation and cultivators along the Savannah nothing but distant memories. Nevertheless, flying into the local airport, field after impounded field, an occasional trunk and canals and levees stretching for miles can be seen, all disused yet evocative of the endless, often brutal, extravagant, costly labor required to reshape fever-ridden marshes and swamps for plantation production.

Hurricanes were wildcards when people gambled with development of low-lying lands along the Combahee and Savannah or on the islands where farmers could (and sometimes did) lose everything to high winds and tidal surges racing up estuaries on which their operations ultimately depended. Savannah River rice growers saw their fields damaged by high water in 1824, John Hamilton Couper reporting the same storm damaged Hopetoun, his plantation near the mouth of the Altamaha, where the crop "unusually forward and promising was destroyed by the very severe hurricane of the 14th September which broke and ran over the banks and flowed [over] all the fields."

Contemporary newspaper accounts speak of devastation on St. Simons Island, and severe structural damage and deaths in Darien. "On Sapelo Island, Mr. Spalding lost all his out-buildings, crop and one Negro. The overseer's father, two sons and four or five Negroes perished."[15]

Following the destructive hurricane of 1804, storm towers of brick or tabby patterned after towers from the lower Santee where the enslaved and livestock might take shelter had been built along the Altamaha. If similar structures ever existed in Beaufort County, nothing tangible survives. Newspaper reports do record that Savannah and Charleston experienced major coastal storms in 1831, 1846, and 1852, the last event unusually severe. Sea Island planters were well aware that weather factored into the success or failure of cultivation. Records they kept concerning meteorological conditions are mostly lost. The journal of Thomas B. Chaplin (1822–90) is an exception. For the period 1845–57 it documents extremes—periods of drought, heavy rains, high temperatures, winter frosts, occasional (very occasional) snow, and one hurricane. Striking St. Helena Island September 7, 1854, it uprooted trees, destroyed crops, the highest tide Chaplin had ever seen flooding Tombee's yard to a depth of two feet. The storm's eye apparently came ashore well south.[16] Damage was bad but not nearly as catastrophic as damage caused by the great "Sea Island Hurricane" of August 27, 1893. With sustained winds estimated around 120 miles per hour, it was one of the strongest and most violent storms ever recorded for Beaufort or its islands.[17] The Charleston *Sunday News* (September 3, 1893) reported: "lost lives are counted by the hundreds and hundreds, and there is no money with which to buy even the strict necessaries of life. All is gone! Home, crop, and often family." St. Helena Island reported two hundred fatalities (probably an underestimate); there were eighty dead at Coffin Point (the Red Cross later reported fifty). Body counts given by the coroner from Dathaw—where the population stood at one hundred—indicate thirty-seven persons drowned.[18] Mortality was high elsewhere, including forty-seven dead listed from Warsaw Island; twenty-five at Coosaw; and twenty-six from the Pacific (Phosphate) Works, Chisolm Island. On Lady's Island twenty-seven perished, and twenty-four more died on Parris Island.

How many structures were lost or damaged is not stated. St. Helena Island was "a mass of ruins," and wrecked houses, along with dead bodies, human and animal alike, that "rested in the trees."[19] Fortunately walls of Brick Church, where many sought shelter, held, though not the roof covering, which blew away. Tombee's tabby piers and the house they supported also escaped major damage. Seaside Plantation survived too, and Coffin Point's big house kept its roof frame even though timbers cracked, racked, and skewed as winds blew in from St. Helena Sound. Did waves driven by the same winds sweep over and completely inundate old slave cabins still inhabited nearby? My guess is that, given the high mortality figures cited for Coffin Point, this was indeed the case.

Along the Combahee what remained of rice cultivation was wiped out seemingly forever. To the south fields along the Back River were inundated, the *Savannah Morning News* (August 28, 1893) reported, noting: "Nobody will suffer heavier losses by the storm than the rice farmers. . . . The rice crop of this season was said to be very fine and the harvesting had just begun." Always in danger when overwhelmed by excessively high waters,

Savannah River plantations counted, it was said, 130 dead on this occasion. Commercial shipping saw losses. Off Hunting Island the SS *City of Savannah,* with a crew of forty-seven and twenty passengers, ran aground, having been set adrift after its pilothouse, smokestack, and sails were carried away by huge seas. Today the wreck remains one of St. Helena Island's favorite fishing places. How high was the storm surge? Several individuals saw waves twenty feet high passing along the Beaufort River; others estimated (perhaps more accurately) the wall of water was seventeen feet tall. Destruction in Beaufort, especially along Bay Street, where boats were tossed onto the highway and old live oaks leveled, was widespread, the apse of St. Helena's parish church collapsing to the dismay of its congregation. Mercifully there were but two in-town fatalities and for, the most part, minor injuries.

Whether the tabby-built plantation house formerly owned by Lewis Reeve Sams on Dathaw Point (Dataw Island) overlooking the Morgan River was undermined and drowned by the same storm or an earlier event is uncertain. What little can be seen during low tides shows the structure has been overwhelmed by the river, walls and chimneys breaking up and settling into marsh mud. Erosion, subsidence, or a combination of both is invading adjacent sites formerly occupied by slave dwellings and a schoolhouse.

Haig Point's lighthouse keeper found the sea had advanced more than 150 feet inland overnight, eating away the bluff where the now-restored Daufuskie light stands. In an era of global warming and rising sea levels, these should be cautionary tales, reminding us that neither coastal processes nor the environments they create are static, that islands can quickly change shape or disappear, that historic structures standing near the sea are vulnerable to destructive forces of nature. Past disturbances and marine regressions and inundations have not inhibited modern development extending into wetlands and unstable dune lines. How unrelenting population growth and irresponsible zoning will impact Beaufort County's natural and heritage resources over the long term cannot be predicted. Chances are that environmental and cultural attrition will continue, making HABS drawings, photographs, and histories painstakingly collected by several generations of dedicated field workers increasingly valuable records of structures too easily lost.

1

Old World Antecedents and Their Diffusion

Traditional tabby fabrication techniques described in La Rochefoucauld-Liancourt's account of Beaufort (1796) surely evolved over lengthy periods, yet few authors writing before the mid-twentieth century showed much curiosity about how tabby first entered common usage in British colonial North America or from where knowledge of tabby making ultimately derived. Thomas Spalding was an exception. Aware of the material's close relationship with European rammed-earth traditions, he cited other antecedents when explaining tabby's appearance at Frederica, Georgia, just before 1740. Not knowing perhaps that the town's founder, James Oglethorpe, must have viewed tabby under construction while staying at Fort Frederick on Port Royal Island in 1732 and 1735, Spalding wrote: "General Oglethorpe, doubtless borrowed this method of building from his neighbors the Spaniards, but adopted it to the means he had in this country." A letter dated July 10, 1844, adds: "Gen. Oglethorpe was a gentleman and scholar, as well as a soldier, and probably had either seen these buildings in Spain or in other countries along the shores of the Mediterranean; or was sufficiently acquainted with them to determine him to adopt them, and the Spanish in the adjacent Province of Florida afforded him the means of getting men to instruct his own people."[1]

Earlier Spalding mentioned more ancient forerunners, which he knew from "Grey Jackson, late Consul General for Holland at Mogadore [Essaouira] in Morocco," informing an unnamed correspondent in 1816: "[tabby structures] are the cheapest buildings I know of, the easiest in the construction, may be made very beautiful and very permanent. They are the buildings of Spain, the boast of Barbary where some have stood these many centuries."[2] It is possible Spalding also read *Travels in Barbary* by the learned Dr. Thomas Shaw, D.D. (1694–1751), chaplain to the English Rectory of Algiers, who observed: "most of the Walls of Tlemsan [Tlemcen, Algeria] have been made in frames and consist of a mortar made up of sand, lime and small pebbles which well-tempered [tamped] and wrought together hath attained a strength and solidity equal to stone."[3] Published in 1738, this perceptive remark was reprinted verbatim by John Armstrong, "Engineer in Ordinary to His Majesty" (1756), and Lieutenant Colonel Thomas James of the Royal Artillery (1771), both authors drawing analogies between what the "ingenious Dr. Shaw tells us" and medieval tapia walls still standing at Gibraltar.[4] Writing in 1740 ostensibly from Minorca (where he encountered huge tapia cisterns), Armstrong said:

15

"the Moorish Castle of Gibraltar is the most noble Specimen of this kind of work I have seen."[5]

Coincidentally—or otherwise—reference to the latter place was made in 1741 by Henry Myers, one of Frederica's earliest settlers, who explained that "as the bricks were dear and much labor for young beginners, we have fallen upon a much cheaper and better way of making houses, of a mixture of lime and oyster shells (of which we have vast quantities) framed in boxes, which soon dries and makes a beautiful, strong and lasting wall."[6] Myers continued: "they have beautiful houses in Gibraltar of that sort, though they have plenty of stone," a statement for which military informants seem the source since two companies of Frederica's garrison (some of whom spoke Spanish) were drawn from Gibraltar, an intermediary between Europe and Africa where, along with the melding of disparate human populations, cultural exchange had gone on for centuries.[7]

Linkage between tabby and Iberian building tradition is supported by recent authorities. William M. Kelso suggested South Carolinians probably borrowed tabby construction from Spanish settlers, reiterating Stoney's conclusion published in 1938 that: "the nearest thing to an architectural heritage the South Carolina plantations had of the Spanish is the system of building in concrete, called tabby."[8] Sickels-Taves and Sheehan (1999) also conclude that tabby found along coastal areas of the southeastern United States is most likely derived from the tapia of Morocco and Spain, arguing how moveable or reusable formwork both distinguishes and characterizes this construction type irrespective of mixes employed in any given place at any given time.[9] So broad a view offers parallels between Old and New World practice, parallels that are explored below through examination of materials called *tâbiyah* by Arabic writers and *tapia* (sometimes *tapial*) by Spanish ones, these terms here being used interchangeably according to context.

Discussion does not exclude the possibility that individuals working across the Americas drew on experience gained from other cultural contexts or made their own particular contributions to the tradition. Indeed the likelihood of multiple influences and local innovation is very strong although difficult to isolate with any precision. Where or when oyster shell—the most distinctive if not necessarily defining ingredients of North American tabby—was first used as an aggregate is uncertain. St. Augustine or another place under Spanish colonial rule is probable, but positive confirmation has not yet surfaced. Without better research into coastal West Africa's vernacular building history, it is equally hard to say whether rammed-mud and cob-like fabrics of the region predisposed enslaved individuals originating there to recognize and perhaps refine tabby making when involuntarily resettled across the Sea Islands, though considering the ubiquity of these materials in Senegal, Gambia, and neighboring countries, I believe this could be true.

◦ Tâbiyah and Tapia ◦

Before proceeding with descriptions of Afro-Iberian tâbiyah we should be clear about what is meant by the term *tâbiyah*. The best modern definition was given by Martin Malcolm Elbl, who stated: "*tâbiyah* is inherently not just 'rammed earth' or 'tamped earth.' It is any kind of mix containing in different proportions earth, clay, aggregate (sand, gravel, ceramic fragments, ceramic tile and brick rubble) and added stabilizers (e.g., lime, chalk) shaped in situ using removable box forms." Significantly, Elbl added, tâbiyah "is more a technique than a specific type of invariant low grade material," a conclusion borne out by various fabrics given this name in historic sources and recent architectural studies.[10] Among early informants the renowned Arab historian Abd al-Rahman Ibn Khaldûn (1332–1406) is the most explicit, his *Muqaddimah* (Introduction to History) written at the Castle of Oran, Algeria, about 1377 relating the following about formed-earth construction:

> One builds with it by using two wooden boards, the measurements of which vary according to local custom. . . . They are set upon a foundation. The distance between them depends on the width of the foundation the builder considers appropriate. They are joined together with pieces of wood fastened with ropes or twine. The two remaining sides of the empty space between the two boards are joined by other small boards. Then, one puts earth mixed with quicklime into [this frame], the earth and quicklime are pounded with special mixers only used for this purpose, until everything is well mixed throughout. Earth is then added a second and third time, until the space between the two boards is filled. The earth and quicklime have combined and become one substance. Then, two other boards are set up in the same fashion and the earth is treated in the same manner . . . until the whole wall is set up and joined together as tightly as if it were one piece. This construction is called *tâbiyah* and the builder of it is called *tawwab*.[11]

Ibn Khaldûn said nothing about material proportions required for tâbiyah mixes; neither did he discuss countless practical details, knowing from experience gained when supervising royal building projects in North Africa and Andalucía that operations would differ according to location or resources at hand. And it should not be assumed that all tâbiyah was made exactly the same way with identical ingredients throughout the wide geographical area where it was commonly used. Nevertheless Ibn Khaldûn's account prefigures much later accounts, the craft he described being strong enough, flexible enough, and, above all else, useful enough to persist for a millennium or more, traveling from continent to continent during periods of conquest, migration, and colonial expansion.

Whether or not Ibn Khaldûn's tâbiyah represents a survival of Roman construction modes is debatable. It is certain that earth tamped or rammed between timber boards with—or without—addition of lime has a historic distribution extending right across more arid regions of the Old World, appearing in China as far back as the Shang period if the seven-kilometer-long wall near Cheng-chou (Honan) is correctly described.[12] Certain traditions—those of Central Asia and Yemen—could be discreet. Others distributed around the Mediterranean basin, south into the Maghreb and north through western Europe, are, Glick concluded, products of diffusion over millennia.[13]

Variant kinds of tâbiyah occur in every conceivable building context along Morocco's Atlantic littoral and among its successive inland capitals (i.e., Marrakech, Fez, Meknes), formed materials still being utilized for rural structures (farmhouses, walls, and stores) across hinterlands of the High Atlas. Similar, identical, or nearly identical construction methods typify Hispano-Islamic architecture, knowledge of the material likely having crossed the Straits of Gibraltar into what became al-Andulus with Arab and Berber incursions starting with an expedition led by Tarik b. Ziyad, who took possession of the great rock in 711 A.D. Subsequently craftsmen traveling with royal armies or competing military forces moving back and forth across the same straits brought architectural uniformity over an area that, under Islamic rule, extended from the Pyrenees, through much of Iberia, south into the Sahara.

Morocco and Andalucía remain rich in medieval *tâbiyah*, with examples dating back to the ninth or tenth century A.D. Because research into and documentation of these building traditions is very uneven, I have described representative Northwest African and European sites separately, my treatment—primarily chronological with respect to Morocco, more thematic when discussing Iberia—reflecting disparities among relevant sources.

⁕ Northwest Africa ⁕

Tarik fortified Gibraltar, but nothing of his work survives, early fortifications still standing there including the Tower of Homage (Calahorra)—one of the largest medieval towers of al-Andalus—being the work of later rulers. Today, despite multiple reconstruction episodes, ramparts enclosing the historic core of Marrakech, with their many (perhaps 150) square interval towers giving views over serried groves of olives, oranges, and date palms toward the High Atlas, illustrate early "Moorish" defensive building. Started at an astrologically propitious hour by the Almoravid prince Ali b. Yūsuf (1107–42), whose rule extended over most of northwestern Africa, the original circuit was more than five miles long, fronted by a wide ditch and interrupted by numerous gates.[14] Spalding's informant, James Grey Jackson, observed that these "extremely thick walls" were built of "a cement of lime and sandy earth, put into cases and beaten with square rammers," a footnote adding, "this cement is called Tabia by the Moors."[15] New walls were built inside old ones, and the city's ramparts were extended southward by the Almohads (puritanical Berbers who displaced the Almoravids after 1130) to enclose a Qasaba housing palaces and administrative quarters. Today, this Qasaba is built over, its once-guarded enclosures—now broken and incomplete—appearing and disappearing among multiple blind alleys of densely populated urban quarters. Unlike the city's outer circuit, walls of these palace compounds are not prettified by newly applied pink stucco. Rather they present something of their original state, exhibiting clayey, calcareous mixes containing abundant small pebbles and broken brick and pottery all cast in 2'8" high vertical increments using smooth-faced horizontal form boards. If dating from Marrakech's first major expansion period (1185–90), construction is attributable to sultan Abū Yūsuf Ya'qūb (1184–99), whose immediate predecessors occupied much of southern Spain, establishing what became a brilliant court in Seville, renowned for philosophical, poetic, and architectural innovations.

Expansion across al-Andalus required secure staging posts where armies might be assembled, fleets built, and material stored, activities explaining Ya'qūb's foundation of

Ribat-al-Fatah (Fort of Victory) in 1195.[16] Located astride an ancient strategic corridor linking Fez and Marrakech, commanding views of the Atlantic and fertile lands bordering the Oued-Bou Regreg estuary, this site (now Rabat), with its sister port of Salé on the opposite bank, was destined to become the most notorious haven for "Barbary" pirates during the seventeenth century. Covering over eleven hundred acres, the new *ribat* was enclosed along two inland sides by "concrete" (i.e., tâbiyah) walls strengthened with seventy-four towers, all cast using an exceptionally rich lime-and-clay mix containing quantities of pebbles. Much still stands, carefully integrated into early modernist town plans imposed during the French protectorate period. Most towers are hollow and linked by continuous walkways running along the top of intervening walls. About twenty-four feet high, these walls are eight feet thick and were cast in thirty-five-inch vertical increments.[17] An applied coating covers all exterior surfaces. Moroccan workers interviewed in May 2016 were preparing what they called a "melange" of an almost blood-red clay, rotten lime, sand, and minimal quantities of water to patch spalled finishes of the southern section, a traditional activity differing little—if at all—from medieval practice.[18] Four early gates penetrate the tâbiyah circuit. One, the Gate of the Winds (Bab Rouah), is an exceptionally grand monument fabricated of cut stone enriched with geometric and interlaced floral carving probably designed in the same atelier as Bab Agnou, the lavish slate-faced ceremonial entrance into Ya'qūb's Marrakech palace where severed heads of his enemies were displayed.[19]

Masterworks of Almohad art, both gateways are relatively well preserved. Not so the same ruler's Hassan Mosque, commenced in 1199 but never finished. Excepting its truncated yet superb stone minaret (now Rabat's most iconic architectural artifact), this mosque—Islam's second largest when built—fell into ruin soon after the founder's death, its materials, including enough high-quality cedar to refurbish a Muslim fleet, being carried away. Despite subsequent earthquake damage and partial appropriation by the present royal family for sepulchral purposes, tâbiyah walls of an enclosure measuring about 600' north/south × 456' east/west can be traced. Occupying elevated ground above the Oued-Bou Regreg, this space was divided internally by rows of stone columns into twenty naves, the minaret centered on its north side.[20] Along the west side, tâbiyah wall fragments stand about twenty-seven feet high. Building techniques used here match those employed by the Almohads in Andalucía. Rows of rectangular cavities (measuring 5" × 3" or 4" × 3") spaced at more or less regular thirty-two-inch intervals along pour lines attest use of timber "needles" to support what Elbl calls side-shutter assemblies (*luh* in Arabic) measuring about thirty-four inches in height made up from horizontal plank. To achieve the desired wall thickness, opposing shutters were distanced 4'6" apart or slightly more if we allow for centuries of surface erosion. Removable vertical staves mortised into the needles to brace formwork as it was filled have left no trace. Evidence for the length of individual forms is also lacking. It can be observed that needles were uniformly staggered row by row to line up only after every second pour. Very rich in lime, the tâbiyah mix incorporated dark-red clay probably excavated near the site, broken rock, and water-worn pebbles all pounded into the luh in relatively thin layers. Old photographs show that, unlike the enclosure's other three sides, the south face was buttressed externally by eleven

Hassan Mosque, Rabat, Morocco, detail of enclosure wall
(all images from the author unless otherwise stated).

small "towers" apparently of mixed stone and tâbiyah construction, this wall having special significance for assembled believers, *mihrabs* near its center marking the direction of
prayer.

Commanding highly disciplined armies, naval squadrons, untold numbers of enslaved prisoners, and treasures captured during his Iberian campaigns, Abū Yūsuf Ya'qūb
possessed ample resources. Nevertheless Ribat-al-Fatah's unfinished minaret suggests that
accomplished stonemasons who fashioned its two lower stages were relatively few and
perhaps thinly spread among construction projects scattered from Marrakech to Seville.
By contrast expeditions entailing fortification of innumerable sites across southern Spain
produced large cadres of men skilled in military building. Since tâbiyah was by far the
commonest material employed, these craftsmen surely included numerous tawwab capable of executing royal commissions—ramparts, palaces, or, in this case, a mosque enclosure of unprecedented size—with speed and economy.

While Marrakech remained crucial to Almohad authority, Fez (which came under
their rule ca. 1145) was refortified in 1212, the work—one of this dynasty's greatest (and
last) architectural projects—ordered on the very day then-sovereign Muhammad an-
Nāsir was defeated by Christian armies in Spain.[21] Following contours rising steeply above
Oued el-Fās, new walls enclosed dense nets of winding alleyways, innumerable markets,
workshops, houses, and *fonduks* and religious buildings, including the Qarawiyyin—a
pivotal center of medieval learning. Following establishment of Fās al-Djadid (New Fez)
as an administrative center in 1276 by the Marīnid sultan Abū-Yūsuf (1258–86), defenses

Fez, old ramparts at Bab Guissa, Fez Album
(courtesy Fine Arts Library, Harvard University).

protecting the old medina (Fās al-Bali) underwent renovation, and predominantly tâbi-yah walls were erected around the new city. Heavily restored, circuits defining Old and New Fez constitute one of the most extensive medieval form-cast installations extant. Fortunately for architectural historians, it was photographed by commercial studios before evidence of phasing and the nature of original building methods became concealed beneath modern reconstruction work. Among the best of these records, Harvard University's "Fez Album" preserves numerous images of the city taken by an unknown photographer in the 1920s or 1930s, the example given here showing ramparts near Bab Guissa, an Almohad gate built about 1207, now much restored.

Emulating their Almohad predecessors, Marīnid rulers waged jihad in Spain, mounting six or more campaigns between 1275 and 1299. Footholds were gained at Tarifa, Algeciras, Ronda, Malaga, and eventually Gibraltar, where Sultan Abū'l-Hasan (1331–48) erected new tâbiyah defensive walls decorated to simulate regular coursed masonry.[22] Political gains were ephemeral, yet exchange of craftsman between North Africa and Andalucía unleashed streams of creative genius among artisans, enhancing royal building projects. Naval power played essential, often critical roles in Marīnid incursions, Abū-Yūsuf installing or restoring shipyards and an arsenal at Salé, where an immense stone water gate, tall enough to let ships through, still stands, although the channel it spanned has silted up and disappeared. After the sultan died at Algeciras, his body was carried back to Morocco and buried outside Rabat within the compound known as Chella, an already-sanctified outpost (ribat) for warriors of the faith. Located beyond the old Almohad city southeast of the Hassan Mosque on the far side of a deep valley sloping down toward the Oued-Bou Regreg estuary, Chella appears fortified, which may be indicative of an earlier medieval function. Archaeology confirms that the present enclosure occupies Roman-era Colonia Sala, the oldest strata, dating back to the third century B.C. Marīnid rulers were doubtless attracted by the site's historic associations, strategic position, abundant spring-fed water supply, monumental *thermae,* and spectacular *nympheum* installed during Trajanic times. Existing facilities were resurrected and enhanced by successive sultans, starting with Abū Sa'id (1310–31) whose successor, Abū-'l'Hassan 'Alī (called the Black Sultan), inscribed his own name above the present enclosure's main gateway (Bab al Gharbi al-Kabir) along with the exact date, July 8, 1339 (739 A.H.) when work was completed.[23] Crenellated and strengthened with twenty rectangular towers, tâbiyah walls define a pentagonal compound covering about fifty-seven acres, which became the burial place for Marīnid sovereigns, their families, and several popular saints.

Flanked by two canted stone towers with cornices supported on *muqarnas* (stalactite pendentives), Chella's main entrance is monumental, an exceptionally high "bent" passageway emerging through a second grandiose gate into the enclosure. Material usage follows established hierarchies. Fine-grained limestone blocks laid with near-invisible joints were reserved for flanking towers and the two (i.e., inner and outer) portals. Masons vaulted the entrance corridor with thin, Roman-style brick. Tâbiyah was employed for the gate's back wall. Stonework of the inner portal was keyed into this back wall, lift by lift, as casting proceeded, but not into the enclosure's outer wall, indicating that the gateway was perhaps inserted into preexisting construction.[24] Whatever the case, beautifully finished stonework and delicate carving contrasts with the tall (20'–22' high) and thick (5'4") tâbiyah circuit it cuts through, bland new stucco now concealing serried rows of horizontal pour lines and rough holes left by formwork supports visible before 2017. Several scraps preserved in situ indicate that lime plaster was originally applied over the tâbiyah, then scored to imitate ashlar, techniques seen elsewhere—notably Gibraltar and Granada.[25] Was this finish primarily aesthetic, symbolic, or designed to deceive by making fortifications appear stronger or the ruler's resources greater than they actually were?[26] Among princely milieus, where functional, formal, and poetic attributes became blurred, it is possible all these meanings came into play simultaneously.

Basset and Lévi-Provençal called the tâbiyah "mediocre."[27] Their judgment is over-harsh considering that much of Chella's enclosure remains intact despite severe damage caused by the Lisbon earthquake of 1755 and subsequent despoliation.[28] Typically individual pours of the northeast outer wall measure twenty-six inches in height, the tâbiyah consisting of ochre-colored clay, lime, quantities of broken stone, small water-worn pebbles, pieces of pottery, and smashed tile probably derived from the Roman ruins. With twenty-seven-inch lift heights, tâbiyah of the western side (which steps down a precipitous incline) is essentially similar, except ceramic inclusions are far less common. Here, as along the northeast wall, cavities distanced between twenty-three and thirty-one inches on center indicate that timber needles supporting the side shutters were both rectangular (measuring 6½" × 6") and circular (measuring about 2½" in diameter) in section. Most wall surfaces are too weathered to gauge the exact length of individual pours—at lower

Chella, Rabat, exterior of main entrance gate, state before restoration.

Chella, Rabat, exterior view
of enclosure's west side
before restoration.

levels they appear very long. Shorter forms were employed above. These were struck and moved horizontally once the cast mix hardened. Along the enclosure's exterior, erosion has revealed stepped tâbiyah foundations. During the 1920s investigators found that towers themselves incorporate two stages plus parapets and merlons, the lower spaces roofed with poorly built brick domes supported on pendentives.

Deep within the enclosure a *zawiya* (Sufi monastery), *hammam* (baths), latrines, and a spring house were erected near Abū-l'Hassan's small funerary mosque. That glazed tile and tile mosaic were applied with equal dexterity over tâbiyah and fired brick backgrounds is demonstrated by what is left of the zawiya. Two (or possibly three) stories high, the building had thick tâbiyah exterior walls at the lower level (typically thirty-three-inches wide cast in twenty-four-inch vertical increments), construction switching to brick above. The narrowly rectangular courtyard about which the plan was organized contained a deep pool fed by low fountains at its two opposite ends, with arcades supported on slim marble columns linking small student cells arranged along the courtyard's two long sides. Cross partitions separating individual cells (backed up to the building's tâbiyah perimeter) were mostly brick, tile, and tile mosaic, concealing differences in underlying materials. Patterns of interlocking stars fabricated from cut and specially shaped glazed white, turquoise, deep-blue (originally black?), and yellow tile remain evocative, recalling radiating *shamsa* designs emblematic of divine light embellishing Islamic manuscripts. Outside the main building communal latrines were similarly treated, with floors and ruined walls displaying remnants of elaborate tile work and decorative plaster. Tâbiyah enclosing this facility was cast over earthenware pipes carrying water channeled from nearby springs for cleansing purposes, the same springs irrigating restored gardens imaginatively designed to emulate lost medieval precursors that harbor—or so popular legend relates—malevolent *djinn* drawn by the site's quiet solitude, pure water, broken spaces, abandoned tombs, and buried treasure.

Later Marīnid rulers abandoned Chella's royal necropolis for Fez, building their tombs high above the old city (Fās al-Bali).[29] Heavily decayed these structures further illustrate the way medieval Moroccan masons integrated stone, tâbiyah-and-brick construction in response to structural, aesthetic, and perhaps economic demands all within the confines of relatively small (if highly prestigious) commissions. At first sight the best preserved appears little more than a broken masonry cube, having lost its roof and surface enrichment. Closer inspection reveals the logic governing masonry usage. To safely distribute tensile stress, the principal facade—featuring a tall horseshoe-shaped arched opening—is mostly brick built, the relatively thin, well-fired units expertly laid. Vertical timber inserts above the arch were fixings for an applied decorative or inscription band. Above there is an attic story of brick pierced by a single window opening. Other pockets left by timber fixing pieces and scraps of carved plaster show that the entire frontispiece was heavily decorated with applied materials—stucco and perhaps *zillaj* (glazed tile mosaic)—all now gone. Supported on low brick plinths, the rear and two side facades are principally tâbiyah, keyed round by round into the brick frontispiece. Vertical lifts (each approximately thirty inches high) are interrupted above the fifth pour level by a band of brickwork pierced by splayed window openings (spanned by thin timber lintels) regularly spaced around three sides of the building. Above this line tâbiyah construction resumes, another four lifts raising walls to eaves level. The entrance facade and side walls were built together, bricklayers and the tawwab closely coordinating their respective activities. More vertical fixing pieces visible on the interior rear wall above the line of window openings suggest another type of timber enrichment—cedar friezes, intricately carved with vegetal motifs and pious inscriptions usual among contemporary *madrassas*. Sadly broken and incomplete, a neighboring tomb chamber shows that interior stucco was applied directly over tâbiyah coat by coat until sufficient depth was achieved to allow carving, pecking still visible on tâbiyah surfaces ensuring good adhesion.[30] Here too polychrome timework—an often dazzling characteristic of Marīnid architecture—is absent, probably robbed along with inscribed marble memorial stones seen by Leo Africanus in the sixteenth century.[31]

Similarly al-Badiʿ Palace (the "Resplendent"), built in Marrakech by the Saʿdid sharif Ahmad al-Mansūr (1578–1603), has been stripped of its once-sumptuous marble embellishment imported from Europe in exchange for Moroccan sugar "pound for pound."[32] Today tâbiyah walls enclosing the building's grandiose central court with its shallow pools, sunken gardens, and royal pavilions are bare except for scraps of broken tiling.[33] Yet this complex is still impressive owing to the symmetry and overwhelming presence of its principal components, mirrored as they are by sheets of water. Water—that ultimate luxury for persons who traveled over steppe or desert to reach the sovereign's presence—must have once been everywhere, channeled down from the High Atlas by subterranean channels and distributed through pipes embedded in the palace's predominantly lime-mortar fabric. Fountains played around the throne, and basins splashed among apartments reserved for ambassadors, with conduits refreshing flowerbeds and citrus trees during the brutal heat of summer.

Remarkably an accurate plan by the oriental scholar Jacob Golius (1596–1667)—who accompanied Dutch envoys to Morocco in 1622, when much of the architecture remained intact—came into possession of John Windus, who used it as an illustration for his account of British diplomatic dealings with the "Emperor of Fez and Morocco" at Meknes in 1721.[34] Along with structures strung out along the court's long sides and four rectangular garden plots, the document shows two pavilions located at opposite ends of the large (approximately 295' × 65') middle pool, each flanked by secondary pools extending right and left. Sunk about eight feet below main ground level, garden plots have, like the pools, been restored, young orange trees planted in serried rows reestablishing the court's once-verdant aspect. The West Pavilion also stands, this hulking form-cast cube entered via openings centered on three exterior facades enclosing the ruler's hall of public audience (measuring 42'6" square) and one small retiring chamber. About 4'6" thick, walls alternate between brick and tâbiyah at the heavily rebuilt lower level, switching to tâbiyah above.[35] Resembling features described from Fez, vertical cavities extending in horizontal bands high about the building interior are probably impressions left by consoles supporting cedar strips inscribed with literary or dedicatory legends picked out with brilliant color, judging from fragments held by museums. Wall revetments are lost, but interior fountains lined with exquisite polychrome cut-tile work (*zillij*) featuring

Marrakech, al-Badīʿ, view of central court and West Pavilion.

elaborately interlaced stars attest decorative schemes that concealed and enhanced visually unacceptable tâbiyah surfaces to produce what the court poet called a garden of delights. Gone too is the marble, mentioned by early chronicles, used for columns Golius saw enclosing this chamber on three sides and its eastern counterpart called the Crystal Pavilion, another predominantly tâbiyah structure designed for more intimate meetings or amusements now reduced down to foundations, only a few water channels and ghost-like outlines of circular fountains reminding visitors of its former glory. Almagro postulated that timber domes sheltered beneath tiled pyramidal roofs covered the two pavilions, citing architectural parallels from Almoravid Spain. He also noted that similar sunken gardens are found in Andalucía, the Alcazar of Seville's Patio de Contratacion installed during the mid-twelfth century being the most notable example.[36]

There are hints of Indo-Persian influence, the main pool's central island reached by narrow causeways recalling emperor Akbar's Fatehpur Sikri, while baroque elements now lost—probably derived via Ottoman Stamboul—are suggested by literary accounts. With its double set of high tâbiyah walls guarding the ruler's person, wealth, and most intimate secrets, al-Badi represented a staggering investment in labor. External enclosures, floors, pools, gardens, pathways, cellars, and prison cells were all fashioned from formed materials—as were the long, blank facades enclosing intimidating passageways leading into the building's interior.

Southward pleasure grounds extended (and still extend), set amid orchards irrigated by networks of open channels and subterranean tunnels. Currently incorporating some 840 acres, this Agdal has two elevated reservoirs of exceptional size. One is relatively modern; the other, first excavated during Almohad times, was enclosed by Ahmad al-Mansūr within a playfully embattled tâbiyah circuit, thirty small towers mimicking in miniature the outer wall of al-Badi. Resting on Almohad foundations, the enclosure is about twelve feet high, the fabric consisting of carefully compacted mixtures containing clay, lime, and small river-worn pebbles poured into timber forms about 6'5" long in 2'6" high vertical increments. With a capacity of about sixty-eight thousand cubic meters, the central reservoir—or "little sea" (*el-buhayra* in Arabic), as early writers called it—is an astonishing piece of medieval engineering, with massive retaining walls of tâbiyah standing over fifteen feet high along its north side holding back wide, elevated pathways besides vast amounts of water.[37] During the early sixteenth century groves of citrus, aromatic shrubs, and roses flourished all around, an elaborate hydraulic system feeding numerous irrigation channels on which the medina of Marrakech and its palace quarter also depended.[38] Today the roses are missing and Ahmad's summer residence built overlooking the tank absorbed into more recent construction. Largely cut off from the ever-encroaching infrastructure of modern tourism, an inherent poetry survives, migrating swallows still hunting over the reservoir, and long rows of oranges and olives with scattered date palms framing distant views of mountain peaks snow clad in spring.[39]

That Sa'did splendor was supported by cultivation and processing of sugarcane is vividly shown by extensive industrial installations located southwest of Marrakech. These required heavy investment in land, water management facilities, specialized buildings for machinery, and above all else labor, provided by slaves captured, along with masses of

Sugar mill, Idaogourd vicinity southwest of Essaouira (Mogador),
crushing (mill) room, 1570s.

gold, from Timbuktu during Ahmad's conquest of the Songhay Empire (eastern Mali) in 1591, an exploit that earned him the title *al-dhahabi,* "the Golden One." Europeans had insatiable appetites for Moroccan sugar, England alone importing tons of refined product via Agadir in exchange for cloth and cash. To facilitate trade Queen Elizabeth I (whose court was a prime consumer of Morocco's best white sugar) chartered the Barbary Company in 1585 and entered into diplomatic relations with Ahmad al-Mansūr, an alliance formulated against Spain, their common enemy, that could have changed American history had both sovereigns lived.[40]

Larger Sa'did mills were water powered, rivers flowing out of Morocco's mountain ranges being tapped via barrages, basins, and aqueducts to irrigate cane fields and drive cane-crushing machinery. Located among relict *argan* forests inland from Mogador (now Essaouira) near the small settlement of Idaogourd, one of the best-preserved installations harnesses Oued el-Qsob several miles before it debouches into the Atlantic.[41] Here an aqueduct some fifteen hundred feet long, built as a solid wall, carried enough water to propel an overshot waterwheel seventeen feet or more in diameter positioned against the exterior face of an enormous (95' × 41') crushing shed. Excavating this building in 1959, Paul Berthier discovered evidence inside for three cane-crushing machines. Finding traces of timber frames (but no millstones), the excavator assumed (albeit tentatively) that these machines were of a roller variety, though whether rollers were mounted vertically or horizontally at this period cannot be said. The boiling room—a long, narrow

Salé, Morocco, aqueduct (built ca. 1341) at its northern entry into the medina,
detail showing lift lines exposed by modern road cut.

structure (measuring about 102' × 33') positioned against the crushing shed—is heavily
damaged, only fragments of its outer walls still standing. Berthier excavated two parallel
lines of boilers and a large cistern—which received cane juice from the mill—located
near the room's northeast corner. Less easily understood are several incomplete ruins
nearby. Masses of broken earthenware littering the ground identify one as a purgery,
where crystallized sugar was transferred into cone-shaped pots and drained of its molasses, marketed in England.

The aqueduct and associated buildings are all constructed of tâbiyah cast in three-
feet-high vertical increments using red clay, small pebbles, and lime well pounded into
moveable timber forms. Terminating in a water chute, the aqueduct stands about thirty
feet high above the tail race, the race itself, an adjacent settling pool, and diversion channels all molded in lime mortar. Walls of the crushing room are about twenty-five feet
high and 4'2" thick—but what kind of roof (if any) they supported is far from certain.
We can only guess whether cane was delivered by mule, camel, or wagon, while the
number of enslaved individuals harvesting and processing is completely unknown. It was
reported by Robert Cecil in 1609 that Ahmad al-Mansūr's revenues from sugar installations "about Marruecos [Marrakech], Teredinid and Mogador" were "yearly worth unto
him six hundred thousand ounces at the least."[42]

Berthier, who first documented Morocco's early sugar industry, confirmed that
pisé construction, which he equated with tâbiyah, was predominant among sugar

manufactories, where employed for hydraulic works and processing facilities alike. The reason is self-evident—suitable clays are available everywhere across harsh semidesert lands southwest of the High Atlas. Gravels and "glistening red slag covered with pebbles" are also ubiquitous—timber much less so, with exploitation of montane cedar (*Cedrus atlanticus*), renowned for its size, scent, and longevity, requiring resources beyond the reach of all but the ruling elite among mostly nomadic societies lacking wheeled transportation.[43] He further noted that aqueducts supplying urban areas follow the same design as those supplying sugar mills, the one bringing water from the spring called Aïn Barka to the medina of Salé, located on the Atlantic coast opposite Rabat, making his point.

This monument is more than twenty feet high where in enters northern ramparts of the old city. Rather than carrying canalized water on arches or vaults common among Roman structures of similar function, it was built as a solid tâbiyah wall (measuring 4'8" in width) with stuccoed exterior surfaces unbroken except for several arched openings intended for cross traffic or drainage. Construction minimized use of timber centering and speeded building processes, utilizing semiskilled labor. Breaches made for modern roads and the Tangier-Casablanca railway make tracing the aqueduct's course difficult, reliable accounts putting its total length near 2.5 kilometers. Cuts near the outer medina wall reveal nine successive horizontal pours, each pour measuring 2'4" in height. Exceptionally dense and well-compacted, tâbiyah contains dark-reddish clay and small stones plus quantities (probably large quantities) of lime, brick, and broken pottery. The water channel was created by casting uppermost tâbiyah pours around long horizontal molds, with several coats of lime-based stucco rendering cavities that were left after these molds were removed waterproof. Once everything had set, a six-inch-deep tâbiyah cap was cast over the top to prevent contamination and evaporation.

Exhibiting exemplary qualitative standards, the work was ordered about 1341 by sultan Abū 'l-Hassan to supply Salé's newly founded madrassa—among the loveliest in Morocco—and adjacent Great Mosque, which, standing on land elevated sixty feet above sea level, must have taxed the ruler's hydraulic engineers. Sugarcane and other crops grown by residents inside the urban perimeter might have been irrigated.[44] What is more important, the aqueduct was high enough, long enough, and solid enough to protect cultivated areas extending north of the city from sudden attack or nomadic marauders during turbulent times. Today, with its primary function lost, new housing developments are encircling the old fabric and garbage collecting on what were once orchards, gardens, or cotton fields. In 1918 Edith Wharton observed how "neglect and degradation of the thing once made" is a recurring theme in Moroccan architectural history, a theme cruelly played out around Marrakech following the plagues of 1597 and 1602.[45]

In June 1598 an English merchant remarked how workers had fled from the sugar mills as contagion spread and rebellions broke out that threatened "ransack and spoyle of the ingines [engines]."[46] Rumors of the ruler's death reached Queen Elizabeth's spy master, Robert Cecil, during the same year, but Ahmad survived until carried off by pestilence in August 1603 despite having set up camp among allegedly healthy fields outside Fez. In Marrakech al-Badi probably remained occupied by al-Mansūr's successors, fratricidal warfare, indolence, and a love of luxury that followed them into their tombs giving rise to another dynasty—the Alawites—after 1659. Assuming power in 1672, Mawlay Isma'il established his base at Meknes, an old city set amid rich wheat-growing areas southwest of Fez. Meknes was transformed, with Isma'il building and rebuilding defensive walls and erecting palaces, gigantic stables, and granaries that epitomize his power, megalomania, and innate savagery. Construction and deconstruction went hand in hand. Al-Badi' was denuded of its marble columns, these being carted off to Meknes, where they were reused in Isma'il's own mausoleum—along with Roman-era capitals looted from the ruins of Volubilis and Sala.

Arriving in Tetuan in 1721 with the British ambassador Commodore Charles Stewart, John Windus investigated local building modes, observing of its inhabitants: "they raise not their Walls as most Nations do, by laying Brick or Stone even upon one another, but their way is first to make a strong wooden case into which they cast the Mortar, and beat it down hard, take the case away when its dry."[47] The majority of houses were two stories high, whitewashed on the outside with flat roofs, these rooftops reserved—then as now—for women and young children. Weeks later, after an arduous journey punctuated by feasts, long negotiations, sundry setbacks, and assurances from nervous courtiers who went in perpetual dread of their sovereign's violence, Stewart and his party were among the few Europeans to escape Meknes unscathed after gaining brief access to the royal presence, private apartments, armories, storehouses, and treasuries. Always the careful observer, John Windus related the following about Isma'il's palace:

> The Emperor is wonderfully addicted to building, yet it is a question whether he is more addicted to that, or pulling down and those who have been near him since the beginning of his reign, have observed him eternally building and pulling down, shutting up doors and breaking out new ones in the walls. . . . This palace is almost four miles in circumference, and stands upon even ground, in form almost square. It is built of a rich mortar, without either brick or stone, except for pillars and arches, the mortar so well wrought, that the walls are like one entire piece of terrass.[48] The whole building is exceedingly massy, and the walls in every part very thick. Some of the squares [i.e., courtyards] are chequered throughout. . . . Others have gardens in the middle that are sunk very deep and planted with tall cypress trees, the tops of which appearing above the rails make a beautiful prospect of palace and garden intermixt. . . . It is reported that 30,000 men and 10,000 mules, were employed every day in the building of this palace, which is not at all improbable seeing that it is built of hardly anything else but lime, and every wall worked with excessive labour.[49]

Construction was still underway. Passing through several courtyards and "long buildings," the ambassador's party saw "Christians upon the tops of the Walls, working and beating down the Mortar with heavy pieces of wood which they raise all together, and keep time in their stroke." This work pleased the murderous old ruler, who, toothless and haunted by dreams of his innumerable victims, did not hesitate to punish any slaves who broke rhythm, having them thrown down or, in a macabre precursor of modern urban legends, buried alive in the tâbiyah mix as it was poured. Today the palace and its ghosts are

Salé, Morocco, Saqala al Qadima, gun platform overlooking the Atlantic, later eighteenth century.

largely inaccessible, Ismail's halls, courts, and colonnades ruined, destroyed, or absorbed into more recent royal building.

Elsewhere about Meknes gigantic structures still stand, one roofless section (with 6'6" thick exterior tâbiyah walls, cast in thirty-four-inch-high vertical increments and buttressed at regular intervals) rising cliff-like above a great reservoir covering about ten acres that stored water for imperial gardens. The ruins are divided into innumerable naves by tâbiyah cross walls comprising identical sets of rectangular piers, which carry somewhat crudely cast arches rising about 6'4" above their springing and spanning 11'8" in the clear. Windus visited this installation "situated about a league from the Town." It must have been larger then than now since he describes "two very large oblong square buildings with handsome arches all around, under which the horses stand without any partition, there being an Arch for every horse; they stand twelve feet asunder, after which manner these stables are reckoned to hold about a thousand horses."[50] Finer details of interior planning are lost and of course the emperor's magnificent equipage long gone, the now extant ruins—more enjoyable for scenic than architectural qualities—presenting perspectives punctuated by massy piers and row after row of arched openings that create diagonal vistas resembling some Piranasian fantasy made eerie by the screams of nesting peregrine falcons. An associated structure is more rewarding—or would be if it could be better seen—with high windowless halls spanned by brick vaults carried on thick mortar walls, lantern-like towers, and deep silos cut into the ground all demanding more study than the dimly lighted interior allows.[51]

Following Isma'il's death (1727) his empire descended into anarchy, central authority not being reestablished until the late eighteenth century, when successive sultans turned their attention toward international trade and defense of coastlines coveted by European powers. Mogador (renamed Essaouira) was fortified in stone with the aid of captured engineers and expanded to replace Agadir as the empire's chief shipping outlet, with other forts—notably Safi, and al-Jadida—seeing renewed building activity. Damaged by repeated bombardment, defenses at Salé were restored, Sultan Muhammad b. Abdallah (1757–90) introducing a new battery on the town's seaward side, this wide, rectangular feature (called Saqala al-Qadima) extending between two stone towers raised high above the Atlantic shore's jagged rocks and relentless surf.[52] Here the platform and parapet consist of an exceptionally hard tâbiyah containing quantities of broken rock apparently cast solid from the beach upward in long, 2'6" high horizontal lifts finished with an equally hard lime-based stucco resilient to storm, tide, and saltwater spray.

Similar fabrics distinguish the huge semicircular tower dating back to the thirteenth century terminating Salé's seawall at its northeastern extremity, which was extensively restored after 1861. These and other late examples—including the delicious two-story pleasure pavilion built by sultan Mohammed IV b. 'abd al-Rahman in 1869 to overlook the Menara tank in Marrakech (another twelfth-century Almohad hydraulic installation)— show that tâbiyah remained indispensable across southeastern Morocco down until early modern times.[53] A mile or two away, walls of the rambling Bahia Palace, begun during the 1890s (but not finished for decades thereafter), are also tâbiyah throughout, their low-grade fabric concealed by stucco on the exterior and heavily wrought (one might say

Taourirt, Morocco, tâbiyah construction, removal of formwork
(Postcard published by D. Millet ca. 1910). Note needles at base of
timber form that supported uprights tied at top with thongs.

overwrought) finishes inside, with decoration of the highest quality, whether it be tile work, timber, or chiseled plaster, representing an effort to revive traditional crafts. Today utilitarian boundary walls, farmhouses, and outbuildings of rammed fabric scattered in large numbers across open countryside west of Marrakesh (near Chichaoua for instance) appear recent if not new.

Along river valleys southeast of the Atlas, fortified collective buildings termed *kasr* (singular) or *ksour* (plural) with formed earth walls two, three, or even four stories high housed scores of individuals, hundreds in some cases. Based on fieldwork completed here during the 1970s, William J. R. Curtis gave an account of construction modes within a kasr describing how earth mixed with "gravel and straw or dung" was pounded into reusable timber shuttering (known as a *louh*) braced externally with uprights tied by leather thongs.[54] An old photograph of masons striking formwork at Taourirt further illustrates how closely their practice resembled Ibn Khaldûn's account of earth building written centuries before.[55] Paradoxically it is doubtful whether any extant kasr predates 1700, the type having generated much speculation but little certainty about its history. The evolved character of ksour and their distribution around oases frequented by caravans bringing quantities of gold and slaves into the *suks* of Marrakech for a millennium or more open the possibility that analogous building traditions traveled this way too, carried by merchants traveling to and from immensely wealthy entrepôts that stretched "south of the Sahara to the shores of the Gulf of Guinea."[56]

⁎ Iberian Peninsula—Andalucía ⁎

Medieval tâbiyah walls defending cities or defining urban quarters were once common across Andalucía, but widespread destruction of defensive circuits occurred from the mid-nineteenth century onward as historic towns expanded and technology advanced. More than twenty thousand feet long and strengthened by 166 towers, walls enclosing Seville suffered near-complete demolition, sections being cut into blocks and sold for domestic reuse in the 1860s.[57] Despite similar losses—in Córdoba, Valencia, Malaga, Jaén, Gibraltar, and numerous other cities—medieval tâbiyah remains common about marcher zones where local rulers battled one another for dominance, with numerous hilltop fortresses distanced only a few miles apart, fortified towns such as Niebla, palatial dwellings near Murcia, and watchtowers commanding high cliffs above the Mediterranean and valleys of Segura de Sierra providing memorable examples of the material's use.

Under the Nasrid dynasty, which ruled from 1230 until swept away by the Christian sovereigns Ferdinand and Isabela in 1492, tâbiyah became almost universal for military, civic, and domestic construction in and around Granada.[58] The Alhambra and Generaliffe, celebrated royal palaces sprawling across foothills of the Sierra Nevada, are well known, though still not fully understood architectural assemblages where tâbiyah is the predominant, if usually overlooked, structural medium. Occupying high ground dominating Granada's urban core, the Alhambra functioned simultaneously as fortress, royal residence, center of government, barracks, and sepulcher, sheltering within its walls bazaars and workshops, baths, kitchens and stables, public audience chambers, and private apartments set amid gardens reserved for the ruler and his harem, advisers, servants, and slaves. Abandoned, despoiled, neglected, and patched up, the site has retained its princely aspect if no longer appearing as isolated, tumbledown, or romantic as it did when surveyed by Jules Goury and Owen Jones during the 1840s.

Oleg Grabar reckoned this palace city's perimeter walls, built almost entirely of tâbiyah, over seven thousand feet long and counted twenty-two towers, some military in character, others enclosing living spaces with large arched windows giving panoramas out toward the city of Granada and snow-capped Sierra Nevada beyond.[59] Enriched with superb applied decoration, these apartments were admired by the wife and courtiers of emperor Charles V who lodged here during 1526. But on closer acquaintance they found their accommodation old-fashioned, uncomfortable, and finally unacceptable, those having sufficient means or rank moving out. Consequently in 1533 the emperor began building an updated, Renaissance-style royal residence inside the old Islamic edifice. Never fully completed, facades were of cut-limestone blocks variously faced with sandstone or marble, rusticated and decorated with figurative sculpture in classical dress—or undress, much, one imagines, to the scandal of Muslim viewers. The break with the Alhambra's former form-cast building traditions was absolute, the juxtaposition of new and old building modes appearing especially incongruous where western Islamic and Italianate aesthetic ideals collide near the southeastern corner of the Court of Myrtles

(Patio de Comares)—a water garden (installed over an earlier garden in 1370) fronting the chief audience hall of former Nasrid kings.

Misplaced and alien when seen against its surroundings, Charles V's work is iconic, the circular courtyard with two tiered colonnades at the scheme's center finding few parallels outside paintings and unrealized architectural projects of contemporary Italian masters.[60] Absent from this most innovative of imperial Spanish building projects, tâbi-yah did not disappear during the later sixteenth century any more than it had after earlier phases of the *reconquista*. Executed first by Muslim masons for Christian patrons, tapia (as it was locally called) continued useful for less-prestigious works long after the last Muslim family had converted or been expelled from Andalucía, surviving across southern Spain and Portugal among vernacular builders well into the late nineteenth century.

◦ Formwork and Materials ◦

As reconstructed by Basilio Pavón Maldonado,[61] formwork commonly used for tâbiyah across the Iberian peninsula was indistinguishable in design from its North African coun-terparts, consisting—like them—of pairs of timber shutters carried on horizontal needles (*aguja* in modern Spanish) into which were tenoned vertical uprights (*costal*). These vertical members, tied at the top with rope, provided lateral restraint to the box-like shutter assembly during filling and tamping operations, the rope when tensioned making everything tight—a necessary precaution since mortar was pounded into these forms with considerable force. Once the mortar mix set, ropes would be released, uprights removed, and the shutters struck, the entire process of assembly and disassembly being repeated for the next vertical or horizontal pour. Needles were sometimes left in place, re-moval for reuse—common in sparsely wooded areas—leaving lines of horizontal cavities (*mechinales*) extending through the finished wall, which might be plugged with mortar or rock.

At Granada several fortified elements of the Alhambra, notably the Torre del Home-naje guarding the Alcazaba's northeast angle built by the first Nasrid ruler Muhammad I (1230–72) and outer circuit's Reservoir Tower, preserve formwork impressions. These in-dicate that shutters were fabricated from three horizontal boards, supported on timber cross members (i.e., needles) 3¼" square positioned twenty to twenty-three inches on center. Formwork was moved horizontally as well as vertically, joints visible on the Torre del Homenaje's exterior indicating that individual forms measured between 6'6" and seven feet in length. Similar practice is widely documented, with tâbiyah walls surround-ing Seville utilizing forms about eight feet long, and those employed for watchtowers (twelfth to thirteenth century) of the Segura de la Sierra (Jaén) measuring between two and three meters in length. To minimize disassociation, joints between horizontal pours were sometimes staggered, level by level, but not invariably, almost lining up from top to bottom at the Torre del Homenaje. Though rare among deforested areas of the western Mediterranean region, where lengthy timbers were (and still are) almost unobtainable, use of much longer forms is known. Describing "framed" tâbiyah walls of Tlemcen (Al-geria), Dr. Thomas Shaw (1738) wrote: "the several stages and remains of these frames are still observable some of them being a hundred yards long and a fathom in height and thickness," a statement that seems wildly exaggerated though more recent authors talked (without being precise) of "grandes assises" here separated by layers of sand or lime.[62] For Seville modern investigators distinguish short and high formwork measuring 31½" and thirty-five to thirty-seven inches high, respectively, though how real this distinction might be is questionable, since examples surveyed show random dimensional variation. I obtained lift heights between twenty-nine and thirty-two inches from Seville's city wall— not the best example since the small section still extant presents a palimpsest of original eleventh-century work executed for the Almoravid sovereign Ali b. Yusuf (1107–43), with medieval or postmedieval repairs and modern finishes. At Córdoba unrestored sections of the tâbiyah circuit probably erected by the Almoravids (eleventh to twelfth century) to enclose an urban quarter called the Ajarquia yielded form heights between thirty-four and thirty-five inches.[63] At the Alhambra corresponding dimensions were standardized at thirty-one to thirty-two inches before 1273 if underlying fabrics exposed by spalling stucco of the Torre de la Vela and Torre del Homenaje are original.[64]

Materials used for tâbiyah vary with geographical location and available resources, the choice made by any particular mason often appearing opportunistic.[65] Earth-based mixes; mortars containing lime, earth and gravel; larger pebbles; ceramics; pieces of calcareous matter or even large, unshaped blocks of stone are all reported.[66]

● Operational Procedures ●

Operational techniques ranged between careless, hasty kinds and the painstaking ones utilized when erecting tâbiyah walls protecting Córdoba's Arjarquia. Strengthened at regular intervals by square towers, this work was raised about thirty inches above present ground level on stone foundations doubtless designed to minimize groundwater penetration into the cast fabric above. Forms (of undetermined length), twenty-four to twenty-six inches high, were filled in three-inch increments, the mortar mix consisting of lime, clay, pebbles, and considerable quantities of unglazed pottery, all well pounded. Filling and pounding operations continued until formwork was within about ½" of capacity. The entire lift was then sealed using lime mortar, rectangular "needles" (positioned about twenty inches on center) being removed, and the forms struck for reuse after the tâbiyah had sufficiently dried. Measuring 14' × 12' in plan, towers were cast as a solid mass, their construction rising level by level in concert with adjacent walls.

For outer defenses at the Alhambra, forms were filled incrementally in layers 2½" deep, the lime-based mortar mix containing finely graded pebble aggregates. Each layer was then tamped until perfectly flat and sealed with red clay, the clay giving the palace its present name, derived from Arabic Qalat al-Hamra, signifying "red fortress." Filling, tamping, and sealing operations produced an unusually dense tâbiyah, while careful leveling provided true horizontal and vertical planes. The same site has revealed techniques that involved spreading earth and lime onto the interior face of forms before casting commenced.[67] Sections taken through the Torre de Homenaje's outer skin graphically illustrate this variant (termed *tapial calicastrado* in recent literature)—showing an inner

Alhambra, Granada, Gate of Justice, detail of facade, Jules Goury and Owen Jones,
Plans Sections and Details of the Alhambra, 1842. (Author's Collection)

tâbiyah core penetrated and shielded by an impermeable (but still "breathable") surface crust of lime mortar.[68]

Despite precautions structural cracking and separation might occur at building angles. Manuscripts prepared for King Alfonso X of Castile (1221–84) illustrate tâbiyah reinforced at its corners with cut stone or brick. Large wall openings required special treatment too. Henri Terrasse wrote, "the Alhambra gates of huge proportions, form deep masses of masonry enclosing vaulted passages with two or even three passageways. . . . They were made of concrete [i.e., tâbiyah] with facade and arches of brick sometimes set off by ceramic decorations," the brick overcoming tâbiyah's lack of tensile strength.[69] Favored by nineteenth-century Orientalist artists, the so-called Gate of Justice (Bab al-Shari'a), dated by inscription to June 1348, underwent considerable patching and restoration over the latter half of the twentieth century. Drawings by Owen Jones (published in 1840) record how brickwork of the horseshoe-shaped arch spanning the exterior entranceway was extended into the tâbiyah "in large stepped sequences (*dentellones*)," these

Tower, el Carpio (Córdoba), detail before restoration showing tâbiyah reinforced with brick at corners and around wall openings.

brick keys matching the height of individual tâbiyah pours, the stuccoed and scored finish imitating ashlar.[70]

Guarding an ancient route along the Guadalquiver linking Córdoba and Jaén, the little *mudéjar* tower of el Carpio (measuring about 36'10" × 54'6" in plan) presents analogies.[71] Generally construction follows conventions already described, the tâbiyah mix containing clay, lime, rounded pebbles, and broken brick and poured into forms thirty-two to thirty-four inches high. Exterior corners received special attention. These were laid up using stone at the lowest level and brick above in alternate long and short patterns corresponding with successive tâbiyah lifts. To equalize stress distribution and maintain accurate leveling, a horizontal brick course was introduced along the length of cast wall sections after every fourth pour. Inside the building large dominical chambers are superimposed one above the other on three floors. Major spaces are vaulted with brick and upper window openings spanned by brick arches. Displaying fully developed structural solutions, el Carpio's fortress is precisely dated, an inscribed alabaster slab built

Alhambra, Granada,
Torre del Homenaje,
detail of exterior showing
construction in tapial
calicostrada.

into an exterior wall (now ex situ) documenting construction in 1363 under "maestre Muhammed" for Garci Mern de Sotamayor "senor de Xdodar," a Muslim master craftsman working here for his Christian overlord.[72] Like many other skilled builders, this Muhammed perpetuated Islamic construction traditions after the *reconquista,* the result being an eclectic, often rich, assemblage of styles.

More than seventy feet high, the Alhambra's Torre del Homenaje is larger in plan (measuring about 40' × 34' overall) and considerably more sophisticated in section. Exterior walls are of *tapial calicastrado* and diminish in thickness with height from 7'6" at base to 2'7" at the uppermost story. Two brick piers rising through five stages both subdivide interior spaces and receive the complex series of brick vaults with which they are enclosed.[73] A sixth-story features an open patio surrounded by four principal rooms intended for domestic habitation, a staircase giving access to the uppermost roof terrace. The two lowest stages probably served as guardrooms separated from living spaces above, which, on occasion, housed deposed royal captives.[74]

Formwork for large military structures involved major expense, especially where wood was scarce. "Molds" could be heavy and unwieldy, making raising and lowering arduous. Formwork was also difficult to fabricate in anything other than rectilinear or polygonal shapes. Consequently tâbiyah defenses can appear primitive, with rounded and semicircular mural towers being uncommon, notwithstanding strategic advantages associated with nonrectilinear features at gates and angles. Still repetition of simple components allowed economic fortification of difficult terrain. Probably built during the eleventh century, a high barrier, some thirteen hundred feet in length, linking the Alcazar of Almería with the neighboring Cerro San Cristóbal, was adjusted to suit extreme slope conditions by varying the length of stepped wall segments while maintaining a common form height. Cast using an exceptionally dense tâbiyah, this feature descends from the citadel's precipitous heights, crosses what to Arab eyes must have resembled an African wadi—where pierced by a single gate—then ascends the opposite hill.[75] Interval towers and paired towers guarding the gate are rectangular and unencumbered with superfluous detail, repetitive merlons constituting the only decorative—yet functional—element.[76]

• Nonrectilinear Plan Shapes in Tâbiyah •

Except for foundations timber wheels raising waters of the Guadalquiver to irrigate gardens of Córdoba's Alcazar *nuovo* have long gone. But an extant section of the enclosure that formerly surrounded this "paradise" has three interval towers (measuring at maximum 22' × 16'9" in plan), each presenting its semicircular face toward the river. Construction is tâbiyah made using forms thirty-three inches high supported by rectangular needles positioned twenty-six to thirty inches on center. Multifaceted shapes rather than circular or rounded ones presented fewer practical problems for larger-scale tâbiyah construction, barbicans and other isolated, or nearly isolated, defensive installations with polygonal plans being more common. Dodecagonal, Seville's Torre del Oro, remains an exceptionally majestic if now-austere Almohad monument despite numerous restorations and its mid-eighteenth-century upper story. Erected in 1220–21 overlooking the Guadalquiver waterfront not far from the ruler's palace (Alcazar), this structure measures about 148 feet in diameter and incorporates three floors.[77] Originally the exterior was clad with gold-glazed faience tiles (*azulejos*), or so the story goes, which, falling away over the course of centuries, left underlying tâbiyah exposed to erosion. Consisting of lime, clay, ceramic fragments, and medium-sized gravel, tâbiyah was cast using forms thirty-seven inches high, the tower's angles being reinforced with carefully cut stone blocks laid in thin beds of mortar.[78] Gradually damaged sections of cast fabric were cut out and replaced with stone. Equally monumental, the Torre de Espantaperros, Badajoz, is octagonal, an old photograph of the then-dilapidated structure showing that lower portions were tâbiyah with perhaps brick and stone at the uppermost stage—construction details currently concealed by renewed stucco facings and modern restoration.[79] Octagonal towers at Caceres (one of dressed stone, another of tâbiyah) and Ecija (of tâbiyah) further underscore the utility of polygonal shapes for fortification.

Seville, Torre del Oro, erected 1220–21, upper story, mid-eighteenth century.

⦿ Finishes ⦿

Most tâbiyah structures have lost their original exterior finishes, or seen them replaced over multiple reconstruction episodes. Many—perhaps the majority—were coated with lime mortar, which, scored, colored, or otherwise enhanced, imitated stone.[80] Less typically several palatine assemblages retain complex interior decorative programs installed over what Oleg Grabar called "hastily and cheaply assembled masonry of mortar and friable brickwork."[81] The Alhambra's Torre de Comares exhibits schemes of unparalleled richness achieved through use of repetitive and standardized finish techniques, its utilitarian construction concealed beneath polychrome skins of applied materials—tile mosaics, marble panels, molded stucco, and timber inartsia. Erected around 1340, about forty feet square, and approximately seventy feet high, this tower—originally tâbiyah, now much repaired and heavily patched with brick—was among the largest of Islamic Andalucía, enclosing one principal chamber (Sala de Embajodres) surmounted by an elaborate, seven-tiered timber dome besides several ornately decorated subsidiary spaces. The product of three or more building phases, the tower gradually lost its primary defensive significance, becoming an audience hall for Granada's later Nasrid rulers, Yūsuf I (1333–54) and Muhammed V (1354–59; 1362–91), whom the historian Ibn Khaldûn served in various administrative capacities. Apart from window openings, exterior facades are blank, with horizontal lines marking casting levels. Metal ties and late brick relieving arches inserted when stabilizing walls riven from top to bottom by earth tremors are flaws distorting the original scheme's smooth plastered exterior facades, which, unless frescoed or color washed, had minimal enhancement.[82]

Inside no greater contrast could be imagined. Brilliantly colored stucco and tile revetments of dazzling virtuosity cover every surface. Purporting to be the voice of the building itself, an inscription opposite the entrance proclaims: "my lord the victorious Yusuf has decorated me [the building] with the robes of glory and excellence without disguise."[83] Describing the same structure, Ibn Zamrak (1333–93), whose panegyric poem celebrating Muhammed V is introduced into the architectural program of this and adjacent apartments, wrote: "with how many a decoration have you clothed it in order to embellish it, one consisting of multicolored figure work which causes the brocades of Yemen to be forgotten."[84] Illusion is complete. Hidden, the Alhambra's utilitarian structure of form-cast material passes unnoticed under thin veneers of applied finishes usually prefabricated or the product of rationalized manufacturing processes. Stucco panels were poured using timber molds, and wall mosaics were made up from small pieces of mass-produced glazed tile cut and assembled into an almost infinite variety of geometric patterns. Painted decoration, predominantly in shades of red, blue, and green, liberally enriched with gold, enhanced marble capitals, timber ceilings, and doors. Besides disguising visually unacceptable wall construction, application of repetitive units and vivid surface embellishment produced splendid yet speedy results, gratifying to royal patrons whose hold on sovereignty was fragile.

Tâbiyah found more humble uses. An illuminated manuscript produced for King Alfonso X illustrates an open belvedere supported on high stone, brick, and tâbiyah foundations. Walls dividing the luscious garden of Seville's Alcazar—where North African and High Renaissance elements owe much of their present character to Charles V—are largely tâbiyah, some being reused remnants of old fortifications. Aqueducts bringing water into the Alhambra and terraces supporting the neighboring Generaliffe's "paradise" garden are also largely fabricated from form-cast materials, frequently renewed over time. Many small rural houses scattered across Andalucía must be similar, stucco and regular applications of whitewash hiding their original fabric.

Alhambra, Granada, Torre de Comares from Court of Myrtles.

Because they were vulnerable to artillery, defensive tâbiyah walls were often faced with masonry from the later fifteenth century onward or, if funds allowed, partially rebuilt. A semicircular gun platform and bastion installed near the Alhambra's Torre del Homenaje in 1589 are of stone raised over cut-down "concrete" (tâbiyah) foundations left over from an earlier tower formerly occupying the site. Decades before (1565), King Philip II reconstructed walls of the adjacent northeast barbican in stone, with Nasrid period work (likely tâbiyah) supporting its lower levels.[85] Some time before 1840, when Goury and Jones published their drawings, the Torre del Homenaje had undergone substantial repair. Weathered or crumbled tâbiyah disfiguring upper levels was patched with stone and the damaged south facade largely remade with coursed masonry.

Nevertheless tâbiyah persisted, finding occasional employment for military construction through the eighteenth century down until at least the 1860s. Lieutenant Colonel Thomas James relates that Gibraltar's defenses protecting the narrow isthmus separating British from Spanish territory were partially rebuilt by Britain in 1730, an "irregular demi (or half bastion) and curtain" being lowered by eight feet and new parapets of tapia "twenty-two feet thick" raised.[86] John Armstrong said this installation was "almost proof against the enemy's shot, which made but little impression upon it, either sticking in the Face of the Wall as some did, or only striking against it, and falling to the Ground."

These attributes were seized on by engineers seeking strength, utility, and economy of means when enhancing, refashioning, or otherwise improving older installations as an odd Victorian-period memorandum confirms. Published in *Papers on Subjects Connected with the Duties of the Corps of Royal Engineers* (Woolwich 1861), describes how describes how British engineers, impressed by the longevity of medieval tâbiyah at Gibraltar, decided to fabricate merlons and "a traverse of the Grand Battery" (measuring 33'10" × 13') using the same material.[87] But Henry C. Jago, the clerk of works, soon discovered knowledge of "tapia" construction had been lost on the Rock, an experiment by local men producing worthless results. Consequently the "Emperor of Morocco" was requested "to send masons to instruct the Engineer's Department in the true composition and manufacture of the material and the method of building." The ruler (probably Mohammed IV b. 'Abd al-Rahmān) obliged, sending two of his best workers, who set about choosing materials, fabricating and filling formwork using methods unchanged since medieval times.[88] Materials found suitable for the tapia mix included dark-red marl "containing oxide of iron, gypsum, volcanic lime sand and some salt as found in the red sand pits of Gibraltar," finely slaked lime, broken stone or loose shingle from the quarries, detritus from destruction of old masonry walls, and refuse from the mortar yard or lime kilns. About twenty-five cubic yards of these diverse materials were mixed at a time with about one thousand gallons of fresh spring water in the following proportions: marl, thirteen cubic yards; shingle, three cubic yards; lime refuse, 2½ cubic yards; and lime, 6½ cubic yards. After "resting" for two days, the tapia mix was rammed into forms consisting of two side shutters each twenty-seven inches high made up from three horizontal timber boards three inches thick "kept together by framework, composed of sleepers, uprights, and caps all of deal 4½" wide by 1¼" thick, mortice and tenoned, placed at a distance of about 2'9" apart." Mindful of the intended professional readership, Jago described the proper positioning of forms and subsequent filling operations in painstaking detail, pertinent paragraphs from his official report reading as follows:

> "In laying the sleepers for the first course of tapia, each sleeper being placed perfectly horizontal, is cased with rough stonework, forming a rough drain, to take the pressure off the tapia of the sill or sleeper, and leaving them free for removal to an upper course.
>
> The formwork having been thoroughly secured, the prepared tapia is thrown in between the planking in layers about nine inches thick, and as many men as can get in the space between the planking (being each provided with a rammer) commence ramming the material. Ordinary field-work ramming is altogether useless. The rammer is raised nearly to the level of the shoulder and brought down with all the force the man is capable of exerting. The ramming is done by word of command, so that all the rammers come down at the same time, the words used by the Moors being *Y Allah, zah!* . . . This is found to be the most laborious part of the work and must be strictly attended to. . . .
>
> After the ramming has been continued some time, the tapia will have been worked into a soft paste; a small quantity of broken stone is then evenly spread over

the surface, and the ramming again proceeded with; this ramming and supplying broken stone is continued until the mass has become tough, and no longer adheres to the rammer; another layer of tapia is then spread, and so on until it has reached the top of the casing or coffer. . . . Compression of the material was about three inches in the original depth thrown in of 9 or 10 inches. . . . Two days should elapse before the frames are removed; an earlier removal may injure the whole work."[89]

Clearly, Moroccan craftsman followed their own traditions when building in Gibraltar, Jago's description corresponding with that of John Windus—minus its disturbing tale of immolation—concerning construction at Meknes that he (Windus) witnessed in 1721. Neither author provided pictorial illustrations. Allowing that under royal patronage or military supervision activities were better organized, a photograph from Rabat probably taken during the early 1900s offers a glimpse of analogous casting activities in progress. Three male figure are seen standing between a set of form boards, each man equipped with a heavy rammer held at shoulder height very much as our informants describe. Formwork is of conventional design, consisting of four horizontal timber boards, supported on exterior faces by timber uprights secured at the top by wooden cross ties, needles ("sleepers" Jago calls them) concealed in this instance by piles of earth, doubtless carrying the entire load. The most striking revelation is that the building technique shown was essentially a folk medium adaptable to a wide variety of programs including installation of simple boundary walls (as apparently photographed in Rabat) or, in Gibraltar, a far more sophisticated piece of military engineering.

To make way for parking and new roadways, Gibraltar's tapia traverse was demolished along with other similar works.[90] Still the record remains impressive, tâbiyah/tapia defenses having protected Gibraltar under Arab, Berber, Spanish, and finally British rule for more than a millennium. And perhaps it is not stretching credulity too far to speculate that military engineers and artificers, becoming familiar with the material here, took this knowledge into South Carolina as well as Georgia—similar patterns of diffusion set in motion by their Hispanic counterparts, carrying formed construction across Latin America.

With the fall of Granada in 1492, the last vestige of Islamic power disappeared from southern Spain. Lands long cultivated by Arabic-speaking populations were colonized by Christian settlers and urban centers forcibly occupied. Mosques were turned into churches, minarets into bell towers, *suqs* into markets, *madrassas* into convents; but the tortuous alleys, narrow streets, and confusing dead ends that characterized Andalucía's Islamic towns survived. And despite Christian domination, the subjugated peoples stubbornly kept their identity, language, and customs. Almost one hundred years later, Spain was still "a maelstrom of competing civilizations whose troubled waters refused to mix."[91] The Spanish crown's recognition of this fact caused mass Moresco deportations between 1609 and 1661, when most "Moors" fled or were forcibly removed from Spain to North Africa.

One group of Andalusian refugees found their way to Rabat (Morocco), where they occupied the sparsely populated quarter established by the Marīnid sultans at the northern end of Ribat-al-Fath. For their own safety, new occupants erected a wall strengthened by twenty-six small towers extending roughly east/west between the old Almohad city circuit (running roughly north/south) and Bou Regreg estuary thereby reducing the urbanized area (medina) to defensible proportions. Segments that collapsed following winter rains in 2016 show that this so-called Andalusian wall was fabricated of tâbiyah cast in forty-inch-high vertical increments. Containing dark-red clay with pebbles but little or no additional lime, the poorly compacted fabric is of indifferent quality and consistent with the poverty of builders who arrived with very few resources. Remarkably it proved strong enough to preserve security after the population took up piracy, exhibiting minimal alteration today except for a circular gun tower (Borg Sidi Machlouf), which the map produced by Richard Simson indicates was added at its southeastern extremity after 1637 when the semiautonomous Republic of Salé and Rabat was blockaded and ultimately occupied by British naval forces under Captain William Rainsborough.[92] Accurate for its time, Simson's map, published in John Dunton's *A True Journal of the Sally Fleet with the Proceedings of the Voyage* (London, 1637), further demonstrates that Rabat's earlier monuments attracted the interest of European raiders, illustrating what he calls a "Roman tower," which in actuality was the great stone minaret of Sultan Ya'qūb's Hassan Mosque. This mosque is also mentioned by Olfert Dapper's *Description de l'Afrique (1686)*, which, except for some wildly exaggerated dimensions, gives a relatively true description of the building, an engraved plate depicting Salé, the Hassan Tower, and New Salé (Rabat) as seen from the sea. Comments concerning construction of the mosque's walls are of particular interest, Dapper, a lifelong resident of the Netherlands, having learned through an unknown informant, these "excessively thick" features were made from what he calls "terre grasse"—a greasy or rich earth and lime (*chaux*).[93] Although the nationality of the informant is uncertain—Dutch and English are both possible—the comment attests that over the seventeenth century, formed construction was becoming known to western Europeans via "Barbary," much as Thomas Spalding suspected.

Across central and southern parts of the New World, knowledge of formed construction was carried by Andalusian migrants including Muslim exiles despite an edict that banned settlement of the latter "because of the troubles with those who have already come."[94] Architectural elements of obvious Islamic inspiration surfaced across South and Central America until the late eighteenth century. Examples include splendid decorative woodwork at San Francisco, Quito (Ecuador); and in Mexico, the monastery tower, which could be mistaken for a minaret, at Actopan (Hidalgo), Cholula's Capilla Real—a mosque seemingly in classical disguise; or most surprising of all, an *alfarje* roof frame of pure *mudéjar* style enclosing the church of San Francisco, Tlaxcala.[95] Whether these building elements were fabricated by immigrants of Christian or Muslim descent is perhaps irrelevant. Following the *reconquista,* building techniques quickly crossed cultural barriers, passing into the wider vernacular building tradition of the Iberian peninsula and far beyond. Braudel noted: "in the sixteenth century Seville, and the Andalusian hinterland, still half-Moslem and hardly half-Christian, were engaged in sending their men to settle whole areas of Spanish America," a process that continued for several centuries as men were exchanged for American silver or gold.[96] Treasure ships bringing improbable piles of bullion coined in places as distant as Potosi, Lima, and Mexico City along with ingots and bars mined at genocidal cost to America's indigenous peoples were unloaded in Seville.

Although it is certain that building materials resembling tâbiyah were transmitted from Spain and Portugal into their New World colonies during early settlement phases, the process is clouded by contemporary writers who used building terms inconsistently, the terms themselves often meaning different things to different authors. The Spanish word *tapia,* derived from Arabic or Berber *tâbiyah,* can indicate rammed earth (alternatively known as *pison*); fabrics containing stone or oyster-shell aggregates (called *tabby, tappy, tapia,* or *tapier* in English sources); even adobe. Semantic difficulties aside, the probability that tapia construction ultimately derived from Iberian prototypes diffused rapidly need not be doubted, the earliest example appearing at La Isabela, Santo Domingo, in 1493 then being used for the residence of no less a personage than Christopher Columbus. Erected on stone foundations, this building's tapia walls consisted of "dark-red clayey sand mixed with lime, gravel and lumps of unfired clay" and were almost two feet wide.[97] Under Spanish rule analogous construction was fabricated around the Caribbean, tapia offering an alternative to stone where protection of European settlers demanded defensive structures. In 1540 the Spanish crown ordered (without apparent result) a fort of tapia at Cartagena, Colombia.[98] Ten years later Trujillo (Honduras) was, through intervention of the Audienca of Guatemala, surrounded by tapia walls. Utilizing the commonest materials, cast materials facilitated cheap and speedy fabrication, important factors considering that Trujillo had been recently sacked and remained threatened, while Cartagena's nearest source of building stone was located six miles away. Subsequently (in 1580) la Yaguana, Hispaniola, petitioned for a tapia strong house, incorporating curtain walls and a central tower.[99]

Central America convents built during the sixteenth century from "tapia con rafas de piedra y ladrillo"[100] were evidently of tamped earth buttressed with brick and stone. Started in 1670 the church of Nuestra Senora de los Remedios, Antigua Guatemala, had similar walls. These still exist, but buttresses mentioned by archival sources were perhaps never executed or have been lost.[101] In the same city, "even so important a civil monument as the Capitanía, the seat of the *audiencia,* had portions of its exterior walls built of tamped earth," these collapsing during the earthquake of 1717.[102] Available funds were never sufficient for complete restitution, the building being patched up time and time again following more tremors, an especially severe event in 1751 causing even greater damage. To save money Diez de Navarro, a military engineer, urged reuse of existing fabric following its reinforcement with "some stone buttresses and courses of brick in order to serve as load-bearing walls for the second storey," these proposals not being put into effect until 1769, when a new arcade of stone was added to the principal facade overlooking Antigua Guatemala's *plaza mayor.*[103] Along with the entire city, the Capitanía was abandoned during 1773, "the most melancholy epoch in the annals of this metropolis," when yet another quake of catastrophic proportions together with several aftershocks left it and most other public buildings (including the nearby cathedral and university) ruined.[104]

Celebrated for depictions of Mayan sites, the English architect Frederick Catherwood (1799–1854) sketched the building as it stood in 1840, his atmospheric illustration with two intermittently active volcanoes, Agua and Fuego, in the background (rising almost fifteen thousand feet, his companion John L. Stevens said), providing an invaluable record of the venerable old palace before later nineteenth-century restoration.

Antigua Guatemala's original grid plan may have been devised about 1541 by Juan Bautista Antonelli, an influential architect and engineer whose career serving the Spanish crown took him to outposts scattered around the Caribbean besides Mexico and Reino de Guatemala, which included much of Central America. Excavation at el Morro, San Juan, Puerto Rico, of work designed by Antonelli and executed in 1589 by Piedro de Salazar demonstrates that military specialists, especially peripatetic ones, were often responsible for tapia's diffusion. Now encased within what Kubler and Soria called "immense masonry carapaces,"[105] el Morro's late sixteenth-century walls were of *mamposteria,* a species of concrete made with lime mortar, pieces of stone, and brick all tamped into timber forms faced with fired brick.[106]

Threatened by Dutch and English privateers, bedeviled by earthquakes and yellow fever, and teeming with "dangerous and bloodthirsty sharks" when Alexander von Humbolt passed this way at the very end of the eighteenth century, La Guaira, port of entry for Caracas, Venezuela, opted for a small tapia fort of irregular shape in 1669.[107] Soon found inadequate, it was replaced by more up-to-date defensive structures of stone. Fortifications built over the seventeenth and eighteenth centuries to defend sea lanes, distant outposts—Isla de Margarita, once famous for its pearl fisheries; Acapulco, far away on the Pacific coast; and Caribbean ports (Havana, Santo Domingo, and Cartagena among the most populated)—further demonstrate that Spanish military engineers valued stone above all material alternatives, notwithstanding the massive financial burden their preference imposed on an often bankrupt or nearly bankrupt Spanish crown.

Earlier Portuguese engineers made similar calculations when establishing trading colonies along North Africa's Atlantic coast to protect the immensely lucrative Lusitanian spice trade with India. In the years 1541–48, Benedetto de Ravenna, an Italian appointed royal engineer by the Emperor Charles V, augmented plans for Mazagan (now El Jadida), a fortified coastal town located sixty miles southwest of Casablanca. Star shaped and measuring about 250 meters × 300 meters overall, the new scheme included four corner bastions, seawalls, moats, and a carefully protected inner harbor, the project exemplifying all the refinements devised by Renaissance military engineers. Construction was of cut stone throughout even though tâbiyah was a ubiquitous and relatively cheap building material used for centuries hereabouts.[108] Florida provides another instance. Hastily constructed, St. Augustine's old timber fort was almost useless by 1586, when discovery of coquina deposits on Anastasia Island prompted consideration of a stone replacement.[109] Not begun until 1672, the new Castillo de San Marcos drained the presidio's slender resources until 1762, British administrators then taking it over still unfinished. Cast materials using mixes containing lime and oyster-shell aggregate (i.e., tabby) were used—but not for walls. Rather the Castillo's internal rooms were given tabby roofs, masonry "bomb proof" vaults subsequently (ca. 1750) replacing these features.

Anastasia's quarries also changed St. Augustine's domestic building. Houses built of tabby predominated during opening decades of the eighteenth century, substitution of oyster shell for coarse aggregate having been made quite early in the presidio's history, though exactly when is debatable. It seems the material was rarely used for dwellings before 1702, when St. Augustine suffered destruction by British forces, who burned and sacked the place. Rebuilding proceeded slowly, tabby not coming into its own until the 1730s for those settlers who could not afford coquina. Manucy calculated that 132 houses with tabby walls stood in 1764, this representing 39 percent of the total housing stock.[110] In 1765 William Bartram observed that "most of the Spanish houses was built of oyster shells and mortar, as well as garden and yard walls. They raised them by setting two boards on edge as wide as they intend the wall, then poured in lime-shell mortar mixed with sand, in which they pounded oyster shells as close as possible. And when that part was set, they raised the planks, and so on until they had raised the wall as high as wanted."[111]

Taking Bartram as his guide, Albert Manucy offers a reconstruction drawing of the technique. Two parallel form boards set some distance apart are shown in the process of being filled with a lime mortar and shell mix. Circular dowels secured with pins keep opposing faces of the form boards in place at the lower level, simple wooden cross ties tenoned into the top of the boards further ensuring that everything stayed secure during filling operations. While Manucy's drawing is convincing, it should be noted the original document makes no mention of dowels, pins, or cross ties, much later sources—notably Thomas Spalding's 1830 account of tabby—having apparently been extrapolated for details Bartram neglected to mention.[112] The latter's 1766 diary does confirm the practice of burning oyster shell for lime. Traveling near St. Augustine, he found Wood Cutter's Creek "very full of oyster banks" and observed that "ye people comes and rakes up what they please brings them in a boat heaves them on shore to dry after which they burn them to lime which is very sharp and much more so than ye lime made of ould shells of which here is prodigious quantities in heaps on the shore heaved either by ye Indians or Spaniards or both."[113] In St. Augustine itself, tabby fabrics cracked and warped over time, causing demolition of more than one hundred houses after 1764, coquina-built dwellings outnumbering tabby houses by approximately ten to one in 1788.[114]

Seventeenth century Guatemalan sources document analogous changes, tapia gradually being abandoned for stone when erecting public buildings. Nevertheless, among areas isolated or frequently hit by earthquakes all across Central and northern South America, use of tapia for vernacular-style houses, outbuildings, and walls survived until displaced by concrete block during the later twentieth century. The small mountain town of Cotacochi, Imbabura Province, Ecuador, is typical, preserving numerous structures of both rammed earth and adobe, the two materials being used together or separately. In central Mexico traditional houses of uncertain date at Calpan (Puebla)—which McAndrew called "a pretty village set among walnut groves at the foot of Popocatepetl"—have external walls built of tapial consisting of yellow clay rammed into timber forms, internal partitions being of adobe.[115] Recent investigators have recorded that 52 percent of about twenty-five hundred houses occupied here in 1995 exhibited similar construction, this structural system occurring, it is said, across an area stretching though the states of Puebla, Tlaxcala, and Vera Cruz.[116]

Along the southeastern seaboard of British colonial America, variant form-casting techniques never gained more than local distribution, nor did they achieve predominance over alternative building modes, again displaying cyclical patterns of usage, first appearing during early settlement periods and gradually losing ground as building skills, industrial production, or distribution of manufactured building components improved. And despite obvious parallels—standardization of formwork height, reinforcement using stone or brick at building corners, and the scarcity of nonrectilinear plans—clear differences exist between tâbiyah, tapia, and tabby construction, driven by environmental conditions, material availability, and local experience. For instance southeastern tabby structures give no detectable evidence that formwork was moved horizontally and filled segmentally along the perimeter of any given building—usually the case with Afro-Iberian tâbiyah. Rather, Spalding observed, timber plank from the Southeast's tall—sometimes giant—forest trees allowed construction of "boxes" going "all around" the intended building, "including even the partitions."[117] Neither does Sea Island tabby give much evidence in the shape of broken- or smashed-oyster-shell aggregates for the painstaking process of repeated ramming, compaction, and sealing layer by layer, which characterizes tâbiyah found across Andalucía. Nor do I know of any North American parallel for the Iberian practice of throwing lime onto the interior face of forms to give walls an economical and lasting protective skin. These and less significant differences aside, it is those shared elements of New and Old World form building that exhibit continuities spanning time, distance, and culture besides telling the incredible history of early modern human migrations whether voluntary or otherwise.

Making its first local appearance in 1733, tabby achieved an apogee of popularity across Beaufort County, South Carolina, between 1780 and 1840, becoming less widespread during the late antebellum period, when new brickyards came into production and steamship service commenced, linking Charleston, Beaufort, and Savannah. Elsewhere cycles of use and disuse played out at different times. In Charleston tabby is documented before 1720, apparently being relegated to minor works by the turn of the eighteenth century. Military building was different, tabby occurring among defenses erected along the Southeastern Seaboard from the early colonial period onward, until Portland cement became widely available near the end of the nineteenth century. Tabby's usage was dictated by the material's relative cheapness, contractors choosing it over stone and brick to offset perpetual financial shortfalls forced on them by difficult sites, unrealistic budgets, bureaucratic delays, and never-ending needs for repairs and replacement. Larger than any tapia structures known from St. Augustine—which suggests that their inspiration came from elsewhere—mid-eighteenth- and early nineteenth-century fortifications of coastal South and North Carolina furnish the best, if now largely effaced, picture of tabby's local evolution.

2

Tabby in Military Building of the Southeast Atlantic Coast

In January 1726 the South Carolina Commons House of Assembly determined that "a good substantial Fort be erected at Beaufort or at the most convenient place about the harbour & to be built with lime shells & sand mixed together as is proposed the new Battery at White Point [Charleston] should be." An attached report gave projections for White Point's labor and material costs, which came to £4,625 in total, including carpenter's work and "overseers to the Negroes."[1] Who furnished these figures—the first of their kind known from the Carolina lowcountry—is not stated although it might be surmised that the irascible (and powerful) Captain William Rhett, appointed a commissioner for fortification in 1707, had say in their formulation. Where he or any other proponent gained sufficient experience to develop so detailed an estimate is unclear, Commons House journals offering no clue about the pathways by which tabby diffused into British North America.

Down until the Revolution, legislators were much occupied with matters concerning defense and fortification pricing, labor allocations, payment schedules, organization, and supervision of contracts all demanding scrutiny and debate. Successive governors entered into deliberations, proffering advice, weighing merits of competing schemes, sometimes demanding action in the name of a distant sovereign. For the period 1755–70, the *Journal of the Commissioners of Fortification for the Province of South Carolina* complements legislative records since it transcribes progress reports, letters, payment authorizations, and other contractual matters pertaining to defensive structures.

After the American Revolution, state records are supplanted by federal files documenting new fortification systems erected along southeastern coasts of the United States after Congress assumed responsibility for its defense in 1794. Implementation shifted to the incipient Corps of Engineers in 1802, Charleston, South Carolina, becoming headquarters for their southern district after 1808. Archived information documenting the corps' activities varies between detailed summaries of work laboriously prepared for government departments and rough notes written by junior officers traveling between construction sites, advising, supervising, sometimes cajoling, always observing. Physically adverse situations—coastal swamps, exposed barrier islands, fever-ridden camps—coupled with the vagaries of suppliers, contract labor, superior officers, and politicians might present near insuperable difficulties. Yet written and graphic evidence indicates that despite frustration, sickness, even death, these problems were overcome by the engineers, tabby being used on a scale greater than obvious from extant military structures now almost effaced by time.

The spontaneity and freshness of manuscript records, scattered and not easily retrieved, is reproduced insofar as my sometimes imperfect copies allow. For Fort Lyttelton—currently all but lost above ground—unpublished archaeological findings generously provided by Dr. L. Lepionka are quoted. Focus is centered on coastal defenses rather than inland ones. Beyond information concerning materials and prices, construction at Fort Dorchester is omitted. Fort Johnson, James Island, is also bypassed since its development was closely linked with defense of Charleston Harbor and the City of Charleston, topics deserving more detailed treatment than attempted here.

◦ Fort Frederick, 1733–55 ◦

Events prompting the building of Fort Frederick—named after Frederick, Prince of Wales—are chronicled by Verner Crane, who described international rivalries played out across the immense wilderness dividing Hispanic Florida and British South Carolina over the first thirty years of the eighteenth century. Among many threats desertion of Fort King George on the Altamaha in 1727 by its garrison of "sottish" Port Royal scouts, who, Governor Nicholson said, were "too lazy and mutinous to fetch good water or make gardens to provide wholesome food," was the immediate cause demanding new fortification.[2] With this outpost lost, almost nothing stood between Beaufort and Spain's military forces or belligerent Indian surrogates except for the town fort, which, perpetually out of repair, had been patched up in 1721 and again in 1724. We know nothing about this building unless John Gascoigne's *Plan of Port Royal in South Carolina* (London, 1729), depicting a square fortified enclosure with corner bastions, gives an actual rather than conventional representation, since William Bellinger never delivered plans showing the fort and "corse of the River on the Front" that was promised to Governor Nicholson in August 1723, his survey crew finding the "weather still hot and Snakes not yett gone"—conditions as trying now as they were then.[3]

Nor is it known where the fort stood, Gascoigne locating it within the boundaries of Beaufort Town overlooking the Beaufort River perhaps near Church Street's south end.[4] By 1726 the structure had fallen into complete disrepair and was deemed incapable of defending either Beaufort or its outlying plantations. Gathered "to consider what is absolutely necessary to be im'ediately done to put this Province in posture of defense," a Commons committee found "the Enemy will be encouraged to invade our Southwards settlements which: as they lye much exposed & as the Harbour of Port Royall is capable to receive many and large Ships of Warr they may therefore be induced to make their attacks first there." Rumors of Caribbean fleets assembling heightened tensions, erection of a new tabby fort replacing Beaufort's old timber one being recommended for an estimated cost of £1,500.[5]

Means were lacking, the great hurricane of 1728 having shattered Charleston's already-decayed military installations, overwhelmed residents, houses, docks, and wharves. Only in 1730 was £5,600 appropriated for building Port Royal's new fort and barracks and another fort on the Altamaha (Georgia). Alarms notwithstanding, Fort Frederick's subsequent construction proceeded slowly. Located overlooking the Beaufort River about 1½ miles south of Beaufort Town, the site allowed observation of shipping, hostile or otherwise, entering home waters from the Atlantic Ocean by way of Port Royal Sound. But time would eventually tell there was little or no natural cover here, and nothing could be done about tidal erosion gradually eating away the site's east (river) side. Moreover inadequate funding, miscalculations, or misunderstandings caused long-drawn-out contractual delays.

Work probably began with the barracks, James Oglethorpe lodging many of Georgia's first settlers in the newly completed building following their journey across the Atlantic in January 1733.[6] One of the company, Thomas Causton, saw the barracks but made no mention of ongoing work involving gathering shell, burning lime, or casting the fort's perimeter walls.[7] Nevertheless the two principal contractors, Jacob Bond and John DelaBere, made considerable progress over the next twelve months, though not enough to satisfy Robert Brewton "from the Committee on Petitions and Accounts," who on January 24, 1734, reported:

> That having examined the Journals of the late Assembly, relating to the Fort & Barracks at Port Royal, find that the House was apprised that the said fort consisted of four Lines & two Bastions: and it appears to your Comm'ee by examining Messrs Bond and DelaBere on Oath, that the Fort and Barracks are completed to the Plan (except the omission of two Hawkers) and that the Wall is five foot high and five feet thick at the top.
>
> Your Comm'ee having objected to Messrs Bond and DelaBere that the Platforms are not made, receiv'd for answer that they never understood it to be in their Agreement.[8]

Consequently the Commons ordered £1,600 paid out of £2,000 raised toward construction, retaining £400 until the contractors "finished the Platforms of said Fort with good & substantial materials with two Inch Pitch Pine Plank and Cedar or light wood Sleepers which Platform is to be laid in the Bastions covering them thro'out and to be laid in the Curtain Line for sufficient traversing all the Guns to be mounted on the said line."[9] When—or indeed if—the platforms were ever completed is uncertain. A report dated June 16, 1738, observed that the fort then mounted four small ship guns even though eighteen cannon and six ships guns had been sent some time earlier.[10]

๑ Plan and Organization ๑

Fort Frederick's original contract documents—plans, profiles, and specifications—are lost.[11] The best verbal description is supplied by Robert Brewton's report, already cited. Apparently reliable it cannot be fully confirmed because, eroded and undermined by the

Beaufort River, the fort's eastern half has mostly washed away. Barracks have disappeared along with a brick magazine for which the Commons House voted (March 4, 1736/37) monies "not to exceed $100." Destructive processes were already far advanced in 1864 although less advanced than now. A survey of Smith's plantation by the U.S. Department of Engineers shows the fort measured approximately 130' north/south × 128' east/west overall, excluding its two bastions. The southeast angle (now lost) was then visible below high-water mark. No bastion is indicated at this point, which supports Brewton's observation that the river bastion was located at the northeastern angle. This places it diagonally opposite the diamond-shaped southwest land bastion still extant. Although damaged the fort's west (landward) line and northwest corner also stand. North and south walls are truncated while the east (river) side is reduced to jagged tabby fragments visible along the shore at low tide only. Taken with the 1864 survey, these fragments indicate that like its southwestern counterpart, the northeast bastion was sharply angled, Beaufort's new fort thus exemplifying the type of small defensive structure first developed during the 1520s, when the Sangallos in Rome and Florence, Peruzzi in Sienna, and Antonelli brothers across the Caribbean region revolutionized military architecture—their projects demonstrating that bastions of "angled, triangular, or pointed shapes . . . were the only forms capable of eliminating the blind spots . . . always to be found in squared or rounded works."[12]

It is possible that reports from escaped prisoners or pirated drawings of St. Augustine's Castillo de San Marcos (started during the 1670s and barely competed by 1700) informed the Beaufort work though there were significant differences between the two buildings. A typical "star" fort, the Florida structure, like the Castillo de La Fuerza, Havana (started 1558), was square with angled bastions at each corner. In Beaufort, as we have seen, the fort was almost square, but only two corner bastions were built—an economical if strategically questionable solution. Similarly the choice of tabby, if less durable, was much less expensive than the stone used in St. Augustine, where construction costs reached over 138,000 pesos by 1695—almost twice the original cost estimate.[13] Size and scale present greater distinctions. Guarding Spanish Florida's principal Spanish outpost, the Castillo de San Marcos—with its coquina exterior walls standing twenty feet high and carefully considered entrance and interior organization—was far more substantial and sophisticated than Port Royal Island's little fort.

How high Fort Frederick's exterior walls originally stood cannot be accurately determined, better preserved sections standing only 4'6" above present grade. Embrasures, or what is left of them, show one "round" of tabby at the top level must be missing. If the same height as lower-level pours (fifteen inches), this would indicate the exterior elevation was about six feet—that is, about twelve inches higher than Brewton's figure. Actual wall thickness correlates more closely with the latter's description, measuring 5'6" just above the presumed penultimate pour level, all walls being slightly battered along external faces. Abraham Reese, in his *Cyclopaedia* (1743) description of pisé, explained how the sloped outer profile was probably obtained, describing how "the head of the mould diminishes gradually to the top, in order that the wall may be made to diminish to the same degree."

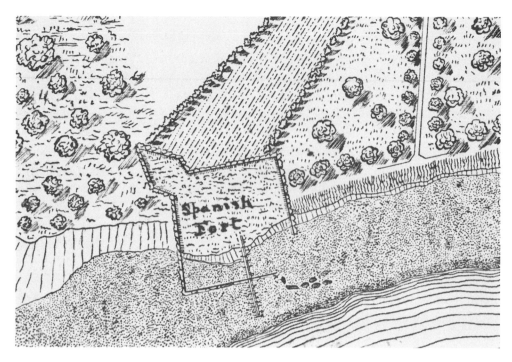

Survey of Smith's plantation, May 1864. Detail showing Fort Frederick,
here called "Spanish Fort" (U.S. National Archives, RG 77 I-33).

Fort Frederick, Port Royal vicinity, gunport.

Interior organization is wholly uncertain. A single opening pierces both north and south lines. Both features are reworked, making it questionable if either is entirely original. All gun platforms (assuming these were installed) have gone along with their supporting sleeper walls. Remnants of six embrasures still pierce walls of the southwest bastion. Typically these measure 7'9" in width on outer wall faces tapering to 3'4" on the building's interior. Nothing is known about the barracks, probably robbed or dismantled for the sake of its materials. Brick scatters encountered during excavation (2015) may represent the magazine, but whether this structure was actually realized can only be confirmed by further investigation. Fort Frederick remained in service from 1734 or 1735 until abandoned during the 1750s, the garrison's strength increasing when expeditions southward were mounted or alarms sounded about enemy intentions, and decreasing to a few men when perceived dangers had passed.

⟡ Structural Failure and Replacement ⟡

Where Jacob Bond and John DelaBere acquired skills necessary for Fort Frederick's execution is not known. There is no suggestion either man had previously undertaken large-scale construction, or even possessed prior knowledge of tabby making. Both were planters rather than builders with ambitions for political life.[14] Moreover Captain William Rhett—probably the project's prime mover—died before construction began. It is definite that within six years of completion, the building was found wanting, portions having decayed and become useless, the garrison "in most miserable and forlorn condition being intirely out of order."[15] As for the defenses, their condition gave cause for serious alarm. Legislators were told: "the new Work (which is esteemed the best Battery and of the most defence) is almost gone to ruin, and all down on the land side. The Repair of which is certainly necessary as the chieftest Part of the cannon are mounted therein."[16] Other faults had surfaced. The magazine proved "leaky," badly ventilated and unfit for storage, so much so that powder was removed to the upper part of one of the barracks, which—not surprisingly—was then deemed dangerous. The barracks themselves "would be more durable were they rough plaistered, and are daily decaying for Want of it. The Weather Boards being leaky and very uncomfortable in Driving Rains, and inconvenient for the sick and ailing Soldiers."

Fundamental design deficiencies existed too, the Commons hearing: "the Fort in general is so very low all round that there is no cover or Shelter for the men in Time of Action, who would be too much exposed to the Fire of the Enemy, if it should be ever attacked: and must be raised if intended for Service, at least three feet on the land side, and four feet fronting the River; and aught to be done after the same Manner as Broughton's Battery is, with Plank framed in. And filled up with Mud or dirt." And, incredibly, no more than one-half pound of musket shot could be mustered by the entire garrison.[17]

Two years later little or nothing had been done even though officials were receiving accurate information about enemy squadrons gathering off the Florida coast. Fortunately for Port Royal's settlers and those few adventurers inhabiting adjacent islands, men commanded by James Oglethorpe beat off the long-anticipated attack when Spanish forces landed on St. Simons Island in July 1742. Thoroughly shaken by events, Lieutenant Governor William Bull sent the Duke of Newcastle a "representation" intended for the king's eyes that described Port Royal's defenseless condition.[18] The danger so narrowly escaped was also grasped by South Carolina's chief justice, Benjamin Walker, who said (October 15, 1742) that the Spaniards were "to have landed 2,000 men at Port Royal and built a strong fortification there; and as soon as they had possession to have sent to the Havannah for ships."[19]

● Projected Replacement ●

Speaking before the Commons on September 15, 1742, Lieutenant Governor William Bull pointed out the urgent necessity of either repairing or replacing Fort Frederick. Two days later the House ordered two galleys placed on station at Beaufort but, distracted with elections, made no reply concerning Fort Frederick until December 4, when it was reported the ruinous fort was incapable of repair. Colonel Wigg agreed. Having "got the plank ready for the platforms at Fort Frederick," he wrote from Beaufort, "in the opinion of most people there, it would be to no Purpose to lay the same unless the Fort was repaired, and that upon a Survey there was a great deal of Rubbish to be cleared away before they could be laid, because a great part of the Wall was tumbled in, that the Wall whereon most of the Cannon were mounted is entirely down and the whole fort is not worth the Money it would cost to finish the Platforms."[20]

The conclusion was inescapable—Fort Frederick was a costly failure and merited no further public expense. Yet solutions posed by its premature ruin did not emerge, the Commons House alternatively hearing opinions concerning a new timber replacement and estimates for making good existing deficiencies. Discussion continued without resolution. In May 1749 the garrison itself repaired the barracks, Lieutenant George Daniel petitioning for reimbursement of £111.10.0 already paid out. Almost one year later, Governor Glenn brought before the House the case of Sergeant Hall, "gentleman" laid low by unspecified circumstances, who Glenn appointed gunner several years before (exact date not given).[21] Hall served about two years without wages before dying in office. Now his widow begged relief stating that £100 "would satisfy her."[22] This request was refused. Glenn himself then gave Widow Hall £65 and another £35 for her return passage to England, a payment not reimbursed until February 1754.

Incapable of authorizing either substantial repairs or any feasible replacement, the Commons referred the entire issue to Governor James Glenn. On May 5, 1752, irritated over what he saw as repeated importunities concerning his own inaction, Glenn issued a blistering reprimand:

> I am most ashamed to answer that part of your message relating to Fort Frederick, or even give it the name of Fort. It is in judicially situated, ill constructed and is a low wall of oyster shells which a man may leap over! And this called a fort . . . a garden fence is full as good a security. Nay it is really worse than nothing. For the name of Fort may decoy people to retire to it in case of danger which will

undoubtably prove destructive to every one who does so, whereas by betaking themselves to their boats or to the woods they may have a chance to escape.

Fort and fortifications, Batteries and Bastions, Ramparts and Ravelings sound well: but if they are empty sounds, they will signify little. Let us therefore not amuse our selves with words, but less take the opinion of persons of experience which of them are good and will prove a real defense in the day of danger: and let such be preserved: but let us not spend our money for what will not profit.[23]

And there the matter rested until March 4, 1757, when, following outbreak of war between Britain and France, local residents reminded legislators how exposed they were to sack, pillage, or worse. The ruinous condition of Fort Frederick and its ill-chosen situation were described again and comments made about the inadequate garrison, which, residents said, comprised "only a Corporal & a few men." Fearing so "enterprizing an Enemy as the French who certainly look with the Eye of Jealousy on the prosperity of this Province," they observed, "a good Fort, upon a proper Situation, of which there are many on both sides of Port-Royal River, would, 'tis probable, discourage any such attempt; or, if made, might render it Unsuccessful."[24]

Wearied perhaps by years of procrastination and indecision, the Assembly granted "£10,000 for erection of new defensive works at Port Royal" on March 10, 1757. No time was lost in securing plans and estimates for a new fort, named Fort Lyttelton, to be built on Spanish Point, slightly north of the old one. This site offered better visibility toward Beaufort Town but, like Fort Frederick, afforded no protection whatsoever to plantations located along the Beaufort or Broad Rivers where indigo production and cattle raising had become important. Perhaps these were considered unlikely targets or not worth defending. Whatever the case construction began on the new site during October of the same year. In November 1757 the Commission of Fortifications instructed "Mr. Rattray and Mr. Crawford to wait on his Excellency the Governor in consequence of a letter which they received of the Superintendent for building the Fort of Port Royal to request His Excellency leave to remove the materials of the old Barracks at Fort Frederick to Fort Lyttelton to be used there."[25] Presumably the superintendent wanted timbers, iron hardware, and whatever else remained serviceable from the old building. Whether leave was given is not recorded, though quantities of broken brick discovered during preliminary excavation indicate that, authorized or not, scavenging took place inside the fort's perimeter.

Despoiled, Fort Frederick gradually fell into complete decay, erosion along one side continuing as the Beaufort River meandered westward. Lands surrounding the fort and the fort itself passed into private ownership, arable areas subsequently being put under cotton. The site saw no more military use until late 1861 or early 1862, when what had become Old Fort Plantation was occupied by Union forces, the First South Carolina Regiment of Volunteers naming their encampment Camp Saxton after General Rufus Saxton, briefly styled "Governor of the Sea Islands." Viewing the camp on December 1, 1862, their commanding officer, Colonel Thomas Wentworth Higginson, saw "a picturesque point, an old plantation with decaying avenues and house & little church amid the woods like

Virginia & behind a broad encampment of white tents." On December 1, 1862, he visited Fort Frederick and an unidentified tabby structure (probably Fort Lyttelton). His journal records: "Had a ride through the plantation to a strange old fort of which there are two here. . . . They are built of a curious combination of oyster shells & cement, called Lupia [*sic*] & are still hard and square, save where waterworn. One is before this house & a mere redoubt; the other two miles off is a high square house, bored with holes for musquetry & the walls still firm, though a cannon-ball would probably crush them."[26]

Near-contemporary photographs show vessels tied up to a pier centered on old Fort Frederick, which had became a staging area for military supplies and perhaps humanitarian aid for "contrabands" now crowded in its vicinity.

• Fort Frederick's Structural Failure •

Apparently no formal inquiry was made into the cause or causes of Fort Frederick's premature and potentially catastrophic structural failure. Thus it cannot be said whether specifications were considered faulty, workmanship defective, or the contractors otherwise in error. Today few significant cracks are visible around the surviving southeast bastion, nor are any substantial settlement evident along land walls. Rather the remains suggest that tabby separated along the horizontal pour line located just above the lowest embrasure level. Damage might indicate faulty casting, mortar mixes contaminated by salt, or improper curing. More plausible, considering walls failed "whereon most of the Cannon were mounted," is that Fort Frederick, like Fort Johnston, North Carolina (ruined before 1766), shattered under stress generated by its own guns fired daily in salute of passing ships, despite inadequate supplies of powder.[27] If so, tabby's vulnerability to sudden imposed load—whether the result of seismic disturbance, hurricane-force winds, or cannon recoil—was not fully understood by contemporary builders, subsequent fabrication of large tabby structures developing by trial and error.

• A Legacy in Georgia? •

Whatever its structural and strategic shortcomings, Fort Frederick inspired tabby defensive works erected by Oglethorpe's officers at Frederica, St. Simons Island, Georgia, a garrison town founded in 1736 to protect British America's tenuous southern frontier against aggressive moves by Spanish forces moving out of Florida or Cuba. A manuscript, "Map of the Islands of St. Simons and Jekyll," dated 1740, gives plans and profiles designed by John Thomas, "Engineer," of a fort accommodating three hundred men flanked some distance eastward by a redoubt and battery made according to captions: "with oyster shells (which are in great abundance) and mortar Ram'd together in a Case which becomes hard as stone when it is dry."[28] "A pretty strong fort of Tappy" and "an extremely well contriv'd building in the form of a square of Tappy work" serving as barracks erected near the town's north end were further described in 1743, the *London Magazine* informing its readers "some houses [at Frederica] are built entirely of brick. Some of brick and wood, some few of Tappy work but most of the meaner sort of wood only."[29] Three years later

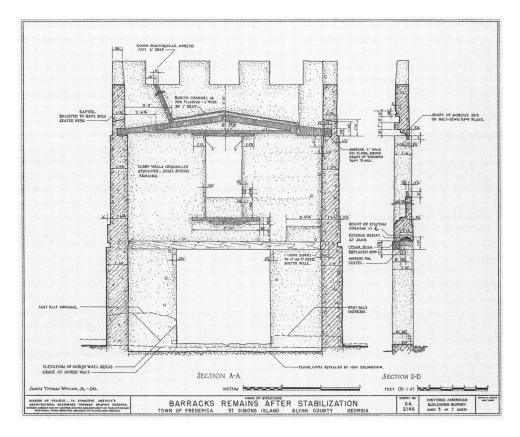

Frederica, St. Simons Island, Ga., barracks,
measured drawing of tabby tower (HABS).

William Thomson and John Lawrence Jr. said the barracks were "ninety feet square, built of tappy, covered with cypress Shingles, and a handsome Tower over the Gateway of twenty-Feet." Eroded and broken, segments of this tower still stand, excavation establishing adjacent tabby walls (mostly lost above ground) were twelve inches thick.[30]

Construction work at Frederica was overseen by Captain (later Major) William Horton, who knew Fort Frederick and Spanish St. Augustine. On mission to the latter in 1735 he was arrested for espionage, the authorities saying he took plans of the town and its castle (Castillo de San Marcos), a claim that was almost certainly true. Threatened with "being sent to the mines" (location not specified), Horton was released after a few days imprisonment.[31] Subsequently he built his own Jekyll Island house and numerous outbuildings of tabby. Elements still standing are described below (see chapter 7). Frederica's domestic architecture has all but disappeared, excavated basements preserving only the footprint of representative structures. The tabby house where Thomas Spalding was born, built, he said, by "my maternal ancestor [who] was an officer of General Oglethorpe," survived the fire that destroyed Frederica in 1758, but its walls were later "sawn up and carried away."[32]

Nothing has been discovered near Beaufort resembling the fortified house called Wormslow erected by Noble Jones on the Isle of Hope, ten miles southeast of Savannah, a building guarded by an eight-foot-high walled enclosure (measuring about 70' × 80') with four corner bastions all constructed in tabby.[33] Kelso has suggested that construction dates from the early 1740s and is probably identical with the bastioned structure shown by DeBrahm's map dated 1752.[34] Other fortified houses of the period, including an example reported from Amelia Island, could mean similar buildings were erected by Sea Island planters exposed to raids by foreign powers and their tribal surrogates during the first half of the eighteenth century. If so, along with most prerevolutionary tabby structures, such fortified outposts have disappeared without record.

⁕ Fort Lyttelton, Port Royal Island, 1757–79 ⁕

Plans for Fort Frederick's replacement were prepared by Lieutenant Emanuel Hesse, a military engineer sent into South Carolina by the commander in chief of British forces in America, John Campbell, Earl of Loudoun. Lieutenant Hesse produced more than one scheme, an undated sketch from the Huntington Collection showing an irregular bastioned affair chamfered toward the Beaufort River rather than the triangular scheme featuring one bastion and two half bastions actually realized.[35] Whether "tappy" work was envisaged from the beginning is not known; neither can we say whether the Commission of Fortifications had reservations about the material's suitability. Whatever transpired out of public view, commissioners were confident enough or cajoled enough to record the following on August 25, 1757:[36]

> Mr. Hesse the engineer laid before the board a plan of a fort to be built on Spanish Point on Port Royal River. The Commissioners made the following estimate for cost of same:

40,000 [cubic feet] of Solid Tappy Work at £2.16.3 the price which Tappy Work cost in Charles Town	£5,125.11.2
65,000 Cubical feet of Earth for the Parapet and Banquette supposing Laborers at 7/6 per day and Overseers wages	.£1,000
6000 cubic feet of Tappy Work in the wall of the Barracks and Provision Magazine	£768.16.3
Carpenters Work & materials for the Barracks. Provisions Magazine, Gates & Bridges, and finishing the Officers Apartments and Platforms	£2,105.12.2
CARRIED FORWARD	£9,000.
10 double Chimneys	500
A powder magazine 12 feet square	£500
	£10,000

Colonel Thomas Wigg, Colonel Nathaniel Barnwell, and George Roupell—all counted among Beaufort's leading citizens, were appointed superintendants.[37] Thomas Gordon, another respected member of the town's mercantile community with commercial contacts extending as far afield as the West Indies, probably had charge of day-to-day activities. Francis Stuart, an indispensable source of credit for local planters, handled financial transactions in Charleston.

Work was well under way by late October 1757. A progress report dated "Beaufort 2 December, 1757" records, "we had nigh fifty Negroes constantly at Work since the latter end of October besides the Tappy Maker and Overseers, and a parcel of Boats and Schooners that have already delivered about thirty thousand Bushells of Shells. . . . The Foundation of the inner Tappy Wall is laid quite round. The Tappy Work is going on & the vessels daily bringing Shells."[38] Up to this date "only £450 of the £1000 brought us by Mr. Stuart" had been spent, but expenditure was climbing. "We shall be obliged to disburse pretty largely on the last Saturday of December," the supervisors informed commissioners in Charleston. Soon thereafter George Roupell traveled with a request for an additional £3,000. This was granted, but by the end of January 1758, only £700 of the latest draw was left. The commissioners therefore dispersed another $3,000, Nathaniel Barnwell and his colleagues reporting that "the Inner Wall of the Fort and Barracks" were "pretty far advanced, and the Boats unloading this day will make up the quantity of Shells delivered Seventy thousand Bushels."[39]

Not much is heard of Fort Lyttelton until July of the same year, (1758) when John Gordon received yet another installment of $3,000 from the state treasurer. By September 6, 1758, supervisors were able to announce: "the Tappy work of the three Water Lines of Fort Lyttelton where the Cannon are to be mounted is advanced. . . . All the rest of the Tappy Work of the Fort & Barracks is finished."[40] A visit by the governor advanced things further although some uncertainly still existed about "what kind of carriages are intended or rather at what height the embrasures are to be from the platforms."[41] The commissioners supplied construction details: "it is our Opinion the fittest Carriages will be the high Wheeled Field Carriages as they will admitt of a higher wall under the Gun . . . and better Cover for the Gunners, besides there being much easier managed than the Truck Carriage."[42]

On November 16, 1758, the superintendents submitted the following return for works done at Fort Lyttelton, which is the most detailed account known for any mid-eighteenth-century tabby building project in South Carolina:

Return of Works done at Fort Lyttelton[43]

A Row of double Barracks for Soldiers Measuring from out to out 136 feet in length, 30 feet 6 inches wide and 9 feet high from the Foundation to Wall Plate of Tappy, in Number [apartments] 16ft by 14 in the clear, with an Angular Chimney in each. The Floors and the middle and cross partitions also of tappy. Flooring Joice laid over head.[44] A Roof to Span the whole width (with a Dormer Window in it over each apartment) which if floor'd would afford good accommodations for a large number of people.—Window in each middle Partition. Door & Windows Shutters to the whole fitted and Hung.

A Tappy Wing to each of the Soldiers Barracks 26½ ft long and 21½ ft wide divided into Four apartments for Officers, Floor'd, Lathed and Plaistered. A Chimney & Closet in each. Floors laid over head in the Garrats. Doors fitted & Hung etc.

The Fort Walls of Tappy carried up Viz.:

From the Salient Angle of S demi-Bastion to the Salient Angle of the North demi-Bastion—the Outer Wall toward the land side 682 feet in length is 10 feet high from the Foundation, 2 ft thick at the Foundation and 1 ft 6 In thick o'top.

The inner wall from Angle to Angle towards the same side 618 ft. long is 8 ft. high from the Foundation, 1ft. 6 in. thick at the foundation 1 ft. 2 in. thick o'top.

From said north Angle to the So. Angle—the Outer wall towards the Water side of the three Lines where the Cannon are to be mounted 502 feet long is 8 ft. high from the Foundation, 3 ft 6 In.: thick at bottom and 2 ft 9 In.: thick o'top.

The inner Wall towards the same side 447 ft 6 In.: long is 6 ft. high from the foundation 2 feet thick at bottom and 1 foot 8 inches o'top. . . .

Note. The Outer Tappy Wall of the Fort is five feet lower than the surface of the earth and the inner Tappy wall is 2ft lower & the distance between wall and wall is 9 feet.

The earth is fill'd & Ramm'd between the Tappy Walls all round to within 2 ft of the present height of the Walls.

The Magazine 24 ft by 16 from out to out raised with Tappy 6 feet high from the Foundation, the Walls 4 feet thick with 2 Cedar door frames set.

A Well 12 feet deep 10 ft. square lined with Frame & 2 In. Plank which furnish'd water for the People and all the Tappy Work.

A small Barrack of wood within the fort 30 feet by 17 with a chimney for the use of the workers and overseers.

A small Barrack of Wood without the Fort 30 ft by 17 for the Laborers and Tools.

Thirteen Cannon brought from the old Fort and two from the Seaside to Fort Lyttelton & mounted on Carriages (sent from Charles Town at present without the fort).

Materials remaining on the Spott:

About two Thousand Bushels of Lime, two thousand Bushels of Shells. Thirty cord of Fire Wood, a parcel of 2 In. pine plank for platforms, Tappy Boxes, Needels and Standards, Wedges, Spades, Wheelbarrows, Hand Barrows etc.

Supervisors now suggested several improvements saying:

It is to be observed:

That there is none of the Earth as yet removed from before the three front Lines & but little out of the space intended for a Ditch. The Earth dug from the Foundation serving to fill up between the Walls to the present height.

That there is a Ridge behind the Fortification at a small distance from it which renders it necessary to raise the Walls about 18 Inches or 2ft higher than they are at present towards the Land side.

That it will be impossible to keep the ditch clear without facing the counter-scarp with Tappy which may be very thin or something to support the earth.

That a Wall ever so slight, of a proper height and at a convenient distance beyond the Ditch, will prevent the Sand and Dust being blown into it and it will in a little time naturally form a Glacis.

That the Tide at high Water will reach the Fort, or very nigh it when the Earth is removed before it which may in time in danger the Foundation if not guarded against.

Fort Lyttelton was only two thirds complete in 1762 when Colonel Thomas Middleton requested another £5,000 to finish construction. A contemporary observer describes the building as it appeared in 1763:

"[Beaufort] harbor is defended by a small fort lately built of tappy, a cement composed of oyster shells beat small with a mixture of lime and water, and is very durable. The fort has two demi-bastions to the river, and one bastion to the land with a gate and ditch; the barracks are very good and will lodge one hundred men with their officers, there are in it sixteen weighty cannon, not yet mounted, the platforms and parapet walls not being finished for want of money."[45]

Gun platforms were finally installed during October 1764 and seven cannon mounted. Despite delays and inevitable cost overruns, results were impressive although the fort's outer walls represented a somewhat makeshift construction system. Rather than being cast solid like Fort Frederick's, these walls consisted of earth packed between inner and outer tabby skins resembling Fort Johnson's "thin case of brickwork very slightly built filled with sand" (see chapter 5), found wanting strength and durability in 1756.[46]

Was Lieutenant Hesse responsible for choosing Fort Lyttelton's construction method, or had it been forced on him by commissioners always mindful of limiting costs? Whatever the circumstance his designs for a horn work in Charleston probably employed an analogous tabby-faced earth-core system, although construction of this kind entailed significant risk if not properly calculated, as total collapse of Fort Johnston, North Carolina, would prove about 1800, thin tabby walls there failing to retain earthen fills.[47]

A drawing presented to the First Council of Safety of the Revolutionary Party on September 26, 1775, when the fort had fallen into considerable disarray, illustrates the barrack block located on its east side. Measuring 136' in length × 30'6" wide, with two wings extending west, this building was the largest freestanding tabby structure erected in Beaufort before 1800. One story high with garrets over, the structure proved that casting very long (and probably slim) tabby walls was feasible especially when enclosing repetitive spatial units. The building's central spine, also tabby, indicates loads transmitted to exterior building skins were minimized, thereby reducing risk of structural cracks and horizontal splits.

Repairs stipulated in 1775 included "5 m of Inch boards" to repair the barracks; "2 tapis walls for the cills under the platform . . . repairing Gun Carriages, Getting the guns out of the sand & up the Bank into the Fort & Mounting the same" and construction of "a Tapis Breakwater," which, with "6 M Bushels of Shells and Labourers," was estimated at £446, bringing total costs to £3,172.5.0. [48]

Providing Beaufort and neighboring islands with psychological security for more than a decade, repaired or not, Fort Lyttelton offered no resistance during the Revolution when British units made an amphibious assault in the vicinity. Instead of risking an encounter with its garrison, they simply bypassed the fort in January 1779, sailing up the Broad River (rather than along the Beaufort River) toward ferries linking Beaufort with the mainland, pillaging and burning settlements as they passed.[49] Reinforcements under General William Moultrie were sent from Charleston to defend Port Royal Island and Beaufort Town for the revolutionaries. They came too late. Moultrie described the ignominious turn of events that preceded his arrival: "would you believe it? The enemy had not more than 300 men when our people took fright, spiked up the guns, blew up the fort and ran away."[50]

◦ Fort Marion, 1811–12 ◦

With Fort Lyttelton shattered beyond repair, residents of Beaufort were again left vulnerable, the intendant telling the secretary of war about this worrying state of affairs in 1807.[51] Beyond military concerns planters and townspeople shared fear of their own "people" fueled by repeated discovery of plots against themselves—real and imagined—formulated among the slave quarters. Arguing for coordinated military building systems some years later, General Thomas Pinckney said: "the whole of the southern states have much to apprehend from the ease from which their black population could be incited to insurrection. . . . Georgetown, Charleston, Beaufort South Carolina and so on south would be useful as a guard . . . to keep the blacks in awe."[52]

In 1808 the U.S. Engineer Department charged Major (later General) Alexander Macomb with Beaufort's protection. Struck by the area's potential suitability for anchoring warships (an unrealized possibility successively promoted since the late sixteenth century by Spanish, French, and English mariners), he suggested that fortifying a small island called Mustard Island, situated near the mouth of Port Royal Sound, would control entry via the Beaufort River. This site was more sandbank than island, "strengthened by living oyster and dead shells washed up by the sea." Macomb noted, with naive (or possibly calculated) understatement, that construction "might be attended with considerable expense in making the foundation."[53] Alternative proposals involving refortifying Spanish Point were more realistic and found acceptance in Washington, construction of a new fort here—named Fort Marion—getting underway before 1811, when it was reported, "this work is of circular form in front, and a straight line in [the] rear. It is at present, only on its foundation four feet high above the ground . . . entirely of *tapier*. There is a comfortable house on the Public ground containing 2 rooms and a kitchen."[54]

D-shaped structures with a semicircular front facing the sea or waterway, guns usually mounted *en barbette* (that is, high enough to fire over the parapet), and an open or closed gorge were particularly favored by Macomb, at least five being erected under his

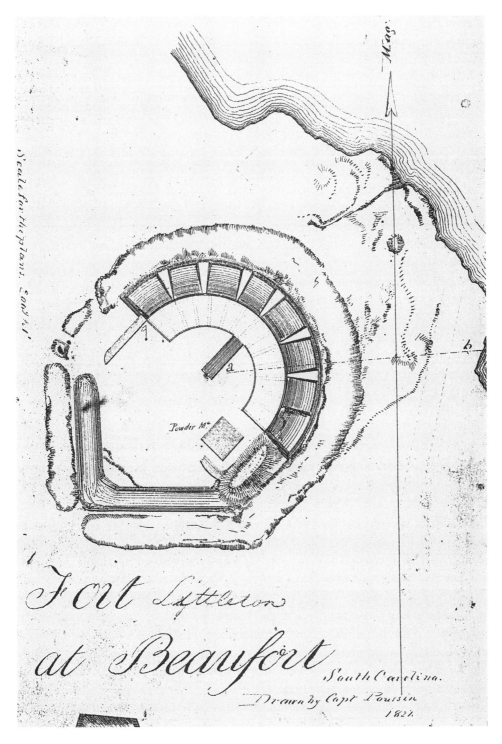

Fort Marion (here called Fort Lyttelton) at Beaufort,
detail of survey by Captain Poussin, U.S. Engineer Dept., 1821
(U.S. National Archives, RG 77 146-5).

direction just before and during the War of 1812.[55] These were loosely based on *demi-lune* forward features or standalone towers illustrated by the marquis de Montalembert's monumental *Sur les Fortifications Perpendiculaire de l'art defensif* (Paris, 1776–78), which rejected the all-pervasive bastioned trace developed over the course of the eighteenth century by Maréchal Vauban (1633–1707) and his Continental followers. To maximize fire power, Montalembert advocated tiers of guns arranged like those of contemporary warships.[56] Macomb presented an analogous, two-tiered casemate scheme for Fort (later Castle) Pinckney, Charleston. This was rejected, a single-tiered, D-shaped brick structure being erected circa 1810.[57] Isolated places (Fort Hampton near Beaufort, N.C.; Fort Winyah, on the Sampit River near Georgetown, S.C.; Fort Marion, Port Royal Island) saw tabby rather than brick construction even though rounded shapes favored by Macomb required relatively complex formwork. Fort Marion's ingenious designer resolved practical difficulties by instituting multiple pours of a rectilinear section approximating a curved outline for the riverfront. Fort Marion also differed from Macomb's other forts in that its guns were fired out of embrasures.

If still extant, documents describing this structure's genesis and erection have not been found. A discontinuous series of letters from the Buell Collection (U.S. National Archives) informs us that the works were overseen by Captain Prentice Willard of the Engineer Department, who in October 1812 was struggling against various problems—not least of which were erratic communication with his superiors—to install gun platforms, the fort's outer wall of tabby having, we can assume, reached completion. An earlier letter (dated September 1812) from Captain Swift at headquarters indicates the casemates were not covered, Swift recommending installation of "a double or treble row of Palmetto Logs on the exterior face so as to raise the work high enough for embrasures or merlons as to be covered from ship top," adding "Capt. Francis Saltus has been directed by me to get Palmetto Logs, Brick, Lumber, Lime, iron . . . do suppose these will be ready for you by 10 of the Month."[58] Swift described Saltus as "intelligent and active," telling his correspondent, Captain McRae, to "call on him for whatever he wanted." McRae was also informed: "Six 18 pds have been sent to Beaufort with. . . . Carriages are ready for them in this harbor, and will be sent." Months later, a letter from Captain Willard, which underscores his frustration over delays, reported: "no very visible progress has been made in the Battery of this place . . . owing to the want of timber for the platform and palmetto logs for the merlons—the hand[s] have been employed in removing earth and repairing the wharf. . . . Captain Saltus has only furnished about half the plank required, whether he has delivered to the amount of his contract I do not know, about 2,000 superficial feet are still wanting—have contracted for 200 logs, 50 of which are cut and made into a raft 1000 bushels of lime are engaged for the magazine."[59] Difficulties notwithstanding, construction continued, Captain Poussin's survey dated 1821 showing that the semicircular tabby wall overlooking the Beaufort River, gun platforms, magazine, and surrounding ditch had been completed, though exactly when is not stated.[60] Poussin's carefully scaled drawing gives the fort's diameter at about 160 feet with an inner radius of fifty-two feet, tabby walls rising about 13'6" above the ditch. There were seven embrasures, one significantly smaller than the rest. The plank-built gun platform was twenty-two feet wide and raised on fifteen radially arranged sleeper walls. Poussin identified an approximately twenty-two-foot-square structure located just southwest of the platform as the magazine. No residential buildings are depicted, an L-shaped feature—perhaps a berm, low earth bank, or even fragment of the earlier Fort Lyttelton—closing the fort along its land side.

Routine inspection in 1815 revealed deficiencies: "the form at Fort Marion is illy adapted to defend the channel and being open at the rear, is consequently exposed to be taken by a descent from a hostile fleet." The same inspector expressed reservations about the mode of construction: "it has been constructed of perishable materials and the effects of decay are already perceptible."[61] Nevertheless the fort survived long enough to see Confederate occupation during the Civil War, twenty-five or more members of the Beaufort Volunteer Artillery camping here in April 1861, while awaiting completion of Fort Beauregard, a sand fort constructed with slave labor near Bay Point Island's southern extremity, which—ignorant of early modern warfare—the Beaufort Volunteers mistakenly thought would bar enemy vessels entering Broad River Sound from the Atlantic. An eyewitness, Lieutenant J. Archibald Johnson, BVA, described works at Fort Marion:

> the six 18 pdrs which had constituted the armament of Fort Lyttelton (then called "Marion") in the War of 1812, were in a state of very good preservation. [Carriages] were specially built in Charleston and sent up on the famed steamer Planter. The large force of hands sent toward by the patriotic planters, soon had the old fortification transformed into a modern earthwork of no considerable strength. . . . Our great engineer General Beauregard, expressed his approbation of the manner in which the work had been executed, as well as the disposition of the armament etc. . . . Fort Lyttelton began to assume the appearance—at least to our inexperienced eyes,—of a formidable little fortress.[62]

Thus transformed Fort Marion lingered into the late 1970s, when overwhelmed by residential development despite preservation efforts by Beaufort mayor Henry C. Chambers, who sponsored archaeological excavation under Dr. L. Lepionka's direction. Today the following observations by Lepionka constitute the best record of an important historic structure shamefully concealed under modern building and landscaping: [63]

> The basic plan of the fort consisted of a semicircular arc facing on to the river, backed by a large building built on tabby foundations. The complete wall had within it six main gunports concentrated in a narrow arc of fire on the section of the river beneath the fort. There may also have been two lateral gun emplacements at the extremities of the front wall for land defense, but this interpretation of Gunport no. 1 may be in error. On the inner side of the wall, was a platform, probably continuous in its original state, constructed of a lower level of tabby and an upper level of wood, which served as a cannon platform. . . . No evidence of a rear wall was found though it remains possible that a wood or earth barricade may have stood there. The rear or west parallel wall of the barracks may have been intended as such a wall but it is dubious that higher tabby courses ever stood upon it because insufficient debris was found. This absence evidence for a rear wall and the

concentration of most if not all firepower on the river indicates that the function as naval fortress had fully prevailed in construction.

Excavation also retrieved information about the unparalleled method of construction:

> As originally built the front wall of the fort was a full half-circle, convex towards the river shore. Somewhat less than two thirds of this wall remains intact. . . . The wall was built in the following stages:
>
> 1. A circular arc was laid out, most simply by attaching a line to a central post. Using the interior line of the wall this arc would have had a radius of approximately 48′5″.
> 2. A trench was dug along this line to an average depth of three and a half feet beneath the surface as it existed at that time. The base of the trench was ca. 8 feet wide, the breadth of the wall, and had slanted walls.
> 3. Within this trench a series of wood forms were built into which the tabby mixture was poured. This was done in a series of horizontal sections and horizontal courses, with a relatively thin lower course (16″ as recorded in one location. Presumably standard for the entire course. Once the first section was poured and had set a new form was built adjacent to it and at a slight angle, using the previous section as part of the form. Thus each section is a rectilinear block, set in relation to adjacent blocks in such a way as to give the impression overall circularity.

Minor dimensional discrepancies aside, comparison between Lepionka's observations and information deduced from Captain Poussin's drawing of 1821 confirms the accuracy of the latter document. Although physical remains are no longer extant or have been smothered by time and tide, surviving drawings from the Engineer Department held by the U.S. National Archives show how closely at least two of the semicircular forts erected along the Carolina coast resembled one another. For instance Fort Hampton, located near Beaufort, North Carolina, was about the same size although more developed in plan. A description dated 1821 reads as follows:

> Fort Hampton. . . . This fort which is of tapia was built to command the "old Topsail" on Beaufort Inlet. Its form is an arc of a circle, whose interior radius is 50 feet, and perimeter measured along the covering line, 48 yards. The gorge is shut by a barracks of tapia and by two small circular flanks, which curve inwards from the two extremities of the great arc: in the right flank there is a proof powder Magazine. The parapet of the Battery is 8 feet thick and those of the flanks of the summit 18 inches. The scarps are 7 feet high and have a very great talus. The Battery may receive 8 guns: its distance from the line of low water 3 feet. The barracks and one of the flanks has been much injured by a violent gale, which was accompanied by an excessively high tide.[64]

The contemporary drawing gives Fort Hampton's plan and two transverse sections ("profiles"). Tapia barracks incorporated two stories and measured approximately 19′ × 78′

overall excluding a ten-foot-wide porch running along one facade. Elevations were carefully considered, the entrance front appearing a spare, symmetrically organized (except for the tall off-center arched entryway), neoclassical essay of some architectural distinction.[65] Perhaps repaired or patched up after 1821, Fort Hampton was swept away by a major storm in 1825. Today the site is entirely submerged.[66]

• Martello Tower, Tybee Island, Georgia, 1812 •

Among early nineteenth-century military structures with nonrectilinear plans, mention must be made of Tybee Island's now-demolished Martello tower, erected during 1812 to protect the city of Savannah and valuable rice plantations bordering the Savannah River. It resembled cylindrical models installed along coasts of southeastern England during the Napoleonic Wars when enemy invasion by sea was anticipated. Construction featured tabby rather than stuccoed brick usual among British examples or the stone employed for several contemporary Canadian and West Indian towers. There was no shortage of good brick around Savannah, but tabby was cheaper. Still the work demanded skill when fabricating forms and casting tabby on the scale required, besides consuming vast quantities of oyster shell. In 1835 Lieutenant Mansfield of the U.S. Engineer Department wrote, "the [Tybee] tower is built of a composition called Tabby or Tapia composed of shells and lime, and is one connected and solid mass. It is five stories high, consisting of a basement which has an outside door, and contains a Magazine and the brick foundation of a hollow column in the center of the Tower supposed to have originally terminated in a well as written on the plan; a 2nd story or entrance story with an outside door; a 3rd story lighted [through?] loopholes and provided with three fireplaces; and a 4th Barbette story intended for one gun on a pivot and provided with loophole machicoulis."[67]

Sketches accompanying Mansfield's description show the circular tower stood over thirty feet high, its floor area diminishing slightly in area with height, measuring 47'10" in diameter at the second stage and 46'3" across at third-story level, the flat roof mounting one gun capable of swiveling through 360 degrees. External wall thickness diminished too—from 10'8" at base to 10'3" at the third stage—which must have saved materials but greatly complicated formwork fabrication. Timber platforms projecting over the uppermost wall level formed the loophole "machicoulis" mentioned by Lieutenant Mansfield, an unusual, vaguely medieval feature for a Martello tower rarely found elsewhere. Platforms, gun, and gunners were protected by timber-framed roofing of conical form covered with shingles "put on within [the last] ten years" according to Mansfield (1835).[68] Overall dimensions and internal organization resembled Martello towers (many still standing variously converted into holiday homes, restaurants, or the occasional museum) erected along Britain's coasts between 1805 and 1808 except for its hollow central column. Housing a hoist, this central column allowed ammunition and water to be drawn directly up to the roof platform, the Prince of Wales Tower, Halifax, Nova Scotia (1796), preserving an analogous feature substituted for the usual solid masonry column favored by British engineers concerned about supporting heavy roof-mounted guns.[69]

Martello tower, Tybee Island, Ga., demolished ca. 1913, (photographer
and date of image not known, Historic Beaufort Foundation).

Lack of adequate funds hampered construction on Tybee; the tower's large size, novel construction, and isolation, as well as supplies and transportation of materials and personnel, doubtless created problems. Progress (or the lack of it) is chronicled by two reports written by Lieutenant Henry Middleton Jr. to Brigadier General I. G. Swift. The first (undated but probably written in late summer 1815) observed, "a Martello tower at Tybee, defending the only anchorage ground to that neighborhood as serviceable as an advanced alarm post is now under construction and will be completed in a few months. . . . The case for containing the *tapia* is almost three fourths filled—the work is still going on, but very slowly being not above forty workers. It appears however to be faithfully executed."[70] By late October of the same year, construction had almost reached standstill, Middleton reporting: "Major Champlain . . . informed me that there are no funds whatever for carrying on the Martello tower which is building at Tybee. It appears from a letter which he received from Isaiah Davenport the person who contracted to build the Tower that two thirds of it were completed about the 18th uto. And that the fourth installment was due on the 28th of the same month, but has not been paid for want of funds."[71]

Whether Davenport finished the work is not known. Middleton's remarks suggest construction may have dragged on until 1825, when roofing was apparently installed over the upper (platform) story.[72] Reference to Isaiah Davenport, "a New England born housewright"—whose own Georgian-style brick-and-brownstone Savannah townhouse remains notable for its fine craftsmanship—is of considerable interest since it could mean other tabby building projects were undertaken under his supervision before 1827, when he died during one of the many outbreaks of yellow fever to visit his city.[73]

Already abandoned by Confederate forces, the Tybee tower was fired on by Union vessels in November 1861. Damage was done, but how much is undetermined, contemporary newspapers reporting that its timber floors were decayed and considered dangerous before hostilities began. Eventually the tower was adapted for commercial and, later, domestic use, fire, which gutted the interior, causing its demolition about 1913.

⚬ Beaufort Arsenal ⚬

Occupying land near the corner of Craven and Carteret Streets, the Beaufort Arsenal was first built between 1795 and 1797, partially demolished in 1852, and then rebuilt. Enlargement followed during the 1930s, and extensive restoration began in 2000. Despite these episodes, the building still encapsulates original fabric built by Colonel Thomas Talbird, who was responsible for several of Beaufort Town's most important buildings including old Beaufort College—an ill-fated enterprise considered in chapter 6. Additionally, excavation revealed extensive structural reuse and late eighteenth-century foundations, walls and other features having been modified then redeployed, this process indicating that Beaufort's antebellum residents retained enough frontier spirit to use and reuse building materials whenever they could. Given the extent of destruction and demolition, it is fortunate that the colonel's nephew, another Thomas Talbird, then keeper, described the Arsenal as it stood in 1825, his account reading as follows: "there are two buildings in size 20 × 30 feet each and 10 feet high built of tabby and covered with tiles . . . one of which is used for the reception of Arms, iron and tin ware etc. and the other for powder and articles composed of leather, wood etc: A shed of 90 feet long extends from one of these buildings to the other and is 12 feet deep—which depth is not sufficient . . . for the cannon, wagons, etc. for which it is intended. The outer enclosure measures 100 feet by 61, and the wall is 10 feet high also built of Tabby work."[74]

Excavation (2001) uncovered incomplete walls belonging to one of the two wings, fragments of the gun shed and early tabby absorbed into the present enclosure wall fronting the building's south side. A "ghost" image of the original east wing (which now supports an added brick story) can just be made out after rain at the present building's northeast corner. About eighteen inches thick and cast in twenty-four-inch-high vertical increments, Talbird's tabby walls—stuccoed inside and out—stand over eight feet high, having lost a parapet perhaps or even merlons. These formerly enclosed one undivided space (measuring 20' north/south × 30'3" east/west) entered from the south and lighted by small windows, the surviving example centered about the east facade. Whether this wing was the armory holding nearly three hundred muskets and bayonets in 1825 or powder store—then containing two hundred pounds of powder—is not known. The younger Talbird's statement that both wings had tiled roofs indicates a pitched roof arrangement. The tiles always leaked, much to the detriment of stored items.

Nothing remains above ground of the west wing. However careful examination reveals remnants of another tabby wall that enclosed the gun shed linking east and west

Arsenal, Beaufort, view from northeast, "ghost" of late eighteenth-century work
indicated by darker-toned stucco at lower level of building (HABS).

wings along its north side. This wall was cut down, trimmed, and reused in 1851 or 1852
as a foundation for the back (north) wall of the present brick building. Original work was
surprisingly sturdy, strengthened by three tabby buttresses placed about 23'6" on center.
Lack of wall foundations to the south show—as might be assumed—that the original
gun shed was open along its south side, this allowing exit or egress of two field pieces
(twelve-pounders) and two ammunition wagons inventoried in 1825. The shed probably
had a pitched roof supported by timber roof trusses supported at their northern extremities by the buttresses mentioned, timber posts carrying loads to the south. Numerous
fragments of clay tile recovered during excavation probably derive from shed's roof, its
excessive weight explaining the sturdiness of known construction.

That the original Arsenal soon fell into disrepair is evident from the 1825 account, the
keeper Talbird finding it "in an unfit state, for the complete protection of the property
deposited therein." The gun shed was "in the most ruinous condition . . . so decayed that
a part of it is in the act of falling, and the whole must [before] long come down." The
enclosure wall also threatened disintegration "in the lapse of a few years." Dilapidation
was so advanced because "these walls have never been rough-cast which is considered a
great preservative & almost indispensable to Tabby work. And it is further calculated to

resist dampness; which these walls are more or less subject to."[75] Recommended repairs were deferred, monies being perpetually short and Beaufort distanced far enough from the state legislature to be ignored. By 1850 moisture intrusion and poor maintenance had rendered the old tabby building unserviceable. In 1852 the Beaufort Volunteer Artillery (founded in 1802) took matters into its own hands, "erecting on the foundation of the old Arsenal a building capable of accommodating garrison of 250 men and a battery of six guns," expending $2,835 on the project.[76] Noted for their parades, dinners, and banquets (which judging from the Arsenal's rubbish pits, involved prodigious consumption of champagne), the Beaufort Volunteers distinguished themselves near the beginning of the Civil War by heroic if ultimately futile attempts to defend Beaufort Town and surrounding plantations from Union invasion.

Tabby Making

Materials and Fabrication

Historic Sources

Tabby making in South Carolina is attested by documentary sources from the late 1720s down until the Civil War, but detailed descriptions of local fabrication techniques are scarce. Aside from the account given by La Rochefoucauld-Liancourt already cited, one of the most comprehensive reports, entitled "Observations on the Art of Making the Composition Called Tapia," was published in 1811 by Captain (later General) Alexander Macomb, styled "Member of the United States Military Philosophical Society" at a time when he was overseeing construction of several tabby forts along the Carolina coast.[1]

"Tapia" (i.e., tabby), Macomb remarked, "is a composition of shells lime and sand in such proportions to make a complete cement which in the course of a short period becomes one solid stone."[2] Visiting Beaufort he found "*tapia* is more practised than in any other part of the United States." Commentary about the composition of mixes, formwork design, and operational procedures follow, drawn from the author's own experience without reference to Palladio, who described Roman concrete, or John Bartram, who had seen tabby at St. Augustine when he traveled through East Florida in 1765.

Notices written by individuals who witnessed tabby making before Union occupation of Beaufort in 1861 give similar information about fabrication techniques. Recalling building activities of his father, Dr. Bernard Barnwell Sams (1787–1855), who owned a flourishing plantation on Dathaw (Dataw) Island, Reverend James Julius Sams (1826–1918) wrote: "the way of construction was to make a box or several boxes according to the length and width of the building, each box so many feet long, say about fifteen or twenty feet, and about one and a half feet wide. These boxes were put in place, filled with the [tabby] mixture, which was packed or pestled down, and allowed to stand until dry. The sides and ends of the boxes were held by moveable pins. When these pins were drawn out, the box would fall to pieces. The box was taken down and put upon the tabby already dry, and so box after box was packed and pestled until the walls were as high as you designed."[3]

Local descriptions are augmented and enlarged by Thomas Spalding. Following experimentation during the early 1800s, he began recommending tabby as an inexpensive building material to neighboring planters and others who solicited information about agricultural subjects. His article "On the Mode of Constructing Tabby Buildings and the Propriety of Improving Our Plantations in a Permanent Manner," published by the *Southern Agriculturalist* (December 1830), claimed to have revived almost forgotten techniques utilized by founders of Frederica, Georgia, during the 1740s, which were abandoned when the town went into decline.[4] Whether this means Spalding did not know or chose to forget that along southernmost areas of the South Carolina coast tabby construction had never fallen out of favor is questionable. Surveys of extant structures show that very few Beaufort planters followed his advice concerning the composition of tabby mixes or his style of formwork. The majority relied on, and sometimes improved, practices developed in the context of their own tabby building traditions, which by 1830 extended back over one hundred years.[5]

Besides, the truth was that Spalding made indifferent tabby with an inadequately sized labor force. Nevertheless his ability to place tabby making into historical perspective and associate the material with fashionable concepts of rural improvement was a unique achievement. Moreover, while frankly promotional, Spalding's accounts are more comprehensive than most, allowing reconstruction of historic fabrication processes not fully detailed by other sources.

◦ Determinants: Costs and Available Materials ◦

Why tabby was favored around Old Beaufort District is explained by local geological history, which has seen multiple progressions and regressions of the sea extending back through the Pleistocene and Miocene periods. During successive episodes sand and silt were gradually deposited over thick limestone beds named the Ocala Series—marine sediments containing late Eocene faunal assemblages. These limestones are buried eighty-five feet below mean sea level at Bay Point Island and forty-four feet deep near Beaufort Town, making commercial exploitation well-nigh impossible. The area also lacks clays, only one or two sites producing fired brick before 1861, the largest enterprise at Brickyard Point, Lady's Island not coming into production until the 1840s. With their thin sandy soils and salt-laden marsh mud, the Sea Islands lack building earths of the kind used for cob or similar building compounds, this deficiency being widespread across adjacent mainland areas.

What the lowcountry had in abundance was shell, timber, sand, and fresh water. Supplies of oyster shell, formerly burned for lime and used as coarse aggregate in tabby mixes, still seem inexhaustible all along the coast, intricate networks of rivers and tidal creeks allowing transportation by flat, bateau, or schooner to almost any tidewater site in antebellum times. Shell's utility for building purposes was realized very early by European settlers. Reporting to King Phillip II in March 1580 from Santa Elena, Spain's settlement on what is now Parris Island, the Adelantado Pedro Menéndez Marqués said that half of its more than sixty houses were "constructed of wood and mud, covered with lime inside and out, and with their roofs of lime. And we have begun to make lime from oyster-shells."[6] Recounting an exploratory voyage to Port Royal in 1666, Robert Sandford also saw potential for the area's great oyster banks and "heaps" of shell, informing the Lords Proprietors how future settlers might build kilns along any river or creek they chose since there would be enough shell to burn for lime long into the future.[7] Residents of coastal North Carolina were making shell lime, too, near the beginning of the eighteenth century, John Lawson writing, "in building with brick we make our lime of Oyster-Shells."[8] Newly arrived in 1745 from New Providence (Bahamas), where he had restored and enhanced Nassau's fortifications, Peter Henry Bruce found residents of Charles Town followed similar practice. "They have no lime," Bruce said "but what they make of oyster and other sea shells."[9]

Forests, which left travelers in awe about their extent and diversity, meant that timber was available in almost any length or quantity everywhere along the southeastern coastal plain, pine being preferred for formwork and lime burning. Likewise sand of pit, river, and drift varieties occurred most places, wells or swales supplying fresh potable water necessary for slaking and mixing even on desolate barrier islands. For builders located where analogous ecological conditions prevailed, these natural resources simply made tabby less expensive than other incombustible products once builders learned how to overcome the structural constraints this composite imposed.

That economies weighed when choosing materials for large building projects is demonstrated by defensive works erected at Dorchester (Charleston County) in 1757. Hearing brick cost £15 per thousand plus freight "at forty shillings per thousand and lime at four pence per bushell," South Carolina's Commissioners for Fortifications declared "tappy work" would be "much cheaper and better."[10] Short of funds and fighting disease and a hostile environment, Captain James Gadsen (who had previously written reports concerning Beaufort's defenses) reached the same conclusion in 1821,[11] proposing to substitute "*tapia* for the brick of the revetment walls of the fort at Mobile Point [Alabama]," tapia then being priced at $10 per cubic yard while brick cost $11 per cubic yard—no small saving considering this project's extravagant scale.[12]

On Sapelo Island, Thomas Spalding calculated that each tabby maker produced 120 cubic feet of wall per day after finding his own materials. A somewhat garbled passage implies the same output executed in Charleston brick would have required 1,560 units at "not less than $15 per thousand" including "lime and labor."[13] With the use of enslaved workers, costs for an equivalent in tabby was no more than $1 per day or $6 per week, Spalding said, adding: "there is not one dollar of actual expenditure to the planter, for he neither buys bricks or hires masons to lay them." But this calculation was ingenuous since it ignored the costs of feeding, housing, or otherwise maintaining his workforce and payments made to the "White Man Superintending," Roswell King, who doubled as Major Pierce Butler's overseer at Hampton, St. Simons Island. E. Gilman was more realistic: "the expense of Tapia building, compared with quarry stone or brick, is from half to two-thirds, depending, of course, on the relative facility of obtaining the materials."[14] Detailed outlays required for domestic tabby building are rarely found. South Carolina's military works during the first half of the eighteenth century are better documented, one of the earliest records concerning tabby (dated 1726) estimating that Charleston's proposed White Point battery, mounting ten guns, would require ten thousand bushels of lime at 2 shillings per bushel (£1,000); ten thousand feet of board for casing the work at

30 shillings per one hundred (£150); twenty Negroes for two hundred days at 7 shillings and 6 pence per day each (£1, 000) and twenty-five hundred feet of cypress plank for gun platforms at $40 ($2,400). In August 1757 the estimate made by "Mr. Hesse the Engineer" for "40,000 feet of Tappy Work" at Fort Lyttleton, Port Royal Island was £5,125.11.2 with labor remaining constant at 7 shillings and 6 pence per person per day.[15]

Lime and shell required continuous delivery once casting operations got underway. An advertisement placed on July 12, 1757, shows the South Carolina Commissioners of Fortifications were then paying twenty pence per bushel for lime delivered in Charleston. At Dorchester the price ranged between twelve pence and eighteen pence per bushel. "Wet lime" fetched by boat from "Mr. Burnham's Landing" was lime putty, Burnham supplying 5,712 bushels at slightly over £47 probably destined for Charleston's Middle Bastion (located between Broughton's Battery and Granville's Bastion) then under construction by Thomas Gordon, who received payment for tabby work there the same day Burnham's account was settled.[16]

Warrants issued in 1757–58 detail quantities, costs, and the pace of shell delivery for construction of Dorchester's magazine and its enclosing tabby wall built overlooking the Ashley River inland from Charleston. Available records are incomplete, but we hear of 1,574 bushels delivered by Thorn and Shubrick (paid £78.14, August 18, 1757); 2,325 bushels of live and 857 bushels of dead shell brought in by Thomas Walker (paid £51.19.2, September 29, 1757); 2,660 bushels delivered by Samuel Stevens (paid £133, October, 13, 1757); and another load comprising 2,665 bushels from Thomas Walker, for which he claimed just over £110 (February 9, 1758). With shell costing as much as eighteen pence per bushel, the commissioners realized expenditure must be curbed: "Taking into Consideration the Extraordinary price given for Shells," they agreed in July 1758 "to receive no more from private persons but the Boats in the Public Service be employed whenever any more are wanted."[17] Savings accrued are illustrated by payment several weeks after new cost controls were imposed of £210.13.4 for 5,056 bushels (roughly ten pence per bushel) to John Holmes, an overseer in the commissioners' service.[18] Off Fort Johnson, Adam Irish, another overseer publicly employed, was gathering shell with "two boats and twelve Negroes" for Charleston's defenses, "tappy-work" at the fort being suspended when building funds ran dry.[19] Hired by the commissioners in May 1757, one of the vessels involved was perhaps Samuel Prioleau's "Canoe Built Boat," which "carried upward of Fifty Barrels of Rice" and drew "but two feet and half Water . . . very handy for bringing Shells and Sand."[20]

Live shells were collected from offshore oyster beds. Dead shells occurred along tidal waterways, Charles Drayton collecting quantities near the mouth of the Edisto River in 1799 probably using a sloop, which, besides shell, carried almost everything needed for his plantation—brick and lime, barrels of tar, rice, lumber, livestock, and wine.[21] Mr. Seabrook (probably William Seabrook, one of the Sea Islands' most prominent cotton magnates) told Drayton that Edisto planters preferred dead rather than live shells because they produced lime of the same strength and could be more compactly stowed on board ship.[22] Dr. G. E. Manigault of Charleston related: "oyster shells were gathered at the mouths of the various rivers and inlets bordering the coast, where they are washed up by the action of the water, and sorted into layers of different sizes by the tides and currents, in such a ways to render it possible to obtain quantities of any uniform size that may be wanted."[23] Other planters robbed Indian shell rings and middens, prehistoric pottery inclusions exposed by erosion in tabby of the Edwards House, Spring Island; the Sams House, Dataw Island; and the Callawassie sugar mill and north slave settlement, Daufuskie Island, all confirming the practice. Thomas Spalding gave explicit information: "the shells I have used were old shells from ancient Indian Barrows, some of them of great extent scattered over our Sea Islands from Charleston to St. John's River in Florida. The drift shells, after the oyster is dead, thrown up along the shores of our rivers, are also used, but the salt should be washed out."[24] Further emphasizing the absolute necessity of salt-free materials, he continued: "no salt water should be permitted, as it produces decay, where the shells have much vegetable mold with them they would be the better for washing." As late as 1900 or 1910, William E. Pinckney was collecting materials, much as planters had done in the antebellum era, quarrying the Late Archaic shell ring at Guerard Point when building his smokehouse and sweet potato store overlooking Okatee River at Pinckney Point near Bluffton—structures that may be the last two tabby outbuildings fabricated using traditional lime-based mortar in Beaufort County.

Addressing the subject of sand, Spalding remarked, "I use pit sand and prefer it, as being free from salt. I like a mixture of fine and coarse sand, but am not particular on that subject."[25] Excavation two feet or less below the surface of almost any South Carolina Sea Island site will expose yellow sand leached clean by rain over centuries. If it seems logical that tabby makers would save transportation costs by utilizing sands exposed near their building operations, Sickels-Taves and Sheehan reported otherwise, noting beach and channel sands were frequently used despite high salt contents.[26] Be this as it may, different sands varied the appearance and properties of finished tabby, while substitutions occasionally occurred. "I have built good Tabby from rough gravel taken up from the bed of the Altamaha, near Darien," Spalding said, observing: "stone broken up by the sledge hammer, if more easily procured, would answer equally well."[27] At White Hall, Thomas Heyward's plantation near Ridgeland, South Carolina, the tabby mix used when erecting the main dwelling's two wings contains lime, an oyster-shell aggregate, and sand mixed with what appears to be clay presumably found nearby.

In order to prepare the tabby mix, having obtained shell—consisting of almost pure calcium carbonate—the tabby maker first burned a portion, this process known since antiquity, driving off carbon dioxide and moisture. Quicklime (calcium oxide) resulted, which, slaked with water, yielded lime putty (calcium hydroxide) and clouds of steam. Lime putty mixed with sand and coarse aggregate produced mortars characterized by excellent plasticity. Where large quantities of lime were required for military or other major projects, it is probable that intermittently burning flare kilns or continuously burning draw kilns were fabricated.[28] On Jehossee Island just north of Edisto, Charles Drayton had John Phalley "bricklayer" and one of his own men (presumably enslaved) called Carolina build a "reverbatory furnace" intended for burning shell to lime. Finished on November 5, 1798, this was an egg-shaped affair, 9½' deep and 6'3" in diameter at floor level with a four-foot-wide "upper orifice," construction having taken ten days. Not fired until May 1801, the kiln then produced 1½ hogsheads of unslaked lime, Drayton soon finding that increasing the ratio of wood to shell ensured better burning and produced less "rubbish."[29] Smaller operations utilized primitive clamps consisting of alternate layers of shell and fuel covered with clay and vented through stoke holes. Otherwise ricks might resemble those now made by the National Parks Service at Frederica, Georgia, and isolated communities in the Bahamas, these comprising nothing more than shell and logs piled up, the whole mass being burned together.[30] Irrespective of method a sustained temperature of about 1,000°C was required for the shell's complete reduction.

Burning near the building site minimized transportation costs. "I would burn them [shells] on the spott [at Dorchester] and save the carting," said John Joor in 1747; "I have already had one shilling per Bushel for burning . . . which I believe no Person will do it for."[31] Over fifty years later, Spalding was burning relatively small quantities of shell, his Sapelo Island crew spending two days every week collecting raw materials and building kilns.[32] Lime produced was used "as soon as convenient after slackin [slaking], without the trouble of sifting or other process," a practice guaranteeing that wood ash—with valuable waterproofing properties—found its way into the product.

How the business of slaking was accomplished is not described. One common method involved placing quicklime into a pit and flooding it with water, this causing violent chemical reactions raising the water temperature near or even above boiling point. Slaked lime was usually (not invariably judging by Spalding's accounts) left in the pit for several weeks, months, or years. It has been suggested—plausibly, I believe—that certain tabby makers used "hot mix" made by adding sand and aggregate to quicklime as it was removed from the kiln, sand being sufficiently moist to effect slaking.[33] This alternative method (which generated a good deal of heat—hence the name) obviated long curing times required by traditional pit slaking. It was also useful when temperatures were low or near freezing, conventional tabby making being discontinued when the thermometer read 40°F or less because lime-based materials gained strength slowly during prolonged cold weather.

◦ Formwork ◦

Fabrication, placement, and filling of formwork involved coordinated efforts by the master tabby maker and his construction gang. Fabricating "molds" (or "moulds" or "boxes," as they were alternatively called) demanded carpentry skills and complete knowledge of the finished building's plan because any mistake or change created difficulties once casting operations got underway. Before the American Revolution, formwork design varied according to predilections of individual tabby makers or constraints imposed by the scope of individual projects. At Fort Frederick (1734), pour lines indicate forms were fifteen-inch-high with circular "pins" securing inner and outer boards. The exceptional width of finished walls (over five feet wide) suggests metal bolts rather than timber dowels were used for this purpose, battered exterior wall profiles requiring adjustment of formwork as the work proceeded. Although thinner (measuring 2'10" at base, reducing to two feet on top) and standing about eight feet high above present ground level, tabby walls surrounding Dorchester's powder magazine, erected 1757–58, exhibit similar details. Outer wall faces were battered and the formwork likely tied by small rectangular iron bars. Incorporating four corner bastions ordered like a pinwheel in plan, the fortified enclosure here, with its numerous junctions, would have presented intractable problems for the tabby maker had he fabricated formwork to match the overall building shape. However, by his opting for clear structural breaks at all intersections whether right angled or otherwise, walls could be cast in discrete segments, an arrangement having the advantage of simplifying construction and reducing costs. No attempt was made to fill narrow vertical voids created between the various wall sections. Rather, as in modern concrete practice, these cold joints allowed for shrinkage when the tabby dried and differential structural settlement on what has proved an unstable site sloping down to the Ashley.[34] In 1758 Messrs Brunet and Hall (presumably carpenters) charged £18.4.0 for making the requisite "tabby boxes." Whether this sum included moneys for metal tie rods is not known. It can be inferred that moving, packing, raising, and repositioning these "boxes" was no mean task since the largest ones probably approached one hundred feet in length and were slightly more than two feet high.[35]

Forms used in domestic building were usually less cumbersome than those utilized in defensive projects, rarely exceeding eighteen inches wide or thirty-five feet long. On Dataw Island during the 1760s or 1770s, exterior walls of the early William Sams House were cast in nineteen-to-twenty-inch-high vertical increments. At the Lawrence Fripp House, St. Helena Island (date uncertain but probably before 1770), impressions indicate use of twenty-two-inch-high "boxes" secured by ¾" diameter "pins."

Two design shifts occurred after the American Revolution. First, formwork height became standardized throughout the Carolina lowcountry at or very near twenty-four inches. Second, use of removable circular timber or metal "pins" fell out of favor. How these changes came about is not known though it seems that tabby making was largely an itinerant occupation, gangs of experienced workers hired out by planters taking fabrication techniques with them as they moved from site to site. Dr. B. B. Sams and Colonel Thomas Talbird were two Beaufort landowners who became general contractors when their agricultural activities allowed. It was commonly acknowledged that Talbird had particular knowledge of tabby, his expertise bringing him important commissions to which we shall return.

Military engineers building tabby forts up and down the Carolina coast just before 1812 played similar roles—spreading practical knowhow about the material and its fabrication among new communities and long-settled ones. Alexander Macomb stands out, since his 1811 paper concerning tapia (i.e., tabby) is illustrated by three rather naive drawings showing formwork design and assembly. As far as I know, these are the earliest North American illustrations to do so, while the accompanying description is unusually comprehensive for the period. Macomb described his formwork as follows:

> the mould is made of any number of short pieces of scantling, about three inches square, of a length suitable to the thickness of the intended wall, having at each end a mortice or tenon hole, and as many upright pieces of scantling, or inch, or inch

and a half plank, or three inch scantling about four feet high, with tenons at each extremity, so made as to enter the mortice easily below, and to receive a yoke above, which will keep them together; then pieces of board, an inch and a half thick, are laid against each side of the uprights, and kept apart by means of a small stick, of the exact size or thickness of wall. The cement being thrown into the mould, it will press it to its proper size, and it must remain there till it is dry enough to stand by itself; the box is then taken asunder and another course is made on top in like manner.[36]

With horizontal form boards supported on sturdy cross timbers morticed to receive vertical staves tied by "yokes," Macomb's "mould" closely resembled "moulds" (*moules* in French) illustrated by European pamphlets circulating near the end of the eighteenth century intended to instruct landlords, architects, and builders about the advantages and practicalities of *pisé* construction. The pocket-sized volume *École d'arcitecture rurale* by François Cointeraux (Paris, *Chez l'auteur*, March 1790) was perhaps the most accessible and certainly most influential of these writings either in its original French language version or English translation. A native of Lyons, Cointeraux described two closely related but distinct formwork styles, found in areas of rural France where *pisé* was commonly employed. In the Lyonnais, side shutters carried on cross timbers morticed to receive supporting timber uprights prevailed. About Grenoble and Bugey (Ain) the author discovered that side shutters were kept in place by timber posts extending from basement to roof set at regular intervals along inner and outer faces of the intended wall, these posts each being angle braced from the ground and tensioned by ropes at the top. Side shutters were moved both horizontally and—as construction rose higher—vertically between the posts. The two systems are illustrated by Cointeraux's Plate X, where designated Fig. 1 and Fig. 2 respectively. Regarding the Bugey system, Jeffrey W. Cody observes "by using the external braces it became unnecessary to insert lower horizontal support at the base of the shutters" thereby obviating the need to patch or fill wall cavities.[37] The labor required to cut mortices and tenons, was also eliminated further simplifying and speeding the whole construction process. Cointeraux considered this method ideal when constructing barns and other agricultural structures, especially in hilly or mountainous areas. Despite his enthusiasm, however, it was mostly ignored by later commentators (notably English architect Henry Holland).

Although the possibility that Macomb adapted François Cointeraux's Lyons style formwork cannot be excluded, a probable North American colonial parallel is concealed in an obscure reference disinterred by William M. Kelso from James Oglethorpe's accounts for wages paid during construction of Fort Frederica on St. Simons Island. Two separate entries, dated February 1740, itemize monies paid Thomas Walker for "making Tappy Boxes and Needles for ye Barracks" (9 shillings, 9 pence), and "making Needles for the Tappy Boxes" (9 shillings).[38] The meaning of the term "tappy boxes" requires no explanation, but reference to "needles" is of considerable interest, Oglethorpe likely borrowing this word from contemporary builders to whom it meant "a horizontal piece of timber serving as a temporary support to some superincumbent weight . . . while the lower part of a wall is being underpinned or repaired."[39] If this interpretation is correct, then Oglethorpe's "needles" correspond in function with Macomb's "short pieces of scantling about three inches square"[40] that constituted the lower part of the exterior frame designed to receive uprights. The "Return for Work at Fort Lyttelton" (Port Royal Island) dated November 16, 1758 gives weight to this argument, making reference to "tappy boxes, Needels, Standards and Wedges."[41] "Standards" were timber uprights (posts) likely designed to keep horizontal form boards in position. As described by Rees in his *Cyclopaedia,* wedges inserted into mortises of the needles kept "the posts and the moulds firmly fixed against the wall."[42] While not perhaps conclusive, the two references cited suggest that when building large tabby forts of novel design, Macomb adopted well-tried construction modes used since colonial times, side shutters supported on "needles" and braced externally providing lateral support for weighty mixes cast on an exceptionally large scale, besides being strong enough to withstand ramming required for proper compaction.

Conversely Thomas Spalding reduced the quantity and weight of tabby made in any one "round." Writing to an unidentified correspondent in 1816, Spalding described the simplified type of formwork he preferred:

> two planks as long as convenient to handle, 2 inches thick and about 12 inches wide, are made to unite and to go the round of your building. These planks are kept apart by spreader pins with a double head. . . . The first head keeps the outer plank in its place, the last with the pin run through the point, keeps the inner plank firm while the workmen are filling in the material and setting it down, either with a spade or light rammer. . . . Then the planks at the ends are let into each other . . . with an iron wire.[43]

Elsewhere he noted:

> the boxes are kept apart by pins, at every three to four feet, which as soon as the Tabby begins to harden are driven out and the boxes saved for another round. This in the summer will be in two or three days. Care must be taken in carrying up your walls, that they be kept straight by a line and perpendicular with a plumbob. It is always well to have a range of boxes to go all around your building; it saves trouble and the Tabby becomes better cemented.[44]

Explaining that there was "little art in constructing these buildings," Spalding further observed:

> The boxes are taken to pieces in a few minutes, at every round of a foot, and are again put together in little more time, because they are kept together and preserved at a proper distance by two means only. First, at the bottom at every three feet, round pins with double heads. . . . The larger head is on the outer side. The planks only require to be put together at the end, by joining into each other like a dovetail, where an iron rod keeps them together; the pins are then inserted, and as

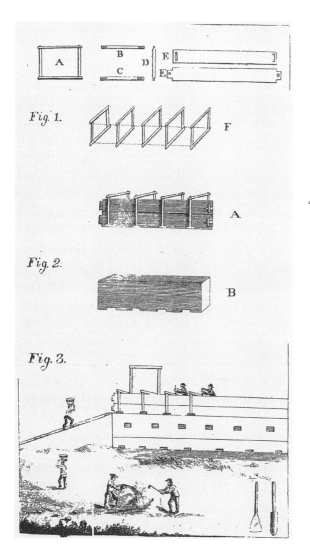

Fig. 1.

Fig. 2.

Fig. 3.

Alexander Macomb, sketch of tabby construction, from "Observations on the Art of Making the Composition Called Tapia," published in the *American Medical and Philosophical Register or Annals of Medicine* 2 (July 1811).

Below: Reconstruction by Richard Wightman of tabby formwork described by Alexander Macomb in 1811, side view (Dataw Historic Foundation, Dataw Island, S.C.).

the boxes descend one inch over the last round of Tabby, the pins keep them together at the bottom, and at the top of the boxes, flat pieces at the same distance.[45]

Despite the obvious technical advantage of formwork so readily fabricated, assembled, and disassembled, few Carolina planters followed Spalding's example. True, it was easier to keep walls plumb using smaller rather than larger "boxes." But assuming comparable wall heights, Spalding's twelve-inch-high forms introduced twice as many horizontal construction joints—always lines of weakness—as opposed to prevailing South Carolina custom. Additionally twelve-inch-high pours extended construction times, each tabby "round" needing several days to cure before casting the next higher one. Consequently examples of Spalding-inspired tabby are rarely encountered in Beaufort County, tabby makers across the Carolina lowcountry setting their casting lifts at or very near twenty-four inches down until the Civil War.[46]

Naturally variations occur within this local tradition. At Bleak Hall (now Botany Bay) on Edisto Island, it seems the owner—John Townsend, who produced the highest-quality Sea Island cotton—was not overconcerned about finer details of casting. Although walls of his "Gothik"-style tabby barn were made (ca. 1840) using standard (i.e., twenty-four-inch-high) forms, the oyster-shell based tabby mix carefully tamped and stucco finishes skillfully applied, there is no sign that pins, cross ties, yokes, or braces held opposing sides of the formwork in place. Rather than investing in any refined structural system, Townsend spent his money on decorative carpentry, which, once installed, must have seemed as delightful to him as it would to us had it survived recent brush fires.

At the opposite end of the spectrum in terms of size, Daufuskie Island's Haig Point House—the largest multistory tabby dwelling known from Beaufort County's Sea Islands, commissioned by an owner prepared to finance work of the highest quality out of his first wife's fortune—exhibited a far more considered structural system. Formwork here was twenty-four inches high with inner and outer faces made up from tongued and grooved horizontal timber planks. Small cavities running through the wall at regular intervals along the junction of successive pours indicate use of 1¼" square timber cross pieces, but exactly how these functioned is uncertain. Considering the huge weight of tabby cast and exceptional thickness (twenty-four inches) of walls at the lower level, formwork must have been braced to prevent bowing. This could mean the cavities were left by removable sleepers designed to receive vertical staves more or less as described by Macomb. Alternatively we might speculate that some other kind of temporary shoring was employed. Either way, despite catastrophic destruction caused by looting and burning, there is certain evidence that exterior corners of the building were laid up in fired brick as casting proceeded, an expensive but highly desirable measure that ensured accurate right angles at vulnerable junctions, that kept forms plumb as they were maneuvered into position, and limited the length of formwork itself. Similar details are visible in photographs of the tabby-built Talbird House in Beaufort Town taken soon after it was gutted by fire in 1907, ruins of an associated privy preserving similar brick corner reinforcement almost intact.

Brick corner reinforcement, privy, Talbird House, Beaufort, S.C. (Jack Boucher, HABS, 2003).

Across the lowcountry, forms usually matched the overall size and shape of the tabby building under construction except where fabrication would introduce unacceptable complexity into the building program (as seen at Dorchester). Spalding noted, "our tall timber trees give us long plank, and instead, therefore of confining our operations to twelve feet in length we construct at once a range of boxes to go all around our intended building . . . including even the partitions," an observation confirmed by La Rochefoucauld-Liancourt.[47] This practice distinguished tabby from European pisé illustrated by Cointeraux, his followers, imitators, and later commentators, who made provision for angled joints along the length of wall pours.

Close examination of near-identical wings added in tabby east and west of the Dataw Island house inherited by Dr. Berners Barnwell Sams allows reconstruction of methods adopted to speed the building process when building on a large scale. Here the owner utilized two formwork sets simultaneously, an arrangement allowing his construction crew to fill one set as newly poured material cured over the course of several days in the other, operations then switching from the east to west wing or vice versa until the job was done. A contemporary, George Edwards, adopted similar time-saving tactics when adding new mirror-image wings north and south of the old tabby house he owned on Spring Island. Whether Edwards and Sams were ever in communication or shared construction workers —as their respective building operations might suggest—is one of those tantalizing unknowns created by Beaufort's loss of historic records.

To keep costs down and avoid mistakes when setting out, tabby buildings were generally rectilinear. Nevertheless exceptions occurred, military engineers devising means of erecting very large semicircular forts before 1815. Unfortunately none survive. Neither does Tybee Island's tall, circular Martello tower, contemporary notice of its "great case almost finished" perhaps indicating that Isaiah Davenport, an accomplished "housewright" working under contract for the U.S. Engineer Department, departed from customary practice when making this formwork. Old photographs show the tabby outer wall was built up round by round though not necessarily in twenty-four-inch increments— indeed the height of each pour may have been significantly more or less—it is difficult to say because cavities left by cross ties normally associated with formwork of the period are absent. Rather exterior wall surfaces, uninterrupted except for gun ports at the uppermost floor level and the entrance doorway, appear heavily eroded, all form-board impressions and exterior finishes (if ever applied) having weathered away.

Two minor tabby structures from Fort Johnson, James Island, offer clues about nonconventional formwork design, besides demonstrating the skill builders applied when faced with complex casting operations. If the interpretation of Dr. Stanley South is correct, these are partly subterranean cisterns designed to store rainwater collected from the roof of barracks erected during the 1790s. Not quite identical, both are circular in plan with twelve-inch-thick tabby walls forming an approximately twelve-foot-diameter cylinder surmounted by a shallow saucer-like dome of tabby pierced by an opening just large enough to permit entry when interior cleaning or repairs were required. Walls stand about eighteen inches above present ground level. The larger southeast cistern is about 6'3" deep below the springing. Its dome—which tapers from twelve to eight inches in thickness—is flared slightly on the exterior, producing an agreeable bell-cast profile. The other roof is less considered but still remarkable considering the practical difficulty of fabricating a domical shape in tabby. Rather than horizontal boards, formwork for supporting walls utilized vertical boards (four to six inches wide) resembling barrel staves—an effective means of achieving the desired radial shape without steaming or bending—this simple method conceivably offering an analogy for procedures followed by Davenport on Tybee. Likewise the cisterns display no evidence for ties joining inner and outer faces of the formwork. Neither is there sign of any structural break between the dome and uppermost wall pour (measuring about twelve to eighteen inches in height), indicating that both elements were cast together—this singular procedure being without parallel.[48] Tabby itself consists of shell lime, sand, and an unusually rich concentration of oyster-shell aggregate. Broken and fractured valves show that the mix was vigorously pounded or rammed to produce an especially dense material capable of holding water without excessive leakage. Internal finishes comprise one or two coats of lime stucco applied directly over the tabby and two finish coats of a pinkish-brown Roman cement.

Tabby cistern,
Fort Johnston,
James Island, S.C.

Conversely, fabricating formwork for an octagonal building required no special expertise beyond knowledge of basic arithmetic. Drawings and descriptions of Spalding's Sapelo Island sugar mill—"forty-one feet in diameter, of tabby and octagonal"—were influential, this building's sixteen-feet-high exterior walls housing cast iron crushing machinery driven by horses or mules.[49] The proprietor of Elizafield near Darien, Georgia, followed Spalding's model, having either contacted the owner directly or obtained his "Observations on the Method of Planting and Cultivating the Sugar-Cane in Georgia and South Carolina," published by the Agricultural Society of South Carolina (Charleston) in 1816. The same influence apparently penetrated into Northeast Florida, disassociated tabby segments at White Oak Plantation, just over the Georgia/Florida state line near Yulee, outlining another octagonal structure almost identical in diameter (41'4") to Sapelo's mill house. Standing very near St. Mary's River, it was probably built by Zephaniah Kingsley Jr., whose Loyalist father traded merchandise out of Daniel DeSaussure's former Beaufort store during the American Revolution. Dr. Daniel Schafer identifies White Oak's feature with the residence occupied by John S. Sammis, an overseer (who eventually became Kingsley's son-in-law). However, I believe it more likely, given its close dimensional relationship with Spading's mill and the fact that Kingsley himself corresponded with the *Southern Agriculturalist* regarding manufacturing and clarifying sugar in 1830, that the now badly broken-up foundation formerly supported a framed building housing milling machinery purchased for White Oak in 1829.[50] No functional counterpart has emerged from South Carolina, but tabby foundations of the McKimmy House, an octagonal dwelling, might easily be mistaken for a mill building if it were not for indisputable cartographic evidence indicating its domestic character. Erected beside the May River (now Palmetto Bluff, Bluffton) during the 1770s, its foundation width (one foot six inches) could mean this building's lower story was tabby. Alternatively upper levels might have been timber framed, an ancient oak growing inside the ruin making full investigation impractical. Limited excavation has determined each of the eight sides was about thirteen feet long (measured on the outer face), their stuccoed and scored exterior faces imitating laid stone.[51]

● Choice of Mix ●

Apart from regular maintenance, few factors affected the durability of tabby more than the initial choice of mix. Thomas Spalding mixed lime, sand, water, and shell in equal proportion, reporting: "10 bushels of lime, 10 bushels of Sand, ten bushels of shells and ten bushels of water make 16 cubic feet of wall."[52] His recipe has been repeated uncritically by numerous recent commentators, yet practical experience with tabby conservation indicates that across Beaufort County mixes invariably contained higher proportions of shell, an observation confirmed by Dr. John A. Johnson (1819–93) whose formula comprised "one part, by measure of lime, two of sand and four to six of shells; quantity of the latter depending upon the strength of the lime and quality of the sand."[53] Conversely the amount of water Spalding recommended was excessive, experiments duplicating Spalding's material ratios separately conducted by Dr. Janet Gritzner and myself producing weak, friable products. Denser, stronger, and more durable tabby resulted when using just enough water to make the mix workable.

Writing in 1811 Alexander Macomb observed that the formulation for an unidentified structure near the mouth of the Cape Fear River (probably Fort Johnston) consisted of three parts shell, three parts lime, and three parts sand. Macomb himself preferred different proportions. The "strongest *tapia*," he said, was made with one part sand, two of lime, and three of shell. This was not a hard-and-fast rule since he continues, "the proportion of sand must be adapted to the quality of the lime, more or less according to its strength or weakness." In 1827 the same caveat was repeated, Macomb now reporting the favored mix along the Carolina and Georgia coasts was one part lime, two parts sand, and four parts shell, or "in the country, where good lime is to be found and no shell the proportions are 1 of lime, 4 of sand and 5 of stones."[54] The latter statement confirms that, like Thomas Spalding, Macomb did not define tabby as necessarily containing oyster shell.

Macomb "communicated" this "mode of building" to E. Gilman, whose pamphlet *The Economical Builder: A Treatise on Tapia and Pisé Walls* (Washington, D.C., 1839) recommended mixes composed of stones of all shapes and sizes place between timber edge boards "cemented in lime mortar." For Gilman the operation was simplicity itself. The novice builder was directed to "place a layer of stone on the ground as closely together as you can, and with a rammer, tap them all lightly, to adjust them in their places, then pour in a layer of mortar . . . over the stone; it [the mortar] should be made so thin that, with a little stirring, it will fill up all the vacancies between them; then another layer of stone, tapping them as before; then the mortar, and so on, till the mould is full."[55]

Regarding the quantity of material required for larger domestic buildings, basement walls of Haig Point House, Daufuskie Island, alone required approximately 5,010 cubic feet of tabby, with an estimated dry weight of about 576,000 pounds. The size of the labor force employed here is unknown—doubtless it was larger—much larger—than the six men and two boys who after collecting their own shell and burning their own lime each produced thirty cubic feet of tabby per week when building Spalding's Sapelo residence in 1805, completion of this single-story house taking two years or more. Military projects demanded far speedier results, achieved by large gangs of predominantly enslaved workers. Commencing tabby work at Fort Johnson, James Island, in April 1758, a Captain Williams said he would not have "occasion for more than Seventy Labourers," an unspecified number of additional men then being laid off.[56] In 1757–58 Fort Lyttelton, Port Royal, saw "fifty Negroes constantly at work . . . besides the Tappymaker and overseers."[57] Concurrently at Dorchester the workforce included both enslaved men hired from local landowners (including Walter Izard who was paid £295.7.6 for his "Overseer, Boat and Negroes") together with an unspecified number of "French laborers" (probably near destitute Acadian deportees) previously employed in repairing, enlarging or augmenting Charlestown's perpetually defective defenses.[58] Hired "Negroes" were also working at Fort Johnson, Messrs Fesch and Guignard (merchants and planters) earning £300 by their labor.[59] A much later account from Georgia relates how the crew building Tybee Island's Martello tower (1815) numbered forty individuals—a number considered less than adequate by military authorities.

Along with properly supervised labor, tabby making required good weather. Rain and excessive humidity increased setup times, while high, unsupported walls were susceptible to wind damage before they achieved adequate strength thirty days or more after casting. Mindful of hurricanes Spalding stopped all tabby construction in August. Gilman told his readers: "it is desirable to commence this kind of work pretty early in the season to give the wall time to harden. In good weather, the work of one day will be sufficiently hard to allow the moulds to be slacked off the next morning, to carry forward."[60] Security concerns made Macomb continue building fortifications irrespective of season or weather. "In the summer time one course of two feet in height may be made in a day," he wrote, "but in the winter it will take two days, and sometimes three or four days to dry the *tapia* sufficiently to receive another course." This firsthand account continues:

> the method of mixing the *tapia* is simply this: the shells are first laid on the ground, or a floor of boards, the lime on top of the shells, and the sand on the lime, which being well mixed together with water by means of a hoe or spade, is thrown up in heaps to be carried to the moulds or troughs; it is then carried either in tubs or hods to the moulds, and then emptied; it afterwards rammed so as to mix it the better, and force it into all the corners, which at the same time makes it more compact. All the implements necessary for carrying on the work and mixing the *tapia* are, a plumb line, pestles for ramming the *tapia* into the moulds, some hods for carrying the *tapia* to the moulds, hoes and shovels for mixing the composition, and the moulds or case.[61]

With or without knowledge of their supervisor, operatives—military and civilian alike—might throw broken brick into the mix along with the occasional conch shell, bottle, pottery sherd, or tobacco pipe, artifacts now exposed by weathering having become useful chronological indicators.

Tamping was important though not always carefully done, irregular pour lines indicating that formwork was clumsily filled using an overly wet mix and barely tamped—if tamped at all—during building of Haig Point's otherwise high-quality tabby-walled slave houses. Concealed by stucco finishes, this lapse did not become apparent until these dwellings fell into ruin and exterior stucco coatings spalled away. Gilman offered an explanation for such shoddy practice, recommending that forms be filled in layers, each layer sealed by "mortar" made so thin that "with a little stirring, it will fill up all the vacancies."[62]

By contrast foundations for tabby walls surrounding the cemetery and parish church of St. Helena, Beaufort—probably commenced by Colonel Thomas Talbird around 1800—exhibit high proportions of small and broken shell, the mix once poured having been vigorously pounded to produce as dense and lasting a material as possible. In any event the wall (enclosing an entire city block) failed above ground within forty years of installation either through lack of maintenance or because the foundation depth was inadequate (less than eighteen inches below grade) to start with. Gradually upper portions were removed and rebuilt using fired brick bedded on cut-down portions of the original fabric, a process that continues today.

ꙩ Risks and Failures ꙩ

Tabby's history is beset by melancholy incidents of failure and destruction, coastal environments with their storms and sudden shifts in topography adding more uncertainty to an inherently unpredictable technology. Started in November 1762 on Cedar Hammock, located eight miles below Sunbury, Georgia, a defensive battery of tabby provides an early example of near-total ruin by elemental forces, strong northeasterly winds "setting the sea to a very great height," washing the work away unfinished.[63] In October 1800 another late autumn storm smashed into Fort Johnson, James Island, reconfiguring the adjacent shoreline and breaching protective palmetto bulwarks. An Engineer Department survey made soon afterward shows that tabby outer walls erected in 1759 to replace an earlier structure dating back to 1708 were left broken and stranded ninety yards or more below the new high-water mark. Work installed by General William Moultrie was overwhelmed and the "Grand Battery" (built 1793 and repaired at government expense in 1796) partially undermined.[64] Few if any reparations were ever made, the fort remaining unused until abandoned during the 1820s.

Vicissitudes of Fort Johnston, located three or four miles from the mouth of Cape Fear River, North Carolina, were extreme, with revolution, storm, and fatal underestimation of tabby's strength all contributing to repeated destruction. Heavily damaged during the American Revolution, the earliest structure was a "small but regular fort of strong cement or *Tapia* walls, with four bastions, a gateway in the rear and lower tier of guns in the front."[65] In 1793 plans for its replacement were drawn up by Colonel Martine, an exiled French engineer from Santo Domingo, but never implemented owing to high projected costs. Stalled, rebuilding did not get underway until 1799, when a contract negotiated with the U.S. Engineer Department called for another fort, the walls comprising inner and outer tabby skins erected in parallel, compacted sand filling spaces between. About nine feet high, measuring 2'6" feet in thickness at base, and "tapering on the outside to 18 inches at the top," the tabby proved too thin (or so the military later thought), consolidation of the sand by pounding causing near total collapse "as if an Earthquake had thrown the Surface into Hills and Holes."[66] Subsequent restoration was desultory, stymied by inadequate funding and lack of purpose, with bored officers and men finding the little town of Smithville on the fort's east side an accommodating and comfortable alternative to their designated quarters. In February 1810 J. G. Swift inspected a "*tapia* barrack 60 by 20 feet nearly finished," which he thought too low and damp, along with officers' quarters begun in tabby but already falling down. An enclosing "tapia" wall (two feet thick) was also unfinished, the only completed element being the substantial tabby parapet (measuring about 6'6" in width) protecting the battery.[67] Why some work seen in 1810 had quickly deteriorated is probably explained by procedures described by Joseph J. Smith who had received orders in September 1804 to complete the battery of tapia at the site Old Fort Johnston—which had been contracted for by General B. Smith. According to J.G. Smith's diary:

> Soon after this the slaves of General Smith commenced the burning of lime in pens, called kilns, formed of sapling pine containing from one thousand to one thousand two hundred bushels of oyster shells (alive) collected in scows from the shoals in the harbor—there abundant. The pens were filled with 'lightwood' from pitch pine. And thus were burned in about one day—very much to the annoyance of the neighbourhood by the smoke and vapour of burning shellfish when the wind was strong enough to spread the fumes of the kilns.
>
> In the succeeding month of November I commenced the battery by constructing boxes of the dimension of the parapet, six feet high by seven in thickness, into which was poured the tapia composition consisting of equal parts lime, raw shells and sand and water sufficient to form a species of paste or batter, as the Negroes term it.

Noxious fumes which so annoyed Smithville's inhabitants indicate that shells burned in their neighbourhood had not been cleaned. Recently collected from local reefs, the shell must have been far from salt free, a condition that would have contaminated lime produced by burning. Consequently, tabby made of such material was defective from the start, the "paste" or "batter" mentioned by J.G. Smith suggesting the mix or—"composition" as he called it—was further weakened by the use of an excessive amount of water.

Premature failure of Fort Frederick (a topic already considered in chapter 2) further demonstrates that misunderstanding or ignorance of tabby's behavior under stress could, and sometimes did, lead military contractors into serious difficulties. The same was true for civilians, an unusually detailed series of reports describing how in 1842 Beaufort's early Baptist Meeting House showed symptoms of stress and structural incapacity. The first sign of trouble was its sagging roof, Reverend Richard Fuller recounting how an informant "had reason to think that the church roof and north end wall were to some extent giving away." Thoroughly alarmed church authorities brought in three "mechanics" who examined the building. Two of them, John M. Zealy and Abram Cockcroft, found (August 26, 1842) that loads imposed by roof rafters and tie beams were forcing exterior tabby walls outward, thus rendering the church unsafe.[68] Moreover walls themselves were not sturdy enough to support any new or repaired roofing system. Despite emotional appeals it was resolved to demolish the old building and erect a new brick (rather than tabby) church immediately south of the original structure—this handsome neoclassical edifice still serving the congregation.[69]

Proper positioning and leveling of "boxes" was critical. James Julius Sams described how his father, Dr. B. B. Sams, was almost killed when "building an outhouse of large dimension in Beaufort. He found that there was something wrong with its construction. It had been carried up beyond the first storey. While walking around it, it fell well-nigh, covering him with its ruins. The defect had been in one of the boxes. It had not been placed in a direct line, square with the others. In other words it produced a bowing wall and a bowing wall will certainly fail."[70]

Accidents occasioned by inexperience, faulty construction, or lax supervision reminded builders that tabby dictated its own rules. These rules could be learned only through observation or trial and error, no published account, not even Spalding's, providing more than rudimentary guides. And for promotional purposes certain authors downplayed the skills required of tabby makers and tabby's inherent deficiencies. Thus Alexander Macomb reported that the officers' quarters at Fort Johnson, South Carolina, were "built by the labour of Africans, just arrived in the last of the slave ships; a fact which I mention to show that it does not require any remarkable skill to construct a tabby building."[71] Additionally he ignored a George Jenkins, who through an inquiry published by the *American Farmer* in 1828 (following republication of Macomb's article about tabby during the previous year) asked, "is a house built of *tapia* not subject to damp?"[72] Personal experience attests the concern was legitimate, tabby walls acting like giant sponges, sucking up ground water by capillary action, sweating, and making enclosed spaces "unwholesome and unpleasant" in summer, as Jenkins feared. Moreover most contemporary authorities emphasized tabby's supposed durability and strength, which was misleading, critical examination of the material's mechanical properties leading to different conclusions.[73]

Tabby Construction Details and Operational Procedures

The early 1800s saw tabby construction on an impressive scale among military projects designed and overseen by professional engineers, or engineers in training. Regrettably not enough of their work survives to fully appraise the exceptional confidence, skill, and courage these men displayed when building in coastal environments distanced for the most part far from urban centers. Techniques used for contemporary domestic structures can be reconstructed with more certainty from several relatively intact dwellings and numerous ruins scattered across the Sea Islands since tabby—even when abandoned or broken—retains evidence for its making, including formwork impressions left before wall surfaces fully dried and cavities created by formwork "pins" after removal. Pockets, sockets, fixing pieces, and ghost impressions preserve dimensions of lost timber joists, plates, flooring elements, window frames, lintels, and trim items. Tabby outbuildings, slave dwellings, and subsidiary structures were similarly fashioned, generally with less attention to detail. Construction details described typify empirically derived norms established through processes of trial and error begun decades before the American Revolution, itinerant building workers hired out by local plantation owners ensuring that this collective experience was handed on through generations until fatally disrupted by the Civil War.

◦ Mechanical Properties of Tabby ◦

While different proportions of sand, lime, shell, and water alter tabby's mechanical properties, tests indicate that the material invariably possesses low tensile and shear resistance. Sickels-Taves and Sheehan reported: "historic tabby has a compressive strength of approximately 350 psi," basing their figure on samples from Cumberland Island, Georgia.[1] At the Barnwell Gough House, Beaufort (ca. 1780), cores yielded average compressive values of 350 psi, with lows falling below 25 psi. Marked variation of this order reflects lack of uniformity during the original casting process and differential preservation, with moisture penetration, subsequent leaching, and deterioration of tabby having occurred at different rates in different parts of the building. Values for tabby's tensile strength are relatively slight, Fischetti suggesting that a figure between 10 to 15 percent of the material's compressive strength is probable.[2] Casting semiliquid materials introduced other weaknesses. Adhesion between successive pours could be tenuous, causing disassociation and structural separation along horizontal lift lines. Ruins of the Callawassie sugar house and Baynard House, Hilton Head Island, illustrate such failure. Following loss of roof framing, exterior walls moved out of vertical, split horizontally, and then fell into separate segments. Not often apparent, eccentric loading of tabby walls, especially those weakened by age or long periods of neglect, might cause them to split vertically (laminate), a condition manifested at Botany Bay, Edisto Island, by the sites' tabby-built garden shed, where outer wall faces are bowed outward while inner faces remain more or less vertical.

In multistory construction local builders offset tabby's limitations by minimizing mass, wall thickness often being reduced with height. Judicious incorporation of timber supporting or restraining members helped achieve uniform distribution of loads and were capable (unlike tabby) of resisting tensile forces. By the nineteenth century tabby buildings were characterized by a certain structural leanness that allowed them to rise through three or even four stories.

◦ Foundations ◦

No set pattern for foundation design is evident even among Beaufort's larger tabby structures. Following excavation of trenches defining the overall building shape, foundations were often cast without form boards as continuous strips, these strips (usually made wider than upper walls) creating firm support and a level base for erection of timber "molds" needed when casting the outer building skin. Examples include Haig Point House, Daufuskie Island, and Phase II, William Sams House, Dataw Island.[3] In Beaufort external walls of the Barnwell Gough and Saltus Houses were formed without any kind of spread foundation after trenches outlining the building footprint had been excavated downward into yellow sand two or three feet below grade.

Now demolished the DeSaussure House (713 Bay Street) had an usually deep basement, its lowest foundation level lying 5'3" below present street level. Construction involved either digging wide foundation trenches or digging out the entire basement area before construction began. General Macomb remarked: "if it is intended to have a cellar, the excavation is made accordingly, but rather larger than for a stone wall, so as to give room to set the frames and let the air pass around the wall after it is made, in order that it may dry quicker."[4] In the present instance walls below ground necessitated three separate tabby pours (each twenty-four inched high) cast 1'9" wide, which seems undersized considering this dwelling incorporated 3½ stories and must have suffered occasional tidal flooding. Tidal action of more direct kind was better anticipated by James Simons, Charleston's collector of customs, when he sought bids for construction of a timber-framed warehouse "80 feet × 30' one story 14 feet high, with a hip roof" situated at Point Comfort near Fort Johnson, James Island. Published by the *South Carolina State Gazette,* June 18, 1800, his advertisement stipulated: "foundations must be substantially piled, so as to bear any weight that may be put in the building, and to be enclosed by a strong breast-work, made of tabby or other cheap and durable materials, so as to defend the walls from being injured by the spring tides; or undermining the building."[5] Although tabby's use in similar situations was probably widespread, instances have passed unnoticed except for so-called seawalls protecting residential areas of Charleston and Beaufort against flooding.

Significantly modified, two Beaufort dwellings—the Johnson House (414 New Street) and Hext House (Hancock Street)—are both late eighteenth-century timber-framed structures raised over an elevated tabby basement standing at least head height above grade. Commonplace among town and rural dwellings, construction of this type guarded living spaces against flooding or hurricane-induced storm surges, facilitated natural ventilation, and helped keep out vermin—rats, snakes, and raccoons—all nuisances for households located near rivers or marshes. Most raised basements enclosed storage spaces or work areas for domestics, one or two (including the B. B. Sams House, Beaufort) having large cisterns holding rainwater—a convenient and sometimes necessary alternative to public or private wells contaminated by salt or water-borne disease—cholera and yellow fever taking tolls with frightening regularity.

◦ Floors ◦

Mortar mixes containing shell-derived lime, finely crushed oyster shell, sand, and water were cast directly over prepared subsoil to create what are loosely termed "tabby" floors. The practice followed ancient precedent, tabby substituting for "plaster" in horizontal surfaces known from classical—even prehistoric—times. Spanish-period tabby floors are documented at St. Augustine, Florida, as far back as the 1670s. Beaufort's earliest recorded example (no longer extant) is described by returns of work done at Fort Lyttelton before 1758. Two of the town's elite federal-style dwellings—the James Robert Verdier House, built about 1825 (now Marshlands, 501 Pinckney Street), and the Francis Saltus House, built about 1790 (802 Bay Street)—had basement floors similarly made. Rural instances include the Horton House, Jekyll Island, where excavation has exposed fragments of lime-mortar flooring attributable to the early 1700s, and damaged fragments surfacing an enclosed space linking two wings added to the old William Sams House, Dataw Island (ca. 1812). An unidentified ruin (possible cotton house) located off Squire Pope Road, Hilton Head Island, preserves a remarkably intact example, consisting of shell-lime mortar cast four inches thick of over a well-compacted oyster-shell fill twelve inches or more deep.

◦ Wall Construction ◦

Beaufort's earliest houses rarely incorporated more than 1½ stories. Soon after the American Revolution, an unprecedented trend toward multistory tabby construction became evident, the Saltus House, Barnwell Castle, and William Elliott House each having three full stories raised over an elevated basement. Following practice set by contemporary stone and brick masons, local tabby makers reduced the thickness of their walls at each higher floor level, this—if properly calculated—minimizing the amount of shell to be collected, burned, hauled, mixed, and placed. The London Building Act of 1619 and its successors issued through the eighteenth century were influential in South Carolina (although never absolutely followed), these acts codifying permissible relationships between external wall thickness and building height. For the "least sort of house" (incorporating two stories, a basement, and an attic) the Building Act of 1667 stipulated basement and ground-floor walls "were to be 2 bricks thick, the first floor 1½ bricks thick and parapet 1 brick thick." For a "better" sort of dwelling (containing three stories an attic, and a basement), basement and ground-floor external walls were to be 2½ bricks thick, first-floor walls two bricks thick, second-floor walls 1½ bricks thick, and the parapet one brick thick.[6] Spalding (1844) said walls of his Sapelo Island residence were two feet wide at their lowest level, fourteen inches wide at first-floor level, and ten inches thick above, noting: "beyond that I would not erect Tabby buildings."[7] E. Gilman recommended slightly different dimensions, although the underlying principle governing wall construction remained the same: "the thickness of the wall will depend on the size of the building, the height to be carried & c. For a large house of two stories I should advise the cellar or basement story to be twenty inches, the first story above ground sixteen and the second twelve inches thick."[8] Beaufort builders utilized similar ratios during the mid-1780s, preferring somewhat higher safety margins when casting lower wall levels. Tall, double-pile structures obviously demanded the most substantial construction. Basement walls of Haig Point House, Daufuskie Island, an exceptionally large 2½ story T-shaped house, were twenty-four inches thick and remain serviceable, despite partial demolition during the Civil War and subsequent adaptation to support the present lighthouse. Single-pile construction featuring shorter structural spans and less weighty roof frames allowed thinner external skins, individual contractors exercising discretion based on their own experience when determining the appropriate wall width for any given situation. Table 1 compares theoretical recommendations regarding appropriate wall thickness made by various authorities with actual examples realized.

Satisfied that foundation strips (if any) were properly set out and everything appeared level or plumb, the tabby maker began casting walls, which climbed incrementally higher round by round. Normally casting was interrupted when work reached each higher floor level. Formwork was then adjusted to make walls less wide, this creating internal ledges used as bearing for wall plates onto which timber floor joists were set before casting resumed. Dead weight was further saved at the Saltus House by lowering story height floor by floor—a sensible precaution for a narrow site washed along one side by the Beaufort River before its owner (Captain Francis Saltus) installed protective bulkheads. Between first and second floors (now altered), the original vertical measurement was about 12'6". Between second and third floors, this height reduced to 10'11", the third floor—occupied by servants or children—measuring only 7'–5½" between floor and ceiling. Brick chimney stacks were carried upward with exterior walls, porch framing, and floor joists, helping to strut and stabilize the building as construction proceeded.

TABLE 1. Wall Thickness in Selected Tabby Buildings from Beaufort District Compared with Requirements of the London Building Act (1667) and Recommendations of Thomas Spalding (1844) and E. Gilman (1839)

Authority or Site	Stories	Date	Basement Thickness	1st Floor Thickness	2nd Floor Thickness	3rd Floor Thickness
London Building Act	2+ basement	1667	1'6"	1'6"	1'1½"	(9" parapet)
London Building Act	3+ basement	1667	1'10½"	1'10½"	1'6"	12½"
Thomas Spalding House, Sapelo	2½	1844	2'	1'2"	10"	
E. Gilman	2+ basement	1839	1'8"	1'4"	12"	
Saltus House, Beaufort	3½	ca. 1797	1'9"	1'6"	1'3"	12"
Barnwell Gough House, Beaufort	2½	ca. 1790	1'10"	1'5"	12"	
Edwards House, Spring Island, wings	2	ca. 1805		1'3"	12"	
B. B. Sams House, Dataw Island, wings	2	ca. 1816		1'2"	12"	
Baynard House, Hilton Head Island	2			1'10"	12"	
Haig Point House Daufuskie Island	2½	ca. 1833	2'	1'4"	12"	

● Floor Joists ●

Judged by modern standards, floor joists appear undersized in most extant tabby structures, this supposed deficiency indicating owners and merchants obtained top-quality materials—frequently old-growth timber—harvested from their own lands or imported from commercial centers. First and second floors of the Barnwell Gough House are carried by 9" × 3" joists of cypress (*Taxodium sempervirens),* a swamp-loving species growing to impressive proportions all across the lowcountry. Laid eighteen inches on center, these span twenty-six feet in the clear between the exterior and a pair of centrally positioned 9" × 9" pine beams running north/south between front and back walls of the building. At second-floor level there is no intermediate support for similar beams other than light timber framing—their span approaching forty-five feet in length. Joists of the Saltus House are similarly sized in section (9" × 3") but cut from heart pine. Spaced seventeen

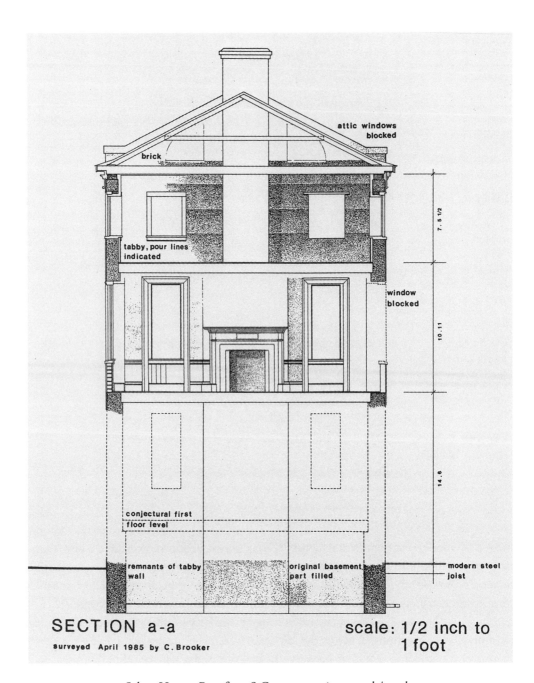

attic windows
blocked

brick

tabby, pour lines
indicated

window
blocked

7. 5 1/2

10.11

14.6

conjectural first
floor level

remnants of tabby
wall

original basement
part filled

modern steel
joist

SECTION a-a

scale: 1/2 inch to
1 foot

surveyed April 1985 by C. Brooker

Saltus House, Beaufort, S.C., cross section, north/south.

inches on center, they span twenty-four feet between back and front walls, again without intermediate support. Tongued and grooved pine boards nine or twelve feet wide (occasionally wider) and one inch or more thick, nailed to joists, stiffened structures besides improving bearing capacity.

⚬ Window and Door Openings ⚬

Window and door frames were placed into position as casting proceeded. The technique was simple. Spalding wrote: "all that is necessary when you construct doors or wi[n]dows, is to drop a short board across the wall between the outer and inner planks and steady it with two poles, to be drawn out at each round and replaced at the next, and so continue until you have reached the height you intend your doors and windows. When you then you drop your Lintall into the Tabby Box, so as to secure the next round of Tabby your wall then becomes an intire whole."[9] Gilman gave similar (if less detailed) advice: "the doors and windows are set as the work progresses, but in an unfinished state. The cills and jams [jambs] only are made, like a box, the thickness of the wall braced out, to prevent the wall from pressing them inwards." He added, "in all thick walls, the frames should of course be flaring on the inside, being handsomer, and giving more light to the rooms." Splayed window openings are difficult to detect in ruined tabby buildings, examples including the unidentified structure already mentioned located off Squire Pope Road, Hilton Head Island, now owned by Mr. Thomas Barnwell; and Phase I, William Sams House, Dataw Island.

Always an accurate observer, La Rochefoucauld-Liancourt noticed that brick was generally used as an infill below windows.[10] This practice, which simplified casting, is confirmed at the Saltus House and Barnwell Gough House, where window openings extend from floor to lintel level, spandrels of fired brick laid in a single wythe infilling space left beneath window frames positioned three or more feet above the floor itself. Dr. B. B. Sams installed spandrels of tabby brick rather than fired units below large windows lighting wings flanking the old William Sams House, Dataw Island. Stuccoed and scored on exterior faces, infill materials have proved lasting and remain almost indistinguishable from stuccoed tabby walls. If porch or balcony access was required, jib doors might be substituted for spandrels, these consisting of a pair of side-hung timber panels resembling solid interior shutters opening inward against relatively deep reveals (typical features of tabby construction) when the lower sash of a double or triple sash window was raised.

Among larger domestic structures, most glazed windows were set nearly flush against the building face and comprised two sashes, the lower sliding vertically, the top one permanently fixed. There is little evidence for use of weights or pulleys among the few buildings where late eighteenth- or early nineteenth-century fenestration survives. Consequently ventilation was not as good as it might have been, and users risked smashed fingers if opening and shutting was clumsily done or small wooden (sometimes metal) props holding an open window in place slipped. Six-over-six pane arrangements were favored for windows lighting principal rooms of better-quality dwellings. Mutins dividing the panes underwent refinement, becoming thinner and more elegant over the Federal period, risking breakage during high winds, as Thomas Chaplin Jr. discovered at Tombee on St. Helena Island.

Insect screens were almost unknown, with "pavilions" (canopies) over beds, as listed by inventories, keeping away mosquitoes and sand flies. Coupled with excessive heat and humidity, these irritants could become unsupportable for island residents. Consequently Beaufort's affluent antebellum families removed themselves temporally to summer colonies (St. Helenaville and Bay Point in particular), abandoning their grander houses for simple timber-framed cottages (strung out along the shore at Bay Point), hoping these would be more "wholesome" than the stifling, ill-ventilated, and, in the case of tabby ones, damp permanent residences they left behind.

Shielding rooms from sun and storm, paired window shutters are indicated by metal "butterflies" set into exterior tabby walls, these S-shaped swiveling devices mounted on spikes holding shutters against the outer building face when opened. Spikes were driven either into tabby walls before they had fully set or into small wooden fixing pieces cast in place as construction proceeded. Civil War–era photographs of Retreat and Myrtle Bank (both on Port Royal Island) show solid timber and paneled shutters, basement windows in the latter case being mere voids furnished with simple wooden frames equipped with horizontal wooden bars for security. Louvered shutters had been installed on several of Beaufort's elite houses by the 1860s, these allowing natural ventilation even when shut. Reflecting social chasms separating owners and their workers, almost all rural and the majority of urban slave houses were sparsely fenestrated, and unglazed windows closed with a single side-hung timber shutter braced and battened for strength, which might— or might not—keep out the weather.

Timber lintels (typically 2½"–3" in depth but sometimes only one inch deep) prevented cracks and splits above windows and exterior door openings, the lintels made wider than frames to give proper bearing. The Barnwell Gough House retains trimmed but otherwise unshaped logs over windows, an altogether primitive treatment considering the dwelling's size and sophistication. Minor dimensional discrepancies around wall openings resulting from defective formwork, poor setting out, or casting mistakes could be masked by decorative timber moldings such as wide federal-style architraves.

• Chimneys •

Fire will burn through tabby. Mindful of safety owners with sufficient means erected chimneys of fired brick when building their own houses. The frugal minimized costs by setting chimneys on tabby foundations, which, if freestanding might take the form of subterranean pads (Haig Point) or alternatively more massy elevated bases. Dr. B. B. Sams cast large, freestanding tabby chimney supports centered in each of two wings flanking the old William Sams House, Dataw Island. A full story high, these rose solid until they reached principal floor levels, where construction switched to fired brick for back-to-back fireplaces and chimney stacks. Brick—considerably safer than homemade lime-mortar units—was an expensive option necessitating shipment from Charleston or Savannah.

Keyed interface between brick chimney and tabby exterior wall, Retreat Plantation, Beaufort vicinity, S.C. (Jack Boucher, HABS).

Excavated ruins of Montpelier Plantation (now Palmetto Bluff, near Bluffton), located beside the May River, exhibit more opportunistic building modes. Here paired central chimneys built solid up to the elevated first floor consisted of a jumbled mass of ballast stone and coral all cemented together with lime mortar and faced with tabby brick laid in two wythes.[11] Coral suggests this unlikely core came from a ship or ships trading with West Indian or West African ports, coral also appearing amid late eighteenth-century fills associated with bulkheads installed in front of his Bay Street house by Captain Saltus. Whether Montpelier's chimney stacks were constructed of fired or tabby brick is uncertain, everything having been thoroughly robbed following the building's ruin. Another solution is presented by Dr. Jenkins's house, located off Seaside Road, St. Helena Island. Substantially ruined this was a timber-framed, double pile T-shaped dwelling, raised over an unheated basement. At ground level two back-to-back fireplaces serving rooms positioned right and left of a central hall were carried by arched tabby supports cast in an H-shaped configuration, the supports projecting from a substantial transverse tabby wall forty feet long dividing front and rear basement spaces. Supports and the wall were poured together without any structural break in twenty-four-inch-high vertical increments. Nothing else remains of the house, lack of fired brick fragments suggesting chimneys were constructed in tabby brick.

When extending his Spring Island house right and left, George Edwards took an entirely different tack, building chimneys—one in each new wing—of fired brick from the ground up. Erected as discrete, structurally independent elements against the inner face of an exterior tabby wall, his procedure—recommended by Macomb in 1827—minimized differential settlement and any consequential cracking, besides reducing the risk of fire consuming the building.[12] Much the same method was followed when constructing the slave tenement located near Spring Island's main house, two symmetrically placed chimneys of fired brick—now totally robbed—standing against the inside face of its north exterior tabby wall.

Brick internal-end chimneys of Beaufort's Saltus House line flush with the outer tabby skin. Again construction was discrete, no attempt being made to bond or key brickwork of these weighty elements into the tabby, through joints so created being concealed by exterior stucco and interior plaster. Stacks rising about 43'9" above Bay Street's present grade had their uppermost two courses corbelled and were stuccoed smooth. Conversely Retreat Plantation's brick chimneys were bonded into adjacent tabby walls as wall construction proceeded. The alternate long and short pattern adopted matches individual wall pours in height. Laid up using an irregular variant of English bond, Retreat's expensive red brick was meant to be seen, the mason creating decorative diaper designs using dark glazed headers, which contrast agreeably with their lighter background.

◈ Roof Framing ◈

Roof framing commenced once exterior tabby walls had reached full height and were deemed sufficiently "cured"—temperature, humidity, and rainfall all entering into calculations made by contractors concerning whether or not work might safely proceed. It was also necessary to complete the business of gathering materials, cutting timbers, adzing them to size, making proper joints, and—judging by roman numerals scratched into rafters—preassembling portions of the roof frame. Large, double-pile tabby dwellings with hipped roofs presented particular challenges. Secondary members were usually carried by heavy (often very heavy) timber trusses, which required lifting or putting together thirty or even forty-five feet above ground without damaging newly cast walls. Attention has already been drawn to solutions exhibited by the Barnwell Gough House, Tabby Manse, and William Elliot House (Anchorage)—all hipped-roofed structures where paired king-post trusses running north/south from front to back span forty-two feet or more without any intermediate support. Nearly identical span dimensions in all three buildings can hardly be coincidental, though whether they were achieved by one group of carpenters moving from project to project or are alternatively the product of an empirically derived local tradition is not known.

At the Barnwell Gough House, each of its two parallel king-post trusses spaced 8'6" apart receive two principal rafters directly and a through purlin secondary rafter system indirectly (the rafters measuring 6" × 2¾"). Bottom cords are each cut from a single baulk of timber adzed smooth and appear slight (measuring 9" × 6" finished) for the very long structural span involved. Ceiling joists extend between lateral (east and west) external walls (where they rest on continuous 9" × 4" timber plates) and the bottom chord of a truss. All carpentry joints are carefully mortised, tenoned, and pegged, showing how

Barnwell Gough House,
Beaufort, S.C., detail of
king-post truss.

labor-intensive, demanding, and expensive installation of this and similar roofing systems must have been.

Captain Saltus realized economies with his gable-ended townhouse by opting for a relatively narrow, linear floor plan, which shortened structural spans. Here long north and south exterior walls are capped by a substantial (10½" wide × 5" deep) timber plate. Plates receive ceiling joists (averaging 8¼" × 2½") cantilevered at their extremities about 8" beyond the building face. Ceiling joists carry false plates (8½" wide × 1" deep) aligned east/west, the two false plates supporting rafters (average size 5" × 4½") pitched at an angle of about 30 degrees. Housed and pegged together, each rafter pair is braced by a timber tie (4" deep × 3" wide), the rafters spanning about twenty-four feet without any intermediate support across the building's short axis. Advantages in terms of weight and monetary outlays gained by this relatively light system were, as already seen, enhanced by reducing exterior wall thickness and ceiling height at each floor level.

◦ Roof Finishes ◦

Although incombustible, slate and tile were infrequently employed for roofing tabby buildings, most owners finding shipping costs too high and both materials too easily dislodged by high winds, which could turn them into lethal missiles. In 1814 the trustees of Beaufort College took extreme measures, stripping the building of its original slate covering damaged during a storm and installing (at the cost of $1,000) timber shingles instead. Slate fragments found during excavation indicate similar replacement occurred elsewhere in Beaufort Town and the Sea Islands, "cedar" shingles generally being the preferred roof finish for tabby and timber-framed structures alike.

◦ Tabby Roofs ◦

The description by Pedro Menéndez Marqués of flat roofs fabricated from oyster-shell lime (*acuteas de cal*) at Santa Elena (Parris Island) is the earliest (late sixteenth century) recorded use of what might be called "tabby" roofing known from South Carolina.[13] Analogous construction can be inferred from later accounts, including Thomas Spalding's mention of drawings of a tabby roof sent to Major Hamilton, most likely then resident on Callawassie Island.[14] Unfortunately the drawings and any structures they inspired have disappeared. If built these tabby roofs followed the pattern of Spalding's own

Tabby roof, dairy at William Sams House, Dataw Island.

house or perhaps flat roofs enclosing double slave dwellings described by Floyd from the Thickets near Darien, Georgia, where riven boards, laid over 3" × 10" horizontal joists, supported shell-lime mortar cast about four inches deep.[15]

Today Beaufort County's only surviving tabby roof encloses the small single-story tabby cold room or icehouse attached to the ruined tabby dairy facing Dataw Island's William Sams House. Asymmetrical when seen from the exterior, this roof defies categorization, appearing gabled at one end and slightly hipped where it abuts earlier construction. Inside all irregularity disappears. Left open to view, it comprises two opposed tabby skins pitched near forty-five degrees. Remarkably, considering the difficulty casting must have caused, these tabby features diminish in thickness with height, impressions indicating use of narrow, horizontal boards for the formwork.

• Brick Vaulting •

Incompletely preserved and heavily restored using salvaged material placed randomly, mention should be made of the King's Magazine at Frederica, Georgia, a large, rectilinear tabby structure erected in the 1740s that had a central gate house (now lost) and two rooms located side by side designed for storage of gunpowder. Both rooms were vaulted in brick, a sketch by Albert Manucy showing the original vault almost semicircular in section, the combination of sturdy tabby supporting walls and brick vaulting recalling Iberian precedent (see chapter 1).[16]

• Internal Finishing •

With roofing complete and everything securely protected from the elements, finishing work commenced. Among tabby houses most internal partitions were timber framed except at basement or ground-floor level, where tabby cross walls might carry and distribute loads transmitted from construction above. Exposed tabby surfaces were usually plastered on interior faces, the plaster applied directly over the tabby. The Saltus House had internal walls lined with timber laths nailed onto timber battens—an unusual treatment for a tabby structure (though normal among contemporary timber-framed buildings), which improved insulation and helped mitigate dampness. Plaster was applied over the lathing in two or more coats.

Decorative plasterwork was always rare in Beaufort Town, a classically inspired roundel of acanthus leaves decorating the dome installed between about 1825 and 1830 over an elliptical staircase winding through the Saltus House and the dome itself having no known local parallels. Fully paneled rooms had become an anachronism by the 1780s, but several townhouses built before 1810 featured this labor-intensive treatment in reception rooms, where it could be admired by visitors—notably at the Barnwell Gough House and Tabby Manse. Since different timbers were used—heart pine for surrounds and swamp cypress for large floating panels—all paneling was painted. Fashionably conceived chimney pieces gracing principal rooms might be enriched with applied "composition" medallions and Adamesque plaques depicting classical figures suitably draped, imported via Charleston from England or the Northeast. Subsidiary spaces received less expensive timber wainscots, molded chair rails, overmantels, baseboards, and cornices, chimney pieces, if classically inspired, being plainer and smaller.

Trim items were usually nailed to small pieces of wood let into the tabby before it dried or set in place during casting. Occasionally more substantial timbers are found, builders of the Chapel of Ease, St. Helena Island, having cast four-inch-square fixing pieces into tabby of the outer walls. These ran completely around the original building, except, of course, where interrupted by windows or doors. Carefully tenoned at building corners, these timbers still retain their fragrance when cut.

• External Surfaces •

Having no desire to see exposed oyster-shell matrices, pour lines, or holes left by form-work cross ties, owners required exterior tabby surfaces be finished with carefully applied coatings. Experience showed that if these were omitted owing to lack of cash or lost through poor maintenance, tabby would disintegrate, moisture penetration into the fabric setting into motion cycles of deterioration accelerated by successive freeze and thaw conditions. Tybee Island's Martello tower (erected 1812) exhibited noticeable surface friability by 1835, having perhaps already lost its external finish (if indeed it ever had one), Lieutenant Mansfield observing, "the nature of the Tabby is such that this tower is gradually wasting away and every wind [leaves a] mark on ones clothes. . . . Fortunately in this case, the walls are thick and sound and will stand for years and answer the purpose intended."[17] Domestic construction was different, margins of safety provided by relatively thin exterior tabby walls leaving little or no latitude for any significant surface erosion. Consequently exterior surfaces of tabby dwellings and all other tabby structures—excepting the most utilitarian—were finished with stucco, which besides protecting underlying fabrics from the elements concealed their inexpensive nature beneath more or less (depending on the owner's taste and financial standing) decorative veneers.[18]

In London the appetite for stucco facings became almost universal after 1770 as great landed estates were redeveloped by several generations of architect-builders who sought to unify speculative housing schemes developed piecemeal street by street and square by square. Cheap stucco was invariably scored and painted to resemble expensive stone, a practice taken up by lowcountry builders, who, like their European cousins, imitated coursed ashlar blocks (usually twelve inches high) and incised voussoirs, even keystones over wall openings. Regular whitewashing or color washing kept painted surfaces fresh and sealed any hairline cracks. Occasionally stucco over tabby would be burnished and frescoed to mimic marble, the Fuller House (northeast corner of Bay and Carteret Streets, Beaufort) being treated this way. Quoting the prominent English architect Henry Holland, the *American Farmer* noted, "plaster proper to serve as a ground for Fresco painting or colouring, is made of one part lime, and three parts clean, sharp, washed sand; this sort of painting has lately been executed with great success at Woburn Abbey, and some other places," these structures, designed by Holland, being of pisé de terre.[19] In Beaufort visual inspection suggests constituent materials included oyster-shell-derived lime; fine, perhaps drift sand; and instead of marble dust recommended by English pattern books, well-pulverized oyster shell, all carefully sifted. Mixes were applied in two or more coats over smoothed and filled tabby surfaces.

Macomb stated, "*tapia* walls . . . may be rough cast on the out and marked off like stone work if desired." Rough casting consisting of lime or lime mortar to which sand or gravel had been added is mentioned by contract documents dated 1799 for the ill-fated rebuilding of Fort Johnston, North Carolina, which cite Baldhead Lighthouse as an exemplar. Colonel Thomas Talbird told trustees of Beaufort College that he would rough cast their new tabby building for 2 shillings and 6 pence per square foot, but, it is doubtful if his offer was accepted. At the Ralph Emms Elliott House, Beaufort, an elevated three-story tabby structure built around 1800, with brownish exterior stucco applied and pebble dashed during extensive alteration about 1900 as part of a major remodeling program by William G. Preston, an architect from Boston, suggests the new finish material was some kind of Roman cement (typically reddish or brown in color) applied over existing lime-mortar stucco or perhaps the original tabby as a means of waterproofing the neglected building at a time when American production of natural cements was at its peak.

Tabby Brick, Wattle and Daub, and Cements

Shell lime with or without added aggregate found numerous uses across the Sea Islands. The principal binding agent in traditional tabby mixes, it might be molded into tabby brick—an altogether cheaper material than imported fired brick—daubed onto timber hurdles (wattles) enclosing slave houses, or simply plastered over chimneys to render them more or less weather tight. Tabby brick was fabricated from the mid-eighteenth century onward but did not long survive the Civil War. Always ephemeral wattle and "tabby" (i.e., shell-lime) daub construction—with a history going back into earliest colonial settlement periods, Spanish, French, and English—also disappeared at much the same time. These changes reflect major social shifts and dislocations following military incursions, the dispersal of formerly enslaved workers, an influx of "outsiders," and paradoxically perhaps the advent of improved distribution systems for new building products.

Among the latter cheap hydrated lime produced out of state from quarried limestone, and cements—either natural or artificial—displaced locally made materials, shell burning becoming a thing of the past. Dr. G. E. Manigault said this trend became evident in Charleston decades before the Civil War began. What may have been true for a major urban and shipping center did not necessarily reflect the situation among small towns much less Sea Islands cut off from highways and railroads, these places remaining hopelessly backward and cruelly isolated from civilization in the view of well-meaning northern missionaries who found nothing about them that accorded with their own progressive ways.[1] It is clear that traditional tabby making did cease—though not at once. Old methods were adapted or modified, formed materials comprising sand and oyster-shell aggregates bound with natural cements at first and Portland cement after 1900, still being fabricated through the 1930s. Conscious revival of historic tabby-making techniques also occurred along the Atlantic coast from the Carolinas to Georgia and St. Augustine, Florida. Although appealing to northeastern industrialists and certain real estate developers, "Revival Tabby" was an architectural dead end and quickly died out.

• Tabby Brick •

Where building clays were unavailable or unexploited, coastal builders who could not justify the expense of importing fired brick chose less durable tabby brick made locally

Chimney foundation of tabby and truncated chimney stack of tabby brick,
Lewis Reeve Sams House, Dataw Island.

from mortar containing shell lime and sand mixed with water poured into molds and allowed to cure, the finished product approximating kiln-fired common brick in dimension though not appearance. Certain islanders added finely ground oyster shell to their mix, Floridians building sugar manufactories along the St. John's River adding coquina shell. There was some art involved, best-quality tabby brick, if less regular than fired brick, appearing evenly sized, heavy, and weather resistant. How closely masons followed conventional bricklaying practice is demonstrated by the truncated chimney of the Lewis Reeve Sams House, now stranded and sunken by storms off Dataw Island's northern tip. Nine courses of tabby brick survive almost intact, the lowest one bedded directly onto a large tabby block foundation resting on pluff mud. Tabby bricks—nominally 9" × 4½" × 3"—are expertly made and carefully laid using English bond—an expensive solution characterized by alternate courses of headers and stretchers. In each heading course, a queen closer is placed next to the quoin (i.e., end) brick, to ensure that joints of the stretcher course above are centered over every alternate header. Purpose-made rather than fabricated from cut brick, closers indicate levels of expertise consistent with workers acquainted with prevailing Anglo-American craft traditions more typical of urban settings than rural ones. The freestanding kitchen associated with the William Sams House on Dataw Island exhibits similar construction. Measuring 15'8" × 22'6" overall

Kitchen, William Sams House, Dataw Island, fireplace
and chimney (Jack Boucher, HABS).

with twelve-inch-thick tabby exterior walls, the outbuilding is now ruined except for its
massive end chimney standing almost intact. This has a tabby base of shouldered form,
the shoulders rising as stepped courses rendered smooth with shell-lime mortar. About
6'7" wide × 5'8" high, the hearth opening was spanned by two parallel timber lintels
(now restored) measuring 1" × 6" and 9" × 8" respectively. Tabby brick facings form zones
of transition between the base and stack. Rising 20'6" above grade, the stack is tabby
brick carefully laid up in English bond, a string course and corbelled cap flaunched with
shell-lime mortar adding decorative notes to an otherwise utilitarian structure. Following

custom, the Sams kitchen was distanced far enough away from the main house to safe-guard residents against fire and penetration of cooking odors into the their living quarters but not too distant for supervision by the mistress, her husband, or his steward, who required timely meals delivered either hot or suitably chilled.[2]

Windows of the original dwelling here were partially blocked with tabby brick when enlarged by Dr. Berners Barnwell Sams around 1810, several courses in situ showing that one end chimney (erected before 1770?) was built of tabby brick raised on a solid tabby base. Similar features occur among later slave settlements, especially those erected by "im-proving" owners, tabby chimney bases—described more fully in chapter 6—invariably proving more resilient than the tabby brick stacks they supported.

Assuming "lime bricks" are synonymous with tabby brick, the earliest Beaufort ref-erence concerns projected building by St. Helena's Parish. Minutes dated August 6, 1753, record that "the Vestry agreed with Mr. Joshua Morgan, to make 20,000 good Lime Bricks, and to deliver them, in good Order, on the Glebe Land near the Spot on which the Parsonage House is to be Built at £6 per m [thousand]."[3] Units of this kind reduced costs significantly, fired brick costing £15 per thousand plus freight (forty shillings per thousand) for supply at Dorchester in 1757. The quantity Morgan supplied was sufficient for the parsonage's entire lower story assuming overall building dimensions (18' × 38') stipulated on January 26, 1753, were actually realized. Or perhaps the building had tabby brick chimneys—extant records give no explicit details.

St. Helena's Parish Church Registers record that "Joshua Morgan of London" married Anne, daughter of Richard and Rebecca Capers of St. Helena Island, in 1740. Under the terms of her father's will (proved October 1754) Anne Morgan received use of "Coopers (later Morgan) Island in St. Helena's Sound," where cattle were probably raised. Whether or not Joshua Morgan was making brick there or on his father-in-law's St. Helena Island holdings during the early 1750s cannot be said. It is known he was an executor of Richard Capers's estate. Historic records cast Morgan in other roles including mariner and part operator of the Port Royal Ferry. Marriage may have been his most lucrative enterprise, Magdalen Albergotti, his second wife, bringing him more lands, her "fortune" and Beau-fort Town Lot No. 305 apparently occupied by a house and outbuildings.

Widely distributed throughout Beaufort County, tabby brick is reported from coastal Georgia, East Florida, and Ellenton, Manatee County, near Tampa, where fired clay brick, probably imported from North Florida, was eked out with tabby brick at the Robert Gamble House (built 1844–56) and wedge-shaped tabby brick used for porch columns.[4]

◦ Wattle and Tabby Daub ◦

Demanding little time and minimal skill, wattle and lime daub constituted the walling material of choice for Spanish settlers erecting dwellings at Santa Elena (Parris Island) during the late 1500s. Consisting of hurdles made from interwoven branches or strips of wood fixed to vertical timber posts, and daubed with lime mortar inside and out, the technique was age old, appearing, disappearing, and reappearing over vast areas of the Old and New Worlds. Because wattle-and-lime-daub construction quickly disintegrates when neglected or abandoned, physical traces of the material are slight in archaeological contexts. Excavation of the Spanish Mission Santa Catalina de Guale, St. Catherines Island, Georgia, exposed wattle-and-daub walls enclosing the seventeenth-century church. The daub consisted of "marsh mud, sand and plant fibers" applied over interwoven "marsh canes" attached to wooden uprights.[5] Near Savannah marines pressed into service by Captain Noble Jones when building his fortified manor house called Wormslow probably occupied wattle-and-daub "huts" in the 1740s.[6] Other examples include chimney fragments from an antebellum slave settlement excavated at Cotton Hope, Hilton Head Island, and disassociated segments of uncertain date from Smith's plantation near Fort Frederick, Port Royal Island. Traveling through Beaufort District in 1833, Reverend Abeil Abbott came across "Major Wigg's plantation," called Okatie, where he found a "curious" storehouse built of wattle and daub. Measuring twenty-two feet square it was elevated on blocks "like a northern cow barn" had six-inch-thick walls made of "oyster shell lime mortar, laid on both sides of a wattled lathing" with exterior faces "smoothed and checked like blocks of Chelmsford granite."[7]

A few miles away stood a temporary dwelling, cabin, or shelter built around 1817 for enslaved workers at Hamilton's sugar mill, Callawassie Island, where boiling operations demanded constant attendance day and night during harvest time.[8] Fallen chunks of mortar with numerous impressions show this small (10' × 10') structure was enclosed by woven wattles fixed to vertical timber posts and daubed both sides with oyster-shell-lime finished smooth. About four inches thick, walls were supported by 1'2" wide tabby strip

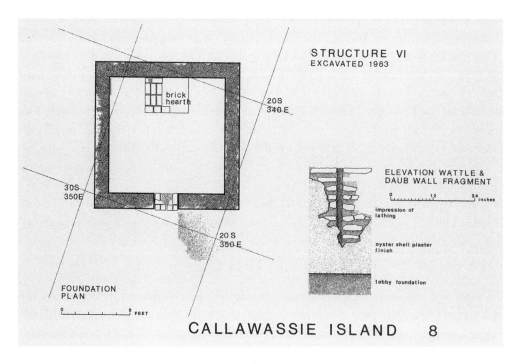

Hamilton sugar mill settlement, Callawassie Island,
drawing of slave house with wattle and lime-daub wall.

foundations. Opposite the single entranceway, excavation revealed a simple brick hearth but no evidence for any firebox or chimney, which suggests smoke escaped through holes in the (thatched?) roof. The size of window openings, or whether windows actually existed, is not known.

A slightly larger permanent slave dwelling of similar character is reported from Behaviour, Sapelo Island. The excavator recorded that it "measured 4.7 m long and 2.5 m wide, with a northwesterly facing doorway located midway along one of its long walls, the walls apparently finished to give rounded corners."[9] Foundations were of formed tabby and the roof probably thatched. Daub was shell-lime mortar not the mud or clay usual across northern Europe. Since analogous late sixteenth-century buildings known from early Spanish settlements along the South Carolina/Georgia coast cannot have lingered long enough to influence Anglo-American construction techniques, Sea Islander's use of shell-lime daub must reflect processes of reinvention, parallel evolution, or secondary introduction.

◉ Cements, Tabby Concrete ◉

Communicating with the *American Farmer* in 1827, Alexander Macomb observed, "*tapia* also makes very good floors for cottages, or other Apartments. . . . By adding to the composition a small part of canal or Roman cement, it will make the floor exceedingly hard and hold water like a cistern."[10] Although no floor of the latter kind has been identified, mention of Roman cement confirms tabby makers were then familiar with this relatively new material. Roman cement was a misnomer, having little or nothing to do with classical antiquity. A hydraulic lime, it was first patented in 1796 by James Parker, "gentleman" of Northfleet, Kent (southeast England), who discovered septarian nodules containing belite (C_2S) collected along beaches of the Isle of Sheppey and other Thames Estuary locations when burnt and ground produced brown powder. Mixed in the proportion two parts powder and five of water, it made good mortar, "stronger and harder than any mortar or cement now prepared by artificial means." Compared with mixes containing hydrated lime, initial setting times of this natural cement were reduced to ten or twenty minutes on average, "in or out of water," Reverend Parker said.[11] Mixing one or two parts sand and one part cement produced an ideal stucco. The British architect John Nash (1752–1835) used similar mixes for facades of Park Crescent, London (completed 1812, rebuilt after bombing during World War II), a chaste neoclassical exercise leading toward Regent's Park, where his giant residential terraces with their showy monumental arches, richly decorated columns, statuary, and stuccoed palace fronts have an "extravagant scenic character," which belies cheap and sometimes shoddy underlying brickwork.[12]

Visiting Savannah in 1833, Reverend Abeil Abbot observed: "a fashion beginning to prevail here is to build of brick or tabby, and to coat the surface with Roman cement, so called, imported from England. It makes a beautiful surface, and the architect assured me it is durable."[13] The English architect William Jay is usually credited with introducing this fashion into the city with his regency-style Owens Thomas House (1819), but he had left the city in 1822 so could not have been Reverend Abbot's informant.

Preliminary investigation of John Mark Verdier's Beaufort residence on Bay Street (built before 1810) suggests local introduction of the material was somewhat earlier. Tabby enclosing the raised basement story has been coated externally with dark-red or brown cement—almost certainly of the Roman variety—scored to imitate ashlar as usual with conventional shell-lime-stucco facings. Cores show no trace of an earlier external finish, which perhaps indicates the owner (whose British mercantile connections are well documented) imported Parker's proprietary material directly from England.

Parker's discovery was duplicated by engineers building the Eerie Canal. White's "Water Proof Cement" (patented 1820) was first manufactured at Chittenango, New York, around 1825. Numerous brands with slightly differing properties soon followed, these products remaining popular down until the early twentieth century. In 1906 the U.S. Geological Survey reported that although the industry had declined, there were sixty-five plants scattered across the United States producing natural cements derived from locally occurring agillaceous limestones heated to over 1,000°C in kilns resembling conventional lime kilns.[14] Resulting clinker was first ground into fine powder and then packed for distribution. Extensively used by canal builders, natural cements were also favored when fabricating so-called seawalls in Charleston and probably Beaufort, these bulwarks often hundreds if not thousands of feet long protecting made ground and other low-lying areas bordering the river systems that generated these two ports. Resembling traditional tabby walls, riparian defenses of "tabby concrete" were cast incrementally using timber forms, oyster shell serving as the principal coarse aggregate in the various mixes employed.

Disentangling building sequences along Charleston's extended waterfront is difficult, early seawalls of shell-lime tabby having undergone repairs, extension, and partial replacement, various cement-based materials substituted for the lime during later construction episodes not being obvious or identifiable without laboratory analysis. Rehabilitation and restoration of "tabby concrete" embanking Colonial Lake—an elongated tidal basin fed by the Ashley River on Charleston's west side—has provided opportunity for extensive material investigation and experimentation. Installed during the early 1870s (and much enjoyed thereafter by small boys and their fathers sailing model boats), the rectangular basin's enclosing walls displayed extensive erosion and deformation in 2015. Petrographic examination revealed that the original "concrete" consisted of oyster shell (whole and broken) and probably dredged river sand bound with hydraulic lime gauged with a natural cement of uncertain origin in the approximate ratio of one part binder to 1.76–2.76 parts aggregate. The *Charleston City Year Book 1887* records the basin was damaged during the great Charleston earthquake of August 31, 1886, an incident explaining deformation along the basin's north side and cast-iron tie rods with pateras plates exposed along the basin's outer face (all now heavily rusted) installed here soon afterward in an attempt—almost certainly ineffective—at stabilization.[15]

Preventing inundation of low-lying property during storms and high spring tides, Beaufort's earliest seawalls were cast using traditional shell-lime tabby, an activity perhaps requiring prior construction of temporary protective or coffer dams. One section still extant protects that part of Black's Point (now the Point) comprising City Block No.

8, Block Nos. 12 and 13. Formerly cut off by tidal creeks and bounded by the Beaufort River elsewhere, the area constituted an enclave inhabited by several of Beaufort's most prosperous planters (including "Good Billy" Fripp) immediately before the Civil War. Probably initiated by shipbuilding activities hereabouts from the late eighteenth century onward, this seawall was first documented by maps prepared for the Beaufort Direct Tax Commission in 1862.[16] Rising about 4'6" above low-water mark, construction is best preserved near the foot of Hamilton Street where Colonel Robert Barnwell's timber-framed federal-style house was built over an elevated tabby basement. The framed superstructure has gone, but the basement survives, having been reused as a foundation for the present house.

Joseph Barnwell recorded how Edmund Rhett (1808–63), when intendant of Beaufort (intermittently during the 1850s and early 1860s), "built brick and concrete (tabby) embankments along the margin of the river on Bay Street." These installations also exist, the best-preserved section located immediately east of Carteret Street along the western extremity of Bay Street rising between eight and ten feet above low-water mark. Made from an exceptionally dense material and remaining serviceable despite daily scouring by tidal processes, it appears harder and more resilient than seawalls fronting Black's Point. If Roman or natural cement was the binder for its densely packed oyster-shell aggregate—as is likely—then rapid setting offered by these materials would have allowed construction between tides.

Seawall, Beaufort River, west end of Bay Street, Beaufort, S.C.

In March 1894 the *Palmetto Post* reported that the Beaufort town council received proposals for building yet another "concrete" seawall one thousand feet long and five feet high measuring three feet at base and one foot at top to embank the Beaufort River west of Carteret Street. W. M. French, the successful bidder undertook construction for $2.35 per linear foot, his mix comprising "1 barrel of Portland cement, 2 barrels of sand and 5 barrels of shell." Mention of Portland cement is important, marking the transition from hydrated lime or natural cements to products that soon became—and remain today—ubiquitous throughout the construction industry.[17]

Developed during the late 1840s along the Thames and Medway Rivers near Rochester, Kent (England), early Portland cement was an artificial cement consisting of chalk (almost pure calcium carbonate) burned with local river mud at high temperature (1,300–1,500°C; 2,400–2,700°F), the resulting mass then being ground and packaged.[18] Compared with natural cements, Portland (especially after addition of gypsum) took longer to set up, had higher tensile strength, and gained strength more rapidly, its whitish-gray color being reminiscent—promoters hoped—of Britain's preeminent building stone quarried at Portland and Purbeck Island, Dorset, a stone imported into North America (notably Williamsburg, Virginia; and Charleston, South Carolina) from the early eighteenth century onward. First shipped across the Atlantic as ballast in 1868, Portland cement commanded ready markets across the United States in areas rebuilding after the Civil War, imports from England, France, and Germany totaling ninety-two thousand barrels by 1878. In 1885, when supplemented by comparable materials developed by entrepreneurs in Pennsylvania, Ohio, Maryland and Virginia, the total reached 554,000 barrels. After an uncharacteristically slow start, North American production grew at an astonishing rate from 335,500 barrels in 1890 to over fifty-one million barrels in 1908, a barrel (usually reckoned four hundred pounds) of Portland typically selling for between $1.25 and $1.50 in 1910. The product's all-around versatility soon eclipsed demand for natural cements, output of the latter across the United States—standing at over seven million barrels in 1890—falling to one million, six hundred barrels by 1908 and declining precipitously thereafter.[19]

It should come as no surprise to learn that, from about 1870 onward, isolated and impoverished communities on St. Helena Island, while clinging to old, well-tried building techniques, opted for the convenience of relatively cheap prepackaged Portland, which eliminated highly labor-intensive lime-burning operations. To this they added locally sourced shell, water, and sand, pouring the resultant mix into reusable timber forms much as traditional tabby had been cast in the old days. Of undetermined date (probably ca. 1900) the two-story, gable-ended Isaac Fripp House (measuring 30'3" × 17'9" overall), with its formed exterior walls, closely resembles a traditional tabby-built farmhouse of middling status except that Portland cement—rather than shell lime—cast in approximately twenty-inch-high lifts was employed for exterior walls. About fourteen inches thick at entrance level, these reduce internally to twelve inches at the upper floor, the ledge produced supporting 2½" × 12" floor joist spaced eighteen inches on center. A pioneering structure—now abandoned among trackless tomato fields—it has not fared well. The three-bay main facade facing west toward Chowan Creek has lost its porch; a

Isaac Fripp House, St. Helena Island, early twentieth-century tabby revival structure
with unreinforced, form-cast Portland cement and oyster-shell exterior walls,
south facade (Jack Boucher, HABS).

spandrel extending between one upper and lower window of the east facade has completely split, and corrugated metal sheet replacing the original (probably timber-shingled)
roof covering is badly decayed, dooming the entire dwelling—despite its National Historic Register listing—to further deterioration and eventual collapse as rainwater leaks
through unchecked.[20]

Generous federal and state funding has ensured that Penn Center, located several
miles away, remains well maintained, the campus established by northern missionaries intent on providing basic education and technical training for freed slaves retaining
several structures of formed tabby concrete, with old photographs from the early 1900s
showing high-wheeled ox-drawn wagons transporting enormous quantities of oyster shell
here for use as coarse aggregate. Shell concrete is clearly visible at Frissell Hall and former
machine shops currently housing museum exhibits. Dating from the 1920s or 1930s, silos
of similar construction reinforced with steel rods or galvanized metal mesh stand nearby,
notably at Seaside Plantation, located off Seaside Road, St. Helena Island. Unusually
the property retains its original prerevolutionary acreage together with an intact early
nineteenth-century timber-framed house of considerable architectural distinction.

Silo of cast
concrete with
oyster-shell aggregate,
Seaside, St. Helena
Island, S.C.

❖ Reuse and Recycling ❖

Reuse of old tabby has occurred down until recent times, recycling of preexisting work offering more cost-effective or convenient solutions than demolition of deeply founded tabby features. Advertizing for sale his lot "in Darrell's Fort" on East Bay Street, Charleston, in April 1787, Job Colcock said that a tabby wall stood on the property "sufficient to build a large house." The wall was probably part of Lyttelton's bastion (erected 1756–57), renamed after Captain Edward Darrell's company occupied the site in 1776. Whether reuse actually took place is not known.[21] In Beaufort excavation leaves no doubt that Beaufort's Arsenal was reconstructed on old tabby walls purposely leveled to make new foundations in 1852.

Haig Point House, Daufuskie Island, foundation footprint as preserved
after excavation with lighthouse built on reused tabby walls
of main dwelling in background (Jack Boucher, HABS).

After the Civil War reuse continued, conventional tabby having become an outmoded structural material, scarcely worth repair. The most notable instance occurred at
Haig Point, Daufuskie Island, where the present timber-framed Rear Range beacon and
light keepers' house, erected in 1872–73 stands on cut-down walls belonging to the Reverend Hiram Blogett's former mansion built overlooking Calibogue Sound sometime
before 1830. Wrecked and burned by scavengers looking for reusable materials during or
soon after the Civil War, the house was one of the Sea Islands' largest tabby structures.

Engineering drawings (dated 1871) show that its ruins were skillfully adapted to support a new timber-framed structure accommodating living space for the light keeper distributed over two stories and a tower at the east end, all organized about a side hall containing stairs communicating with the light proper. The hall and tower above rested on refashioned tabby walls enclosing the central passage bisecting Blodgett's basement. Massive (more than two feet thick) exterior walls of the original building supported the lighthouse on its three other sides. Altogether about one-third of the old plantation house was built over. Everything else was pulled down by federal contractors, who broke up considerable quantities of tabby for fill. Delayed by difficulties experienced from shortage of shipping supplies and sickness, which afflicted the contractor and his workmen, the light was first exhibited on October 1, 1873.

There are accounts of tabby removed for reuse from old to new sites. Spalding reported that Frederica's tabby houses were "sawed up into blocks" and used for construction of the "three first stories of the light-house at St. Simon's."[22] "Newly freed Negroes who were building their own houses" took away brick "by the wagon load" (sold for $1 per load) from the Thickets, McIntosh County, Georgia, along with "much tabby" derived from chimneys on the same site.[23] On Sapelo Island, some of Spalding's tabby was sawn into "twenty blocks" by Archibald McKinley in 1870.[24] Similarly tabby blocks cut from historic structures were reused to support timber sills of several of Spring Island's tenant houses during the early 1900s, similar destruction continuing into the 1970s or 1980s when refashioned tabby from an undisclosed location was used to face a causeway installed over a watercourse crossing the oak alee near the main Edwards House.[25] Chunks robbed from an otherwise unknown tabby structure are also found at Pinckney Point (northwest of Callawassie Island), where they support timber-framed barns erected around 1900.

◦ Shell Paving ◦

Whole or crushed, with or without the addition of lime, shell—usually oyster but sometimes imported coquina—was a common paving material used inside and outside buildings during tabby's heyday. Shell's ubiquity in antebellum Beaufort Town is illustrated by Joseph Barnwell's comments concerning improvements initiated by Edmund Rhett:

> Up to his day the streets were one and all absolutely unpaved, and almost one foot deep in what was called "boggy sand." Carriages and other vehicles could barely pass over them at a slow trot, and in summer and in dry weather to a boy's naked feet (none of the small boys wore shoes in summer) felt like hot coals. Mr. Rhett began treating these roads with oyster shells . . . and soon the streets of Beaufort were in large measure easy to travel over. It is true that when the shell had been ground up by the wheels of vehicles and the feet of horses, and the weather was dry and the wind blowing the white lime was not pleasant for dwellers in the houses adjoining the street, this could be remedied when the weather was dry by watering

carts, but these were not used on the back streets, nor on what was called the "shell road" which led from Beaufort to Port Royal ferry 10 miles away.[26]

Shell-paved roads remained the norm among residential parts of Beaufort down until the late 1960s, contributing much to its then-semirural character. The Point was especially attractive when I first saw it, large live oaks growing adjacent to or even in roadways, and well-worn shell adding a hushed, tranquil quality to the scene. As for the Shell Road, this occupied more or less the same land now traversed by Boundary Street and Highway 21, which roughly follow an early highway linking Beaufort Town and Port Royal Island with mainland routes leading north and south via the Whale Branch Ferry, established in 1733. Despite expensive and sometimes treacherous ferry service, considerable passenger and freight traffic continued to move along the road after introduction of regular steamship services eased the situation in 1816, cutting by hours or days journeys from Beaufort to Charleston and Savannah. Land traffic increased after 1859, completion of the Charleston and Savannah Railroad bringing an influx of passengers into Yemassee Station from where they were conveyed into Beaufort Town by stage. Shortly before (in 1853), Beaufort's intendant, Edmund Rhett, had greatly ameliorated traveling conditions, his Port Royal Land Company surfacing the poorly maintained yet essential artery with multiple loads of oyster shell.

The most remarkable scenes witnessed on the "Old Shell Road" occurred in November 1861 after Confederate forces manning sand forts on Bay Point and Hilton Head were outgunned and ultimately overwhelmed by Union ships under the command of Admiral Samuel Du Pont. Noah Brooks, a northern journalist, described how the entire white population of Beaufort and its neighboring islands then fled in what he scathingly termed the "great skedaddle" writing:

> At once there . . . ensued a great exodus of the . . . citizens of the place and adjoining plantations hastily picking up their most portable valuables and gathering a few of their able bodied slaves, they abandoned all else to the invader and fled, leaving even their dishes upon the table, provisions in the larder and cattle in the stall. Everything locomotive was pressed into service on the long shell road which stretching twelve miles up the island to Port Royal Ferry, and embowered with trees and flowering vines had been the famous sporting avenue of Beaufort, became at once a scene of hurrying confusion, dismay and wild disorder, such as only be imagined, the dreadful cry of "the Yankees are coming" urged them on in flocks and droves, and all day long the rude ferry boat was employed in carrying the flying refugees across to the mainland, where eight more miles brought them to the crossing of the Charleston and Savannah Railroad on the Pocataligo.[27]

6

Tabby Building in
Beaufort Town, South Carolina

◦ Foundation and Early Development of Beaufort ◦

Represented as the most proper place "for ships of Great Britain to take in masts, pitch, tar, turpentine and other naval stores,"[1] Beaufort Town was created by royal charter in January 1711. It was laid out on Port Royal Island near the blockhouse housing thirty men and mounting three guns built during, or soon after, Britain's 1703 expedition against Spanish St. Augustine. The earliest graphic representation of the new settlement is a manuscript plan attributed to the year 1710–11 probably based on surveys of Port Royal made by John Barnwell acting on orders from the General Assembly in 1703.[2] Basically geometric yet somewhat irregular, the town's grid plan was made to fit a great bend of the Beaufort River and incorporated almost four hundred lots. Four were set aside for public use at the intersection of two thoroughfares named Carteret Street (running north/south) and Craven Street (running east/west). Following precedent set by Philadelphia as early as 1682, this crossing is shown as an open square "formed of rectangular corners cut out of the four adjoining blocks."[3] Although designated "Castle" on early maps, it is doubtful that a military structure actually occupied the site, all early eighteenth-century work—if it existed—having disappeared from the scene. So has the square, this space being too small and unambitious a piece of civic planning to function in any formal sense though its location did eventually attract several public buildings: the courthouse (now gone), an equally ephemeral market, and Beaufort's Arsenal, which—much altered—still stands. Twenty-four smallish waterfront lots along the north side of what is now Bay Street were intended for commercial activity. Just beyond the town's western limits, glebe land was later designated for communal benefit of St. Helena Parish, established in 1712.

Beaufort's first land grants were made in July 1717, a second group following several weeks later (August 8, 1717). How much building occurred before conflict with the local Yemassee erupted into open, cruelly violent warfare during 1715 is unknown.[4] It said that most of Port Royal's inhabitants escaped injury by taking refuge on a pirate ship. But their settlements were pillaged and burned, South Carolina's entire southern frontier remaining tumultuous until 1718, sporadic raids and killings continuing well into the 1720s. Subsequent urban development was slow and intermittent. In 1724 Beaufort's parish church was "still unfinished and not fitt to preach in."[5] During the same year Colonel

John Woodward was "empowered" to finish the town fort "with as little expense to the Publick as possible."[6] Whether or not this dilapidated structure was the same fort begun before the town's foundation is uncertain. One year later the General Assembly voted £300 to finish the church, Colonel Samuel Prioleau establishing the first public ferry linking Port Royal Island to the "main" in 1733. After 1735 a new fort, Fort Frederick, erected just south of Beaufort Town put Port Royal Island into a state of readiness against foreign attack—or so settlers hoped. Although these hopes were more illusionary than real, Governor Bull, charged with protecting the southern frontier, was not deterred. "Desirous to promote the better settling of Beaufort Town" in 1737, he ordered "an exact Survey and Plat to be made of the land that was reserved for the Town and Common in order to grant the Lots to such persons as are willing to build Houses and settle there."[7] Always dilatory the Assembly did not authorize this measure until 1740, legislation then requiring every grantee to erect within three years a "tenantable" house measuring at least 30 × 15 feet with one brick chimney. Failure to comply incurred a £2 penalty, fines funding education for poor children.

Disrupted by hostilities with Spain, rural production expanded spasmodically through the 1740s. During the late 1750s and early 1760s profits from cultivation of rice and indigo stimulated new commercial enterprise near the center of Bay Street, including an emporium, shoe manufactory, stores, and slave market, the latter established by Thomas Middleton and partners in 1765.[8] On the bay's south side, "water lots" running down toward the river were allocated in 1763 and 1764, with John Gordon, Thomas Middleton, and Francis Stuart—all prominent merchants—receiving grants. Docks, landing stages, and wharfs were installed, but, by common consent, the waterfront proper was left free of major building until 1796, when Captain Francis Saltus built his dwelling house and started reclaiming land for ancillary structures opposite Town Lot No. 10, a property once occupied by one of South Carolina's most successful merchants, Francis Stuart and Company, which extended substantial credit (much of it insufficiently secured) to planters and traders down until Stuart's premature death in 1766.

◦ Tabby Construction during the Prerevolutionary Period ◦

Beaufort residents were erecting ambitious domestic and more mundane commercial structures in tabby before the American Revolution. Located near the historic center of Bay Street, the best documented building group is associated with Daniel DeSaussure (1735–98), a French-speaking merchant of probable Swiss descent who on February 24, 1774, acquired Town Lot No. 12 along with the water lot fronting it from his former partner "John Delasaye"—presumably the merchant and planter John DelaGaye, who had obtained his title in 1765 from the estate of yet another planter and merchant, the aforementioned ferry operator Samuel Prioleau.[9] The 1774 Act of Lease and Release mentions buildings, but relevant phrases are formulaic, giving no real information about any improvements former owners made. DeSaussure's will, dated May 16, 1778, mentions "the Dwelling house, Kitchen, yard, Stable and other buildings on my Lot Number Twelve." Structures are more fully described by Zephaniah Kingsley, who purchased the property following DeSaussure's imprisonment by British forces for his overt republicanism and complicity with rebel acts during the American Revolution. In sworn testimony—or rather, since Kingsley belonged to the Society of Friends, affirmed testimony—given before a parliamentary inquiry it was stated he had obtained two lots in Beaufort with houses on them for £22,000 Continental papers, which were let for sixty guineas per annum.[10] The property comprised "a lot in Beaufort formerly Dnl. Dessure [*sic*] from east to west in front 60 feet with two good houses in front, one made use of as dwelling house, the other as a dry goods store, a Kitchen, dairy, wash house, smoke house, stables and other Buildings all in good repair built of Tabby and walled with same . . . [plus] directly opposite down by river side, a store built of tabby will hold 1,000 lbs. of rice."[11]

Of the dwelling house nothing remains above ground, the 1907 Beaufort fire leaving it an empty shell, which was pulled down and replaced by what Sanborn Insurance maps called an opera house. In 1924 this too had been demolished and built over.[12] Opened by another fire, the site was partly excavated under the author's direction in 1997, when basement areas of the early dwelling house—filled with mortar and oyster-shell rubble—were found almost intact, tabby walls (1'9" thick) outlining a structural footprint 26' wide × 34' deep in plan barely set back from the public right-of-way.[13]

Remains of Kingsley's dry goods store erected immediately west of the main DeSaussure House are more tangible. Despite alteration, partial demolition, extension, and the addition of a timber-framed second story, this building's original east and west tabby side walls still stand. Built parallel to Town Lot No. 12's western boundary, with the narrow south end facing onto Bay Street, the store originally measured 21'4" wide × 52'4" long. The main floor, approached via brick or stone steps centered on the short south facade, was raised nearly three feet over a full basement. About two feet thick at basement level, exterior walls suggest heavy loads were anticipated during the building's design phase—a circumstance consistent with storage functions.

Between the dwelling house and store, a narrow passageway opened to an open yard paved with ballast stones and partly surrounded by the tabby wall Kingsley mentioned. His kitchen, smokehouse, washhouse, and stables were probably clustered here. Whether they resembled their early form immediately before the 1907 fire is questionable. Sanborn Insurance maps show redevelopment through the 1880s and 1890s, the site both losing and gaining ancillary buildings. In 1884 outbuildings almost enclosed the rear yard—all single story and timber framed except for one tabby structure known only from photographs taken after fire had reduced it to ruin. Incorporating two full stories this one-room-deep, five-bay building might have accommodated a kitchen, stable, or carriage house at the lower level and living space for ostlers, servants, or domestic slaves above.

Additional elements of the layout discovered during excavation included a cistern and brick-lined well—common features throughout the town where supplies of potable water could run short—and tabby footings designed for the wall enclosing the site along its east side.[14] Kingsley's testimony infers that almost the entire lot was enclosed, with walls, fences, and tall gates offering security to most mercantile establishments and private houses on Bay Street and much of Beaufort Town when first documented by itinerant photographers during the 1860s. Stereo images (ca. 1862) of the main DeSaussure House

Daniel DeSaussure House, Beaufort, S.C., oblique view of south front and store (left) from Bay Street (Library of Congress).

by Samuel A. Cooley show that its tabby exterior walls were—as might be expected—stuccoed and lime washed. Accommodation incorporated three full stories over an elevated basement, the roof looking steeply hipped. Organized symmetrically about center doorways, the three-bay south facade was fronted by a raised, two-tier porch, with gently curved brick or stone steps leading upward toward the first-floor entrance. On the second floor, porch handrails broke forward into an ellipse, adding attractive grace notes to this staid yet well-proportioned building. There were other decorative features—cornices of molded or cut brick at eaves level, Tuscan-style columns supporting the porch, and an elaborate, carved wooden doorcase marking the entryway. Whether porch and doorcase were contemporary is uncertain, porch construction having perhaps been replaced while the doorcase remained original. Conversely before post–Civil War adaptation the neighboring dry goods store was entirely utilitarian, enclosing one large space or several smaller ones at entry level, the gable-ended attic having additional rooms lighted by dormer windows.

Occupying the water lot opposite Town Lot No. 12, Zephaniah Kingsley's rice warehouse suffered demolition before 1912. Again we must rely on old photographs when reconstructing this tabby building, which, like the DeSaussure store, appears rather narrow, rectangular, and gable ended, with its entrance from Bay Street centered on the short north facade. Accommodation included two full stories, with an attic over, lighted by dormers. There is no clear indication of any basement—not really an option so near the Beaufort River—although the lower floor was raised slightly above street level. Long facades (east and west) incorporated three bays defined by relatively small rectangular upper window openings. When first photographed, fenestration below was obscured by a timber lean-to running along the length of the building's east facade. This and the two-tier porch facing south shown by Sanborn Insurance maps were probably additions, porch construction perhaps being associated with the 'Negro' restaurant established before 1899 when first shown by Sanborn Insurance Maps.

After the Revolution Kingsley was banished for collaboration with occupying British forces and his considerable estate confiscated. Rescuing what could be recovered of his former property, DeSaussure with several partners repurchased Town Lot No. 12 along with the water lot fronting it for £3,200 in 1783. Domestic life resumed, but the Revolution had taken a savage toll, destroying property and fracturing family relationships. Daniel's sister Mary Elizabeth moved with her Loyalist husband John Kelsall to the Exumas, taking her slaves with her when she left for Cuba after Kelsall's early death, and never saw South Carolina again. In Beaufort John Mark Verdier (1759–1827), who had become Daniel DeSaussure's protégé, traded indigo and rice out of the Bay Street property, also selling building supplies (nails, hinges, and locks), high-quality fabrics (silks and velvets) and more utilitarian weaves (linens, calicoes), shoes, stockings, blankets, spices and sugars, coffee, salt, tobacco, gunpowder, medicines and spirits, china, and plantation goods, all itemized by his account books.[15] Merchandise worth nearly £9,000 was sold in 1785, the figure reaching £11,718 in 1796 despite declining indigo prices, which adversely effected Verdier's trading position thereafter. Entries mention routine maintenance of and repairs to buildings almost certainly located here, including repainting

(October 1785); probable lime washing (May 1786); reshingling (June 1788); papering a room (May 1789); and repair of or alteration to the smokehouse (May 1786). In January 1786 Richard R. Ash, a carpenter judging by his purchase of a saw and "Chizzles and planes," was working on a "shed, salt room, wharf and kitchen." During the following month (February 1786) Edward Barnwell received "9/4d." for a "dogg" (presumably purchased for security) while in March of the same year and in May 1786, "pine poles" were bought "to fill up the wharf." Valued at £200 in August 1788, house and stores were insured by the Phoenix Insurance Company of London, Verdier's agents (Bird, Savage, and Bird) paying premiums as they came due, the store alone being valued at £150 in 1789.

An advertisement placed by Perry Fripp published by the *Charleston Courier* during the winter of 1840–41 may, or may not, refer to Lot No. 12, since it describes a three-story tapia (tabby) house apparently located on the north side of Bay Street. The full text reads as follows:

> The subscriber offers for sale his HOUSE AND LOT, in the town of Beaufort. It is pleasantly situated on the Bay, and is considered one among the best buildings in the place—it is a three story Tapia Building, and has lately been put in thorough and complete repair. It contains seven upright, two garret and two shed rooms; it also has a very large and convenient cellar. The Kitchen and Stable are also Tapia—the former is a two story building, and they are both in complete repair. The Carriage house is of wood, and lately built of the best materials. They are all of a proper size. The Lot is enclosed by a Tapia Wall, and is sufficiently large for any purpose, extending from street to street. For terms, apply to Messrs. Browne and Welsman, Charleston, or to the subscriber, in Beaufort.[16]

This sounds suspiciously like DeSaussure's house and outbuildings. If so, it would supplement our meager information about the property. But there are inconsistencies. The advertisement makes no mention of the store—a valuable source of income—standing next to the residence. More troubling is the statement that the lot extended from "street to street"—an attribute indicated neither by the original land grant nor by later site maps of DeSaussure's property. Neither discrepancy necessarily excludes identification with Lot No. 12. However, Bay Street has seen extensive development and redevelopment, making it possible we are dealing with otherwise unrecorded structures resembling but distinct from those Zephaniah Kingsley described.

Situated one block north of the waterfront at the corner of West and Port Republic Streets, another large tabby house escaped destruction during the 1907 conflagration but ultimately suffered total demolition like many architecturally significant downtown structures. Requisitioned for a guardhouse during the Civil War—when the interior was gutted—this tabby structure accommodated offices of George F. Ricker's "Ginnery" in the 1880s, his steam-operated cotton gins and sorting areas occupying space behind the main building for decades thereafter. Fire during the 1940s and changing technologies coupled with mid-twentieth-century development pressures left no room for the old building, which was replaced by an undistinguished group of offices.[17] Only Civil War–era photographs remain, these depicting an unusually large house with plain, yet powerful,

Campbell House at corner of West and Port Royal Streets, Beaufort, 1865
(Library of Congress, Beaufort County Library, Special Collections).

massing and late Georgian proportions, incorporating two stories raised over the customary high basement with a double-hipped roof and prominent dormers. The five-bay south entrance facade featured a single-tier porch, supported by sturdy Doric columns (probably stuccoed brick) carried on masonry arches—the porch wrapping around the west side. Sanborn Insurance maps show the house was relatively deep, which suggests some kind of double-pile plan probably organized about the usual central hall or passage. Extending westward from the rear of the building, a wing depicted by early photographs no longer existed in 1883, having already disappeared along with the west porch.

Well-informed about architectural matters, Lena Lengnick attributed original construction to Colin Campbell. Tax returns for St. Helena Parish (1824) list one Beaufort lot worth $1,600 under his name, but I doubt this was the Colin Campbell (born 1786) who was then working fifty-four enslaved laborers at Prospect Hill Plantation, Port Royal Island, where he occupied the property's smallish—now completely ruined—tabby house facing east toward the Broad River.[18] Judging by style the Beaufort building was more likely built by an earlier forebear—his grandfather and namesake perhaps—the Reverend Colin Campbell, who left Burlington, New Jersey, for Beaufort in the late 1750s. Alternatively we might speculate that, along with Prospect Hill, the handsome townhouse came into Campbell possession through the younger Colin Campbell's mother, Phoebe Sarah Barnwell, who married his father, Dr. Archibald Campbell (b. 1758) in 1780.

Prerevolutionary origins are mooted for three more Beaufort dwellings incorporating substantial tabby features. Belonging to the late 1760s or early 1770s, the much-altered, double-pile Chisholm House (905–7 Bay Street) originally comprised two stories over a raised basement. Extending about forty-five feet on the waterfront's north side, its five-bay entrance facade was fronted by a two-tier porch until lowering the first floor to street level sometime between 1912 and 1924 caused its removal. Portions of the same facade were briefly revealed by sidewalk repairs in 1997. It then became evident that exterior walls, now completely buried beneath modern finishes, are of tabby, which makes the house one of the oldest and largest of Beaufort's tabby buildings. Lengnick noted, "this is . . . one of those houses which has a drawing room on the second floor."[19] Now almost everything of architectural value inside is cut up or lost except perhaps for the hip roof, which flares out in a concave curve.[20] Behind the main house, an otherwise unknown two-story tabby outbuilding (which disappeared between 1905 and 1912), large enough for stables with servant accommodation above, is documented by Sanborn Insurance maps.

Long demolished the first parsonage house erected by St. Helena Parish deserves mention. Vestry minutes record that monies for its erection were collected in 1726 but that construction did not begin until the early 1750s, successive rectors being obliged (as they bitterly complained) to rent or buy their own living quarters until such time as the parsonage was complete. On May 18, 1753, oversight of the project was placed in the hands of "Coll: Nathl. Barnwell; Coll: Mullryne and Captain Gordon," three vestrymen who, fifteen or so years earlier, had overseen construction of Fort Frederick, Port Royal Island.[21] Appointment of carpenters (August 3, 1752) and an order for one of them to finish construction indicate the parsonage was predominately timber framed rather than tabby built. An entry dated August 6, 1753, mentions "Lime Bricks," which were almost certainly tabby bricks used for part of the work, though which part—basement, chimneys, or possibly outbuildings—is not said.[22]

◦ Tabby Construction during the Federal and Antebellum Periods ◦

The American Revolution brought devastation. Along the Broad and Combahee Rivers and on Port Royal and other islands, many of the area's most productive plantations were destroyed, damaged, or disrupted. By an act passed in October 1781—several weeks after General Cornwallis surrendered at Yorktown—the South Carolina Assembly banished citizens with known, or suspected, Loyalist sympathies and authorized confiscation of their real property. Already facing ostracism, assault, arrest, or worse, members of Beaufort's mercantile and planter communities unable or unwilling to give up old allegiances went into exile, taking refuge in Britain, France, Canada, and the West Indies. Their life often became peripatetic, Zephaniah Kingsley leaving for Bristol, Nova Scotia, then Haiti, and finally East Florida, which had seen an influx of refugees since 1775, King George III himself having wished this province might be "a secure asylum for many such."[23] What had been a steady trickle of refugees became a flood in July 1782, when British forces evacuated Savannah, resulting in about one thousand whites and nearly two thousand blacks pouring into St. Augustine. Charleston's evacuation in December of the same year added about two thousand more whites and over twenty-five hundred blacks if figures given by General Alexander Leslie are approximately accurate.[24]

Across the Sea Islands patriots and those who made peace with the new republican order faced almost insuperable difficulties. Workers were scattered, animals lost or stolen; and escaped slaves, bandits, and hungry, homeless soldiers—all the victims of war—roamed the countryside threatening security and personal safety. Indigo production suffered catastrophic declines, prices tumbling when British and French consumers found new sources of supply and imported better-quality product from their Caribbean colonies and India.

Ironically local fortunes were revived and ultimately enhanced beyond all expectation by Loyalists who sent cotton seed—"black seed" it was called—from the Bahamas and West Indies to family members, former business partners, and friends who remained residents of the United States. Subsequent experiments had extraordinary consequences. After patient selection and crossbreeding, lowcountry planters produced long-staple cotton of a luxurious silkiness that became synonymous with the Sea Islands. South Carolina's first commercial crop was sown and harvested by William Elliott II in 1790 at Myrtle Bank, his Hilton Head Island plantation. Local planters across the coastal plain soon repeated the experiment, their enterprise vastly increasing slave holdings besides accelerating new settlement and opening fresh areas for agricultural exploitation.

Profits from cotton transformed Beaufort Town, stimulating an unprecedented spate of new building, fueling expansion and growth of civic institutions already initiated by revival of rice cultivation. In March 1785 the General Assembly had instructed local officials "to expose for sale in whole or in lots the lands . . . known to be common adjoining the town of Beaufort."[25] Consequently the existing street grid was extended north and west by fifty-two new blocks. Thirty-eight blocks attracted purchasers of one or more lots before 1799, ownership names including planters, merchants, and businessmen whose motives for acquisition were mostly speculative.[26] Despite the appearance of growth, Beaufort remained relatively small, if disproportionally wealthy. In 1796 la Rochefoucauld-Liancourt said there were sixty houses. In 1805 the *Charleston Courier* counted 656 white inhabitants, 944 black inhabitants, and 180 resident students. The same source lists 120 dwelling houses, thirteen stores, nine workshops, four schools, a college, an arsenal, a lodging house, and three churches (Episcopal, Baptist, and Independent Presbyterian).[27] Except for the jail, no governmental or judicial institutions were mentioned, an omission underscoring the fact that Beaufort did not maintain its administrative role after the Revolution, Coosawhatchie becoming county seat in February 1788. This inland place was considered more accessible for lawyers and magistrates, Combahee rice planters, yeoman farmers, and other mainland residents than was Port Royal Island, reached only by boat or expensive, unreliable ferries.[28]

Nevertheless Beaufort Town continued expanding as cotton prices soared, merchants prospered, and prominent planter families became rich. Following formal incorporation (December 17, 1803), the newly appointed intendant and wardens began agitating for boundary changes to enlarge their jurisdiction east and west. These efforts met with little success until 1809, when South Carolina's legislature passed an act annexing neighboring lands, including Black's Point, located just outside the eastern limits of Beaufort Town when first defined by colonial statute.[29] Before the American Revolution and for several decades thereafter, "the Point" enjoyed near autonomy, developing mixed patterns of residential, agricultural, and commercial use including one or perhaps two shipyards. James Black was the best-known shipwright and merchant operating here, his house and boatyard standing on land acquired from Thomas Middleton near the mouth of a tidal creek east of what is now East Street. Black was killed during the Revolution. His former partner, Captain Francis Saltus, maintained the area's distinguished marine legacy, probably using the same yard when building five gunboats for the U.S. Navy in 1808.[30]

Architectural losses obscure the exact progress of development that accompanied Beaufort's expansion from the turn of the eighteenth century onward. Nor is it possible to assess the relative proportion of tabby and timber-framed building any time before the 1880s, when insurance maps first appeared. Fortunately, along with standing examples, early photographs document notable groups of tabby buildings. Through their diverse scale, style, and form these signify an intensive period of experimentation with the material extending over the Federal and late antebellum eras.

◦ Townhouses of the Planter Elite ◦

One of Beaufort's larger tabby structures and probably the oldest urban example still habitable, Elizabeth Barnwell Gough's House (now 705 Washington Street) occupies Town Block No. 42, laid out during the expansion of 1785. This land was purchased by Nathaniel Barnwell II for his sister Elizabeth, her ownership being recorded by Thomas Fuller's map of Beaufort made around 1799. Granddaughter of "Tuscarora Jack," an almost-legendry Indian fighter and pioneer settler of Irish extraction, she was born in 1753, the third daughter of Colonel Nathaniel Barnwell and his wife, Mary Gibbes Barnwell. In 1772 Elizabeth married Richard Gough, planter of Goose Creek. Family tradition relates that the relationship between the high-spirited, beautiful wife and her husband were stormy, ending in separation after the birth of their daughter Mariana (March 1773). Colonel Barnwell was well aware of the marital discord, altering his will to prevent Gough making any claims against Elizabeth's inheritance. Following Nathaniel Barnwell the elder's death (1775), his son and namesake received a fifth part of the estate to be held in trust for the benefit of Elizabeth and her heirs.[31] Where or under what circumstances Elizabeth Barnwell Gough was then living is not known. But it seems the house erected by her brother was finished during the late 1780s.

Elizabeth's household was never large, yet the two-story residence, raised over an elevated basement, with its hipped roof and temple front, is unusually grand for its location, reflecting close connections with Beaufort's leading and most prosperous families rather than the ambiguity of her marital status. It was also among the first Beaufort dwellings built on a T-shaped double-pile plan—"the rear chambers on either side of the central hall projecting beyond the width of the front façade,"[32] an arrangement allowing direct ventilation of rear rooms, though at the expense of long roof spans and complex roof carpentry. Modeled after Miles Brewton's Charleston residence (built 1765–69), which itself reinterpreted designs for the Villa Pisano at Montagnana or Villa Cornaro, Piombino Dese, illustrated by Andrea Palladio's *The Four Books of Architecture* of 1570, Elizabeth

Barnwell Gough House, Beaufort, south facade
(Jack Boucher, HABS 2003).

Gough's house is fronted by its original pedimented portico supported by brick arches at the lowest level, four Roman Doric columns of stuccoed brick at the principal level, and timber ones above. The south facade demanded space for proper appreciation, which explains why land opposite the site (Town/City Block No. 43) remained open, thereby creating an impressive setting more reminiscent of some gracious rural seat set amid parkland than a gentlewoman's townhouse.[33]

Today the first-floor entrance hall gives entry into four rooms arranged in pairs right and left linked by pass-through doors, the two rear rooms projecting about eight feet beyond the stem of the T. Originally both front rooms (southeast and southwest) were fully paneled with heart pine and cypress, an outmoded treatment yet still useful when lining tabby walls, which can sweat in hot, humid summers.[34] The two rear rooms are simpler, having plaster wall finishes and timber architraves around windows and doors, paneled wainscots, chair rails, and cornices. In the back of the main hallway, space widens into a stairwell open through two stories. This unexpected volume, complete with Venetian window, adds considerable drama and sense of parade for anyone entering the building, as no doubt intended. Tucked away behind the staircase, a low doorway with sidelights opened into the raised, single-story back porch (now replaced) accessed from the exterior via splayed brick and tabby steps.[35]

The staircase with its continuous handrails and delicate balusters (wrecked after the house became part of Federal Hospital No. 10 in 1863 and now restored) starts as one flight. At an intermediate landing it turns back on itself, divides into two, then ascends to an upper landing. The upper landing opens into connected rooms of unequal size extending across the dwelling's entire south front, an arrangement common among Charleston's double houses. The larger, so-called ballroom (which, if it followed Charleston precedent, was actually the main dining room) is fully paneled, the paneling painted old rose and deep Prussian blue, rather than the bland white and off-white shades decorators have recently applied believing they are following historic precedent. From here guests might go out onto the south portico through the centrally positioned jib door or, alternatively, move through into the smaller, less elaborately finished west reception area.[36] On the right and left two back bedrooms, both plainly decorated, were also reached via the upper landing. Like corresponding rooms below, these were much enlivened by insertion of tripartite windows during the 1820s. Fashionably wide and tall, new windows replaced smaller ones piercing end (east and west) walls of the two wings. Besides modernizing facades larger openings enhanced cross ventilation making occupants more comfortable when weather conditions turned hot and humid. Back-to-back fireplaces with brick chimneys ensured all spaces except the basement could be heated sufficiently to dry out after rain and stay relatively warm during winter months.

Lack of any staircase between upper floors and the basement, along with the latter's absence of heating, underscores clear functional division made between living and service activities. Food preparation and cooking normally took place away from "family" quarters, a freestanding kitchen positioned behind the house being usual for elite town residences and plantation dwellings—lost here, along with any other early outbuilding—except for an unverified memory that places the kitchen beside Carteret Street. With wines and spirits perhaps locked away in one or two small rooms partitioned off, the otherwise undivided basement may have been too damp for storage of food stuffs—though not perhaps for enslaved servants engaged in household tasks who probably inhabited subsidiary structures or small huts on the property now destroyed.

Exterior walls are load-bearing tabby rising over thirty feet above current ground level. Exterior skins reduce in thickness with height, the tabby appearing expertly cast and carefully compacted, vital factors considering how the long, five-bay north facade approaches theoretical structural limits with respect to stability. Stuccoed and scored exterior wall surfaces imitate regularly coursed ashlar complete with voussoirs and keystones over openings. There is evidence for lime washes, the earliest probably tinted. Inside plaster was applied directly onto the tabby except where paneled or otherwise lined. A few scraps found during restoration suggest that the roof was originally covered with timber shingles rather than more durable, yet weighty, clay tiles or slates.

The Barnwells employed competent craftsmen whose knowledge of local materials and late Georgian design made a handsome if restrained show. But these workers were not inspired enough nor paid enough to emulate the richly carved interior and exterior trim characterizing Brewton's splendid Charleston mansion. Their rather naive, distinctly provincial decorative repertoire is demonstrated by friezes of Mrs. Gough's "ballroom." Rather than any classically inspired motif, these incorporate raised diamond patterns endlessly repeated. After all this was Beaufort, a small, isolated coastal town that might acknowledge metropolitan fashions without being sufficiently populated by skilled masons, joiners, plasterers, metalworkers, upholsters, and the like to influence them.

With temple fronts masquerading as porches and more or less symmetrical massing, groups of tabby houses built on bluffs overlooking the Beaufort River at opposite ends of Bay Street recall Palladio's own precept that "if one can build on a river it will be very convenient and beautiful . . . bringing coolness in the summer and making a more beautiful view."[37] These houses were not unlike villas built for English poets (and a royal mistress or two) along the River Thames above Chiswick, seeing seasonal rather than year-round occupancy and depending on small armies of servants to function. For low-country planters, in-town dwellings provided escape from the discomforts and isolation of island life during oppressive summer months, opportunities for social and cultural interaction, and refuge too during periods of sickness, natural disaster, and political upheaval.

Chronologies are uncertain. Currently called Tabby Manse (1211 Bay Street), the Thomas Fuller House was probably built around 1790–95 after the model of Elizabeth Barnwell Gough's residence. It is quite possible the same contractor erected both houses, using the same formwork set since respective dimensions are almost identical. If so, we do not know his name or anything else about him except that he modified the original scheme, tinkering with its floor plan insofar as formwork allowed, adding decorative touches here and there, adapting the entrance portico by substituting slim, attenuated column supports much less robust than the more classicizing counterpart executed for Mrs. Gough. Minor differences aside roof framing and carpentry details are closely related, showing that while trim changed with shifting architectural fashion, structural framing was ruled by tradition and long-established craft practice.

Nearer the western end of Bay Street (Block No. 123) and known only from early photographs, the tabby-built "Barnwell Castle" (destroyed by fire in 1881) was T-shaped with rear rooms extended laterally into wings rounded at their far extremities. It incorporated three, rather than two, full stories over an elevated basement, the fancied medieval look produced by the two wings giving the building its popular name.[38] Play with similar volumes was taken up by the entrance porch. Semicircular, or possibly elliptical, this raised, single-story affair was approached via an opposed pair of masonry (probably stuccoed-brick) steps wrapped around the base of the porch in almost baroque fashion. More idiosyncratic still these accessed an entranceway featuring two doors, each with sidelights and elliptical fanlight. Paired tripartite windows lighting the second floor above repeated the twinned theme. Joseph Barnwell, who lived in the house when young, describes the building and explained its unusual character:

> The dwelling house which we first occupied had been built by my grandfather
> Robert [Barnwell] together with his elder brother, Edward, who was my mother's

Barnwell Castle (Nathaniel Barnwell House), Beaufort
(Beaufort County Library, Special Collections).

grandfather. It was built entirely of tabby, a material made of oyster shells and lime beaten with water into frames much as concrete or cement is now made. The house contained three stories and a high basement with a high pitched roof, and stood on the bluff in the town of Beaufort where the Court house now stands. It was tenement house divided exactly in the middle, and Robert Barnwell built the western half and Edward Barnwell the eastern half. It was said, however, that the half built by Robert Barnwell cost just one half of what the side which was built by his brother Edward.[39]

The confidence allowing Beaufort's master builders to erect tabby structures of unprecedented height is further demonstrated by the house now called the Anchorage (1103 Bay Street), built by Ralph Emms Elliott (1764–1806), who, among other properties, owned Cedar Grove Plantation, Port Royal Island. Raised over an elevated basement, this three-story T-shaped structure is substantially reworked, intrusive Romanesque-style fireplaces adding incongruous notes to reconfigured interior spaces. Historic photographs taken forty years or so before major alteration began allow reconstruction of the building's early appearance. The clearest image, dated 1862 or 1863, gives an oblique view from the southeast. This shows that the south (riverside) facade incorporated five bays crowned by a central pediment, the latter supported by elaborately carved console brackets. Entry

William Elliot House (the Anchorage), Beaufort, south facade
(Jack Boucher, HABS, 2003).

was made via splayed steps leading up toward the flat-roofed, single-story porch raised on brick or tabby arches. This created an inviting Tuscan colonnade extending along the entire river front very unlike the overscaled, pompous Corinthian-styled double-tiered replacement of Beaux Arts derivation installed around 1900.[40] The original porch roof was enclosed by balustrades and reached from the second floor's central window. Rear rooms projected east and west into canted bays, an arrangement reflecting Charleston's taste for rounded or polygonal room shapes during the late eighteenth century—a persistent fashion (at least in Beaufort) ultimately derived from works by the French architect Ange-Jacques Gabriel and designers he inspired—notably John Soane in England.

All external walls were pebble dashed at the beginning of the twentieth century, giving the building a dull lifeless finish very unlike the bright stucco original and concealing all trace of the underlying tabby. Consequently the fact that this is house ranks among the largest and tallest tabby domestic buildings ever built goes unnoticed. Early carpentry details are also obscured or destroyed except in the attic, where the original hipped roof frame survives nearly intact. Resembling structural solutions exemplified by the Barnwell Gough House and Tabby Manse, secondary roof members are received by two parallel

Fuller House, corner of Bay and Carteret, Beaufort
(stereoscope image, 1860s, Beaufort County Library, Special Collections).

king-post trusses spanning approximately 42'6" between front (south) and back (north) exterior walls, an altogether impressive feat considering the height at which these weighty elements were installed and that any slip when maneuvering them into position might have cracked or split tabby walls below. Work entailed when cutting, hauling, jointing, and erecting timbers for the roof frame was clearly prodigious, this coupled with the huge volume of materials required for wall construction contradicting the common view that tabby was a cheap material.

Although incompletely known, two tabby dwellings lost during Beaufort's "great fire" of 1907 were important. Who built the tabby house occupying the northeast corner of Carteret and Bay Streets is not established though one might assume it was either Thomas Fuller Sr. or his namesake son since the lot is designated "Thomas Fuller's" by survey maps dated 1861. Several Civil War–period photographs show this double-pile, raised, two-story, predominantly tabby house incorporated two parts, the carefully balanced main block being flanked eastward by a wing rounded at its far extremity. A raised two-tiered entrance portico (three bays wide) incorporating familiar Tuscan-style columns imposed discipline on the south-facing (river) facade, an enriched pediment completing the composition. Roof junctions between the main house body and wing are obscure, but roof frames appear hipped throughout. Dormers lighting the main roof space were symmetrically disposed about the pediment, which contained an excellently proportioned lunette window. Details invite comment. First, the difficulty and expense incurred by fabricating nonrectilinear formwork proved an obstacle when making the east wing's rounded extremity. Photographs taken after the 1907 conflagration show that instead of tabby, fired brick was employed for the curved shape required. Earlier images indicate external walls were stuccoed, the stucco scored then painted, "stone" blocks so delineated enhancing an illusion that construction was of freestone or marble. Nothing is known about the rear facade, two isolated brick gateposts facing Carteret Street being the only architectural elements of this once-impressive development still extant.

Occupying the same block toward the eastern extremity of Bay Street, the Reverend Stephen Elliott House, with its carefully modulated proportions, axial symmetry, hipped roof, tall end chimneys, and gravitas pictured by early photographs, would not seem out of place on the Ashley near Charleston or even James River, Virginia. Incorporating two principal stories raised over an elevated basement, the rectangular plan may have been only one room deep with interior spaces arranged symmetrically about central hallways. The south facade featured seven bays, the three central ones surmounted by a pediment. Rather than a temple-like portico, the house was approached from Bay Street via a raised, single-story porch (three bays wide) enriched with slim Tuscan columns. Sanborn Insurance maps show a rear porch extending along the entire length of the building's north facade. Joseph Barnwell stated Stephen Elliott inherited the property from his father, William Elliott.

◦ Mercantile Establishments—the Saltus House and Stores ◦

The Captain Francis Saltus House and outbuildings were the first structures of any consequence erected on Bay Street's south (river) side. These caused consternation among leading citizens who worried about limitations placed on boat access and feared encroachment over existing north/south street endings previously left unobstructed. Consequently lawsuits were initiated, which, however disagreeable for the captain, left considerable contemporary documentation describing his building activities. Saltus himself was a merchant with trading connections stretching from South Carolina to New York, Bermuda, the Baltic, and Russia, who gradually amassed wharves, houses, and rental units along Charleston's waterfront besides cotton lands in Prince William's Parish. Central to his fortunes were successive partners involved with the coasting traffic bringing cotton and other staples into Beaufort and Charleston for redistribution. An element of speculation can be detected, fostered perhaps through friendships or associations made among Beaufort entrepreneurs including James Black, best known for his shipbuilding activities. Partnerships and shared business affairs ensured that Beaufort District remained an area of interest for Saltus throughout his life.[41] He was not alone in seeking to develop Beaufort's waterfront, another erstwhile partner, John Rhodes, having similar ambitions. Little can be determined about the wharf, sheds, or warehouse Rhodes actually built on his water lot. Conversely an impressive memorial placed before the South Carolina Senate dated 1796 details Saltus's project.[42] This document states that his waterside property—a relatively small parcel measuring 150' × 170' located opposite Town

Saltus House, Beaufort, north facade, 1864 (Samuel A. Cooley,
Library of Congress, Beaufort County Library, Special Collections).

Lots Nos. 9 and 10—was bought in good faith.[43] Having no doubt about his title, Saltus proceeded "to erect buildings for a Dwelling house kitchen and stores," expending "great sums of money on improvements." Work had stalled since he (the captain) was "forbidden to make any improvements on part of the said lott until an enquiry could be made into the validity of the original grant," an inquiry that Saltus believed threatened not only his personal interests, but those of other citizens whose real estate titles had long gone unchallenged. Individual legislators agreed, yet the General Assembly took no action until 1800, when another petition submitted by Saltus was considered.[44] Former grievances were repeated, additional information being submitted about contested improvements:

> Your petitioner in a legal manner possessed himself of certain low water lotts in
> the said town of Beaufort, and not doubting that in common with other citizens
> of our country he had an unquestionable right to place such improvements on the
> same as he thought condusive to his interest, there being very little high land to the
> said lotts but a good front, he has been at considerable expence and labour in mak-
> ing land to erect his outbuildings particularly on the ground so much complained
> of, having put his kitchen and other small buildings on said improved land, your
> petitioner begs leave to refer to a plan of lotts in which you will see he has erected a
> wharf, leaving a passageway between his store and dwelling house from Bay Street
> to the Wharf, which the public has access to, should your petitioner be deprived of

the land alluded to he has no place to put his kitchen and other buildings, unless immediately in front of his dwelling house.

Saltus urged relief, to which the General Assembly ultimately agreed. By Act No. 1749, passed December 20, 1800, earlier provisions were rescinded if persons restrained from "ever building more buildings thereon" made no further improvements other than wharfs and left street endings clear. What Francis Saltus thought about this compromise is not recorded. Still legal maneuvers ceased, and the General Assembly was now better informed about Beaufort's mercantile improvements.

Regrettably the map Saltus submitted with his petition cannot be found, and much late eighteenth-century site work has been destroyed, overbuilt, or substantially altered by numerous redevelopment phases. In fact it is surprising the main house still stands since both front and rear tabby walls were demolished at lower levels to facilitate commercial access. In 1954 the *Charleston News and Courier* reported:

> Beaufort's historic Pink House [Saltus House] . . . is now the home of a department store [Belk Simpson, Inc.] whose management listened to the appeals of members of the Beaufort Historical Society and not only preserved the characteristic features of the old, but enhanced them. . . . The old Pink House, which last year had, as tenants a restaurant a liquor store and a fish market, was cleaned out from the rear to the front, roof and side walls. Huge steel I beams were placed in support of floors, windows and wherever support seemed necessary to the mostly tabby construction. . . . The first floor of 1½ stories height, was extended back over 100 ft. Beyond that there will be a parking space.[45]

Huge or not steel beams were inadequately sized for the job, deflecting to an alarming degree under load. Consequently floors and roofing of the original house sagged, and tabby walls split, leaving the entire building precarious.[46] Interior trim is now missing, removed by the current owner, and the old building more or less unoccupied, Belk's additions housing restaurant and retail facilities. Other losses include the original kitchen, store, and wharf, all swallowed up by repeated "improvements." Despite inappropriate alterations, unsympathetic interventions, and neglect, early building programs can be reconstructed, albeit incompletely, from documentary sources, old photographs, site investigation carried out in 1995, and what survives of the much-abused dwelling.

Thus the petition of 1800 remarked on the site's lack of elevation and implied the captain's dwelling house occupied a narrow strip of higher ground extending along the water lot's Bay Street side. This circumstance excluded any T-shaped or "double-house" scheme. Instead a linear, one-room-deep, gable-ended central-hall plan measuring about 44'6" east/west × 24'2" north/south was adopted with living spaces distributed over three principal floors, the first floor (destroyed soon after the Civil War) raised 3'6" or more above present ground level over an undivided basement. Featuring five bays, principal facades north and south reflected the building's simple, bilateral symmetry. Currently blocked, double-hung windows of the south facade gave unobstructed views along the Beaufort River as far as its confluence with the Broad River, allowing Saltus to see and

Saltus House, Beaufort, staircase looking down from third floor
(Jack Boucher, HABS, 2003).

possibly signal his own ships coming in from or going out into the Atlantic Ocean. The west facade had single double-hung windows flanking the end chimney right and left at each floor level. Paired demi-lune windows (now blocked) piercing the two gables lighted roof spaces.[47] Exterior embellishment was restrained. Early photographs show that the Georgian-style doorcase centered about the street facade followed models illustrated by William Paine's *Practical Housebuilder* (London, 1789). Above a timber eaves cornice enlivened by pyramidal consoles and carved panel depicting olive branches still remains in situ. As usual exterior stucco finishes were scored in imitation of regularly laid and coursed stonework.

On entry visitors passed into a lower stair hall, doorways right and left accessing east and west rooms, the small "back hall" behind the staircase leading out into the south-facing porch known from early photographs but now completely lost. Second and third floors repeated the same basic plan arrangement with minor variations. The third floor's lack of fireplaces, plain moldings, and utilitarian board-and-batten doors indicate its two main rooms probably provided sleeping accommodation for children or servants. By contrast, although compressed into an extremely small area (about 8' × 8'), the staircase is a minor masterpiece of vernacular carpentry. Before destruction of its lowest level, it rose on an elliptical plan through three stories, with continuous handrails of understated elegance, slender rectangular balusters, reeded-timber dados, and a shallow elliptical plaster dome above complementing the composition. Cut, trimmed, and otherwise modified floor joists show that the stair hall was reworked during an undocumented secondary building program, the reconfigured space being enclosed by timber studs covered with lath and plaster. Reeded-mahogany and molding profiles suggest alterations were completed around 1825, the present stair replacing a less fashionable square or almost square one. Similar designs characterize structures attributed to immigrant carpenters who built numerous innovative houses around Milledgeville and Augusta, Georgia, during the first quarter of the nineteenth century.[48] While no example provides an exact parallel, the staircase at John W. Gordon's House, Jones County, Georgia, built around 1822 (which features a semicircular plaster dome over its well), and an elliptical stair, four stories high, spiraling through Augusta's Ware House (ca. 1818), are—it seems—more or less dependent on the same unidentified prototype.

The Saltus site combined commercial and residential functions. On the west the house was abutted by an ell containing one main story with habitable attic, all raised over an elevated basement that boasted cobbled flooring, barrels, casks, or other heavy items presumably being loaded and unloaded here. Further west stood a detached store resembling the commercial structure owned by Daniel DeSaussure before the Revolution. Much altered, Saltus's long, narrow, gable-ended tabby building standing at right angles to Bay Street incorporated retail space with heated living areas above and attic spaces lighted by dormer windows.[49]

South of the main dwelling house, near the Beaufort River, no construction was possible without "improvement." Saltus decided to make ground, installing timber bulkheads and rafts or cribs on which he subsequently erected several small buildings including his kitchen. Chinese export porcelain, English stoneware, and ballast blocks of Caribbean coral recovered during excavation of associated fills indicate how far the captain's commercial interests stretched. It was this kitchen that caused public outcry, repeated reference to encroachment on street endings suggesting it blocked or otherwise interfered with the south end of Scott Street. Between his house and stores, a wharf is mentioned by Saltus that later became the customary berth for steamships run by his son-in-law, Dr. Henry Thomas Willis Lubbock.[50]

Captain Saltus died in 1831 and was likely buried near his principal plantation at Old Sheldon Church, Prince William's Parish. An affectionate tribute paid by his grandson, General Francis Richard Lubbock (1815–1905), who became governor of Texas (1861), recalled Francis Saltus as "a rich cotton planter . . . a shipowner and wharf holder in Charleston" and "so jolly and good to me that I loved him very dearly."[51] Whether the Beaufort property was sold before or after the captain's death is uncertain, local tradition associating it with the Habersham family (which might explain the dwelling's reconfigured stair), well-to-do planters and merchants of Beaufort County and Savannah. Relevant records are lost, making it impossible to either confirm or deny the long-held Habersham association. It is documented that Daniel Mann held title when the house and wharf called Pier No. 2 were requisitioned by the military for a commissary following the Battle of Port Royal (November 1861) and Beaufort's subsequent occupation by Union forces.

Photographs by Samuel A. Cooley include an image of the building's new military guise taken in 1864. The original street entrance was intact, but one window (to the

View of Beaufort waterfront ca. 1862, showing side of Saltus Store (*far right*) (Library of Congress).

east) had been made into a subsidiary doorway approached via stone or timber steps, and another window enlarged. These alterations may have been carried out by military authorities since stucco around the openings looks broken. Alternatively Daniel Mann had perhaps converted lower-floor areas into commercial space just before the hostilities began. Upstairs the house looks deserted. Second-floor windows are shuttered. On the floor above, window glass is missing. Roof flashing had become loose, resulting in moisture penetration followed by disassociation of stucco along the building's east gable end. Structural elements unknown except for Cooley's image include a single-story porch extending along the dwelling's south elevation. To the southeast stood a detached, two-story brick dependency with gabled roof and end chimneys. To the west the ell built against the main house appears intact, although locked, shuttered, and perhaps out of use.

◦ Contractor ◦

Structural elegance exhibited by the main Saltus House indicates considerable prior experience by its builder. The name of this person is not recorded. Nevertheless stylistic similarities link the house with yet another of Beaufort's lost tabby dwellings located on Hancock Street where the Talbird family had held property since the late eighteenth century. Though it was destroyed by fire in 1907, photographs of its burnt-out ruin illustrate a gabled, five-bay, federal-style building with internal end chimneys of fired brick, comprising two full stories raised over an almost-full-story-height basement.[52] The plan was one room deep, two principal spaces on each floor presumably being divided by central hallways. Paired windows of rectangular form-piercing gables carried up in tabby show the attic provided additional living or sleeping accommodation. No firm information exists about this dwelling's construction date. It is known that several Talbird family members were planters and building contractors from the late 1760s (when Henry Talbird was rebuilding St. Helena's parish church) down until the mid-1850s, when Franklin Talbird erected Brick Baptist Church, St. Helena Island. Colonel Thomas Talbird was particularly active, his works including Beaufort's arsenal (1795); repair of St. Helena Church (1798); building a new parsonage house for the same parish; and what is most important, erection of Beaufort College (1802–4) described below. Considering his long experience, it would come as no surprise to find him building a tabby house on the Talbird family lot or a similar dwelling for Captain Francis Saltus during the 1790s although neither house can be positively identified as Colonel Talbird's work.

◦ Raised Tabby Basements ◦

Tabby basement walls, dating as far back as prerevolutionary times, show the material was considered durable enough for subsurface work and strong enough to carry timber-framed buildings of considerable size. One example is described by an agreement with Beaufort College (recorded August 21, 1801) for erection of the president's house, two contractors named Mulheron and Flynn undertaking construction measuring "40 feet by 20" . . . raised on a solid tabby foundation seven feet high . . . partitioned off, and

Berners Barnwell Sams House, north facade, tabby basement
with brick above (HABS, Jack Boucher, 2003).

provided with fireplaces and finished for offices."[53] Accommodation above incorporated
two stories (presumably timber framed), each measuring "10 feet in the clear with two
rooms and a passage on each floor." A porch "10 feet wide in front and at one end of
the house with a small room at one end of the *piazza*" was also specified, the contractors undertaking to complete all work for £750 sterling and finish the job by May 1802.
Around the same time, John Mark Verdier used tabby for the lower, unheated story of
his Adamesque house on Bay Street, an exceptionally stylish building with a well-carved
doorcase, exterior rope moldings, and carefully considered interior enrichments borrowed at some remove from Robert Wood's splendid *Ruins of Palmyra* (London, 1758).

Similar tabby raised basements were built throughout Beaufort Town down until the
Civil War. One 1850s example, the Berners Barnwell Sams House (201 Laurens Street),
marks an architectural transition. Tabby encloses the building's raised basement, but exterior walls of the two upper stories are fair-faced brick showing no sign of any surface
finish. Robust, double-height, monumentally scaled Doric columns of stuccoed brick
that capture some of the grandeur of their antique prototypes carry the two-tiered south
entrance porch. What architectural model did B. B. Sams have in mind when he ap-
proved these supports, which look so attractive from the green still left open in front of
his house? The "massy Greek columns three feet in diameter" Robert Mills employed
for his County Records Building in Charleston (1822).[54] plantations remembered from

Louisiana, old engravings of Paestum? Or were they perhaps the hallmark of an itinerant mason responsible for other grandiloquent, but anonymous, Beaufort porches such as the one fronting Abraham Cockcroft's mid-nineteenth-century brick-built house on the waterfront? Regrettably relevant records are missing so the talented designer or designers involved cannot be identified.

When enlarging his father's Dataw Island house thirty or so years before, Dr. Berners Barnwell Sams employed expensive fired brick only for porch supports and chimneys. His extravagant use of the latter material for his own townhouse and similar usage by several other privileged planters reflects increasing commercial exploitation during the 1840s and 1850s of clay deposits found on Lady's Island, and of course, exceptional profits made from Sea Island cotton.

⦿ Urban Outbuildings ⦿

Numerous outbuildings—kitchens, stables, privies, and storage sheds—were once scattered through Beaufort Town, but time, neglect, and gentrification have destroyed almost all these utilitarian structures. Antebellum census returns indicate slave dwellings were commonplace, Joseph Barnwell remarking that three elite mansions and their yards located toward Bay Street's western extremity east of Carteret Street "contained one hundred Negroes. In the first place there was the head waiting man and his assistant, and then the seamstresses and washerwoman, and maids and the cook and coachman and his assistant and wives of all the men and the younger children, and the minders of the younger children," boatmen, carpenters, stable lads, gardeners, oystermen, and the like swelling this enslaved population. Regrettably their names are erased from memory and personal histories lost along with their dwellings except for a few mutilated remnants—tabby foundation strips belonging to a hut situated behind Henry McKee's former town residence (511 Prince Street), and single-story quarters built in tabby by Dr. Bernard Barnwell Sams for domestic slaves behind his Laurens Street mansion.

Conversion into rental accommodation makes determination of what is original about this L-shaped service range uncertain, features exposed during repairs indicating each unit had its own brick-built fireplace and chimney. There was also a communal kitchen and privy. Windows are now glazed, but this was probably not so during the antebellum period.

Coincidently another much larger privy stands behind the site of the old Talbird House (Hancock Street). Now roofless and badly damaged, it was freestanding, measuring 26'2½" × 10'2" with ten-inch-thick tabby walls cast in twenty-five-inch vertical increments and corners reinforced with fired brick laid in alternatively long and short patterns, their height corresponding with successive tabby pours. Slots in lower exterior walls and small windows above provided ventilation. Separate entrances indicate interior division into two discrete spaces divided off by a timber-framed partition (which has not survived), one space presumably intended for males, the other for female members of the family. Below the floor four cesspits enclosed and divided by tabby walls received waste. More tabby belonging to a ruined kitchen remains incompletely preserved at the same site. Similar structures are attested by contract documents agreed by Beaufort College in November 1816 "for the building of Tabby kitchen & other outbuildings," all now lost above ground.

• Civic Structures •

During its brief existence, Beaufort College was one of Beaufort's largest public structures, incorporating about fifteen thousand square feet of enclosed space distributed over three floors—evidence enough that its contractor had successfully mastered technical problems of building large and high with tabby. It has the distinction of being among Beaufort's few documented early nineteenth-century civilian structures although the account given by the *Journal of the Proceeding of the Trustees of Beaufort College* is limited.[55] We learn that an ambitious plan—overambitious it transpired—was devised for establishing higher learning in Beaufort Town, Colonel Thomas Talbird, himself a trustee, proposing on December 23, 1800, "to build, complete and finish, an edifice of tabby one hundred and five feet long, forty four feet wide and three stories high, with eight private rooms and one public one in each story—to cover the building with tile and to find all the materials for the same except paint and roughcasting for the sum of three thousand, nine hundred and sixty five pounds . . . to rough cast the building at any time after it shall be completed at the rate of 2/6 per square yard."

Knowing Talbird's capabilities, trustees with practical experience of tabby making—including John Barnwell, Thomas Fuller, John Mark Verdier, and Stephen Elliott—adopted his proposal with only minor modification. Their finding the building "too narrow," an additional sum of 150 pounds was authorized (January 12, 1801) to increase its width by six feet. Later (August 4, 1801) St. Helena Parish contributed a prime site for the nascent institution, comprising twenty acres of glebe land "bounded by the West Line of the town of Beaufort, to be included in a line to be run parallel to Hamar Street at the distance of 525 feet from the said Street, beginning at the bluff on the River."[56] Construction started November 2, 1802, near the site's southern extremity overlooking the Beaufort River. Concerned about the increased building width, Talbird proposed "to run two counter walls across the body of the College as high as the foundation for a support to the two principal girders, offering "to build the same for £35." Grudgingly this variation was agreed. Tabby exterior walls reached second floor level before August 1803. Timber roof members were raised on December 28, 1804, slate shipped out of New York then being substituted for roofing tile originally specified. Lack of funds and several unforeseen calamities prolonged construction. Hurricane-force winds damaged the building before it could be opened, Talbird's death in 1804 causing further interruption. Nevertheless successful conclusion was eventually achieved within the original budget, though whether or not rough casting of exterior facades was ever completed is not recorded.

Not much is heard about the building until September 1811, when another Thomas Talbird (the colonel's nephew) repaired more storm damage. Subsequently (November 22, 1814), it being found that replacement of missing slates was too difficult or expensive, it was "resolved to strip off the slate and shingle the roof of the College for the sum of one thousand dollars." Work dragged on until July 1816, the college apparently standing empty during the interval. If the premises were ever reopened, contemporary observers agree they were closed again in 1817 following an outbreak of yellow fever that sent the institution into crisis, as students sickened and died, enrolment fell off, staff resigned, and trustees removed themselves from the board.[57] By May 15, 1822, financial problems had forced drastic measures, the treasurer being authorized "to advertize the College buildings or any part thereof for Rent and that in the case he shall not be able to rent them in the course of two months he shall be authorized to dispose of the stables and fences for whatever they may bring."[58] Twelve months later (May 1823) the main structure (now storing lumber salvaged from fences once surrounding the site) was boarded up. After four or five weeks everything was advertised for sale. No buyer came forward until 1833, when "Thomas Talbird offered $400 for the college building." His offer was accepted. By April 1836 the college had been "torn down and removed," timber, brick or other reusable items probably being salvaged. No longer needed for educational purposes, former glebe land was then returned to St. Helena Parish. This disastrous history has left no information about the college's appearance, but mention of "a cupola" (either damaged or unsafe by 1815) suggesting it was of some architectural consequence. Nor has it been determined how thorough the younger Thomas Talbird's demolition might have been, though, considering the excessive labor dismantling and hauling tabby away entailed, foundations were probably left intact.

Vestry minutes preserve scraps of information concerning contractual procedures for one more lost structure designed to replace the earlier St. Helena parsonage previously mentioned. On July 7, 1801, it was resolved "that the proposals of Col. Talbird for buildg. a Parsonage House be agreed to, with paymts. to be made in the follg. manner Three Hundred pounds at the Commencemt [of] the Buildg.—Three Hundred pd. when the building is raised, three hundred pounds when the Buildg. is Compleated, Three Hundred pounds twelve months after."[59] On April 5, 1802, vestry ordered "that the Church Wardens do pay Col. Talbird the sum of Three Hundred pounds as 1st payment towards the Parsonage House so soon as the first Box of tabbey is made."[60] Like Beaufort College this building has disappeared, leaving little or nothing behind.

⦾ Congregational Buildings ⦾

Almost from inception the parish of St. Helena was capable of raising sufficient funds to build the body of its Beaufort church of good-quality fired brick imported from Charleston. Tabby was relegated to subsidiary projects—the parsonage already described—walls around the churchyard and foundation work for church extensions made during the late 1840s. Attracting several wealthy planters, patronage did not encourage extravagant building programs when a Baptist mission was established. The earliest meeting house was modest, measuring 30'2" north/south × 36'4" east/west overall excluding any porch or porches. Enclosing walls about fifteen inches wide were of tabby stuccoed with a fine lime stucco on exterior faces. Over time the congregation outgrew this space, existing work being first modified then enlarged. Matching the original in width (36'4"), new construction produced a building measuring 65'8" from front to back. The north wall of the original meeting house was apparently demolished down to its foundation to make one uninterrupted interior space. Almost nothing survives of construction above ground except for traces of the new floor preserved by joist sockets along the addition's north tabby wall. These indicate that 10" × 3" timber joists spaced about twenty-four inches apart were set on timber wall plates approximately 3" high × 6" wide as wall construction proceeded. Spanning north/south joists were probably supported at their opposite (south) end by cut-down remnants of the original building's demolished north facade.

The meager structural record is amplified by Dr. John Archibald Johnson, who writing after the Civil War recorded that Beaufort's earliest Baptist community "erected a small but convenient place of worship . . . in consequence of the great distance from their homes" of the mother church on the Euhaw [and] the intervention too, of Broad River, not infrequently impassable in open boats." Johnson related that "a large and commodious building" was built of tabby during or soon after 1804, a tower and bell being added about thirty years later.[61] Overloaded the tower and exterior tabby walls began failing in August 1842, two local contractors then finding the building unsafe. Despite appeals membership decided on partial demolition and erection of the present brick-built neoclassical church to its south. Old tabby walls were cut down but otherwise left standing and the area occupied by the original church site divided into burial lots, Reverend Richard Fuller (minister) and other prominent church members, including Lewis Reeve Sams and Joseph Hazel (both planters), receiving divisions. Thus: "the site of the old church was fixed" and remains visible today.[62]

Enlargement of St. Helena's parish church in 1842 also necessitated demolition of an earlier building, but here more complex programs involving reuse of old materials and introduction of new work were embarked on. Brick walls of the church, built by Henry Talbird in 1769, were first cut down and then used as foundations for colonnades supporting galleries lining the interior of a new congregational space created by additions extending fourteen feet north and south beyond the original building's confines. New exterior brick walls were supported by freshly made tabby strip foundations (measuring 2'6" in width × 2' in depth).[63] Whether the latter were seen as a means of saving money is not known, although cheap or otherwise the foundations described remain in excellent condition.

◦ Minor Construction—Tabby Walls ◦

Far more durable than timber fencing, tabby proved especially useful for walls enclosing public and private spaces. Reference has already been made of tabby walls surrounding Daniel DeSaussure's Bay Street lot, which secured the property and its valuable stores from theft. Behind what is now 807/811 Bay Street, portions of another tabby wall about seven feet high still stand. Cast in two-foot-high increments, this probably enclosed a garden or yard associated with an early nineteenth-century tabby house fronting the south side of Bay Street. Owned by an unidentified Fripp family member before the Civil War, this relatively small, two-story-high, gable-ended dwelling was demolished soon after 1880.

Samuel Cooley's circa 1863 image of the Campbell House shows a surprisingly substantial tabby wall (now destroyed) bordering West Street immediately south of Campbell's residence, which probably enclosed land extending behind the Chisholm House,
one of Beaufort's more significant dwellings, now much changed. Still standing despite
inept buttressing, the enclosure surrounding St. Helena's churchyard likely preserves fabric belonging to the tabby wall that the vestry asked Colonel Talbird "to consider building" in February 4, 1800.[64] Whether Talbird actually undertook the job is uncertain, but
it was eventually completed, the entire block on which the church stands being enclosed.

Tabby in the Domestic Architecture of the Sea Islands before the American Revolution

• Cattle and Indigo •

During most of the eighteenth century, cattle ranching was an economic standby for coastal landowners, with fresh, salt, or pickled beef along with rendered fats and hides finding steady markets at home and abroad.[1] Moreover ranching required little or no land clearance, drainage, or enclosure. Returns never matched those of merchants and speculators who sank fortunes into opening up wetlands for rice cultivation. Nevertheless the lease dated March 6, 1746, of St. Philip's—one of Beaufort County's barrier islands—by Richard Capers to James Dawkins, a prominent planter and wealthy landowner with estates in England and Jamaica—for ten shillings per annum suggests money could be made running cattle on such unpromising property as this tract, with its dense woodland, serried ranks of old dune ridges, wide marshes, mud flats, and open exposure to the Atlantic Ocean.[2] Mr. Henry C. Chambers reported that a small tabby structure believed to be used by cattlemen was standing here, but I am unable to confirm this information, the sea having overwhelmed the structure around 2016. Capers's will (1755) does prove he owned tanning vats "located next to Mr. William Chapman" of St. Helena Island, features that, if still extant, have not been found. Dabbs put the Capers homestead on Little Capers, a marsh island off St. Helena, where she saw tabby house foundations and traces "of the old family cemetery."[3] Field survey in 1999 revealed that whatever remained of the dwelling had been ploughed out. Except for the fine double tomb commemorating Charles Capers (1728–98) and his wife, Anne (died 1793), the cemetery has also disappeared, scatters of oyster shell suggesting it was once enclosed by tabby walls.[4] No photograph of the Capers site before its destruction is known, local landowners recalling that the lost house resembled early dwellings documented from Spring and Dataw Islands described below.

Tabby's utility for isolated coastal holdings is better demonstrated at Jekyll Island, Georgia, where Major William Horton settled and built his residence before 1738 over the site of an old campground inhabited intermittently by indigenous peoples well into the colonial period. Shielded from Atlantic gales and approached by the navigable, winding waterway penetrating the island's lee side called DuBignon Creek, after a later owner, this house—about which nothing certain is understood—was deliberately destroyed by Spanish raiders in reprisal for their losses during the Battle of Bloody Marsh on St. Simons Island in 1742. Sometime before 1745 Horton rebuilt his residence and out-buildings in tabby. The major was well acquainted with the material, having overseen construction of Frederica's barracks in 1740. We know from an interview recorded by Lord Egmont that Horton (who was then in London raising recruits for Oglethorpe's regiment) adamantly opposed slavery, observing "Negroes if introduced would be the absolute ruin of the colony [i.e., Georgia]."[5] This could mean that men drawn from Frederica's garrison were employed for rebuilding his Jekyll Island structures.

Comprising about five hundred acres, the holding raised cattle and horses, grew vegetables, and brewed beer to supply Frederica's residents who, Oglethorpe remarked, would have starved or abandoned the place without Horton's industry. Horton spent time in Beaufort too, finding cash and securing credit for Georgia's newly founded garrison town, the trusty Major, grown weary of his own plantation and constant bickering among Frederica's civilian inhabitants securing election to the Sixteenth Royal Assembly of South Carolina in 1747 as representative for St. Helena Parish. This position was cut short by death the following year. Thereafter, his island estate was maintained by another of Oglethorpe's officers, Captain Raymond Demere, a French Huguenot of considerable fortune said Thomas Spalding.[6]

Visiting Jekyll in 1753, Jonathan Bryan found "a handsome dwelling house of about forty feet long by twenty wide, neatly finish'd and glazed, a good house for his Overseer about thirty by twenty, a Malt House of eighty or one hundred foot long by thirty, all of these of Tabby." There were still fruit trees and gardens near the house, horses and herds of cattle ranging the island.[7] Occupying the same site delineated by early plats of Horton's property, the present Horton-DuBignon House (measuring 41'6" × 18'2" overall) nearly matches dimensions of the dwelling described by Bryan. Exterior walls are tabby, two stories high, with principal facades organized into five bays and an internal chimney centered about one end wall. The ground floor probably incorporated two spaces of unequal size, the larger forming an old-fashioned through hall. Nothing indicates the entry was raised more than several inches above ground, however, considerable destruction, reconstruction, and alteration make the building's original appearance, height, and organization uncertain.[8] The issue is further clouded by the 1768 survey of Horton's property by G. McIntosh depicting a one-story house with single end chimney and smaller structure (the overseer's house?). This could be artistic convention. On the other hand, rather than an icon we might have an actual representation—making it the earliest illustration of tabby dwellings extant.[9]

Horton experimented with unfamiliar crops, hoping they might flourish along the Georgia coast, inexperienced settlers believing subtropical, even tropical products might suit. In 1738–39 a Captain Dymond noted that Horton planted cotton on St. Simon's Island, though of what variety or from where his seed was obtained is not said.[10] On Jekyll in 1742 John Pye saw a plow going with eight horses and eight acres under indigo. If Eliza Pinckney was the first to grow indigo commercially near Charleston between 1741 and 1744 with seed sent by her father from Antigua, then Horton's was a pioneer achievement. His efforts—like those of André Deveaux, who helped popularize the crop across South Carolina—were probably aided by Caribbean knowhow, Pye mentioning an Englishman and Spaniard working the place who thought their indigo was "as good as that made in the Spanish West Indias."[11]

Encouraged by an act of Parliament passed in 1748 allowing "a bounty of sixpence per pound on all indigo raised in the British American plantations," many coastal planters cultivated the plant before revolution, war, and foreign competition destroyed the trade.[12] Most indigo cake from the southeastern coastal plain was exported to Great Britain, though not all, La Rochefoucauld-Liancourt finding it was used to dye clothes worn by "nègres."[13] Agricultural commentators agreed the plant needed dry, well-drained soils, pine barrens located near open water with protection against late frosts proving ideal if properly fertilized. The majority of Sea Island plantations had land meeting these requirements. On St. Helena Island 473 acres fronting Chowan Creek belonging to the William Chapman mentioned by Richard Capers was bought in 1753 by Peter Perry, who transformed it into an indigo plantation called Orange Grove.[14] The present Orange Grove preserves traces of early settlement, but these are scanty except for tabby chimney and wall fragments standing near a cemetery, enclosed by low tabby walls, containing burials of the Perry family. Two miles or so away, off Battle of Britain Road, badly damaged ruins exist of the Laurence Fripp House. Now reduced almost down to its foundations and filled with household rubbish dumped by nearby residents, this tabby structure retains quantities of fired brick apparently used for leveling between wall pours, with remains of a large tabby corner chimney standing at the building's far end. Details suggest construction during the later eighteenth century, when returns from indigo could sustain the high-quality work seen here. Coincidently or otherwise Dabbs reported indigo still grew wild in adjacent fields during the 1940s.[15]

Lands bordering the Broad River were extensively developed for the crop. Obtained by Colonel Thomas Middleton around 1761, Laurel Bay Plantation on Port Royal Island is among the better-documented examples. An inventory of 1776 records that "3 sets of indigo vats" valued at £60, a "pump" valued at £4, and indigo seed valued at £24 were then on hand, the plantation's workforce numbering fifty-nine enslaved laborers.[16] Thomas Middleton owned more vats than most of his contemporaries, though not as many as his father-in-law, Nathaniel Barnwell, who, having become Beaufort's largest indigo producer, left six sets when he died in 1775.[17] His rather shaky signature in one of their account books indicates that along with numerous local growers, Barnwell brokered his "indico" through the Beaufort firm of Daniel DeSaussure and Verdier, headquartered on Bay Street.[18]

Overlooking the Broad River at Laurel Bay, Port Royal Island, foundations of Middleton's house remain. Joseph Barnwell (whose father owned the property until dispossessed during the Civil War) reported that "the house was constructed of brick and probably was the finest dwelling house in the province of South Carolina west of the Ashley River. . . . It has been the tradition that the house had every improvement which could be found in the province at that time."[19] With financial interests in Charleston, the Atlantic slave trade, and extensive land holdings, Middleton was wealthy enough to choose durability over convenience, and prestige over practicality, opting for brick construction instead of cheaper tabby or timber despite substantial additional costs incurred. Along with certain speculations, this decision perhaps precipitated the financial ruin that ultimately consumed his property and reputation. At Old House, St. Luke's Parish, Daniel Heyward made similar calculations, building at least the lower story of his dwelling on Euhaw Creek (a small tributary of the Broad River) in brick after 1757. Heyward grew rice and experimented with tide-driven mills, agricultural ventures making him a very rich man whose holdings eventually comprised some fifteen thousand acres besides townhouses in Charleston and Beaufort. Old House was raided by British forces during the Revolution and later replaced by a new brick dwelling erected on another site close by called White Hall.[20] When enlarged by Daniel's son, Thomas Heyward, whose spendthrift ways became notorious despite his declining fortune, construction methods changed. Wings (each incorporating one full story over an elevated basement) were added in tabby instead of fired brick, the tabby stuccoed and scored to imitate stone. Whether the existing brick house was also stuccoed cannot be determined, only two or three courses remaining above ground.[21]

Located three or four miles outside Beaufort Town just north of Battery Creek, Retreat is Beaufort County's only habitable plantation house with probable prerevolutionary origins. Tradition asserts that the tract where it stands was first granted to Jean de la Gaye, variously called John de la Geyey and John de la Gayé, an *émigré* merchant who settled hereabouts during the 1730s, 250 acres warranted him "in family right" by the South Carolina Council on July 1, 1738, probably referring to this property. Erection of the existing tabby house is attributed to the same individual before 1769, when he sold his extensive business interests and left Beaufort for Nimes (Languedoc), France. The existing 1½ story, gable-roofed, predominantly tabby-walled structure has a wide porch (doubtless rebuilt) raised only slightly above grade along its five-bay entrance facade. An internal chimney fabricated of fired brick laid up in irregular variants of English bond still dominates each gable end, the brickwork keyed into the tabby as wall construction proceeded. The exterior face of each chimney exhibits an elongated diamond pattern outlined by vitreous headers subdivided into several smaller diamonds, also defined by vitreous headers, an unusual treatment for Beaufort District, the only comparable example flanking the east entrance of Prince William's parish church, Sheldon—by far the grandest planter place of worship south of the Combahee.

Similar patterned brickwork occurs in England, where it appeared during the late sixteenth century and continued popular down until the mid-eighteenth century among domestic buildings of vernacular character. The tradition probably reached Philadelphia by the late 1600s, enjoying an exceptional florescence throughout southern New Jersey, where over one hundred examples erected between 1720 and 1760 are recorded.[22] English and American bricklayers often incorporated their patron's initials and the construction date into these decorative schemes. Workers at Retreat did not, even though whoever was responsible for laying up brick walls of Sheldon Church followed long-established precedent by incorporating the date "1751" into that building's east facade.

Contracts are entirely wanting for any aspect of Retreat's construction, making it impossible to know if the chimneys described were fabricated by John Kelsall, who, Dr.

Above: Retreat Plantation, Port Royal Island, Beaufort vicinity, south facade (Jack Boucher, HABS).

Retreat Plantation, Port Royal Island, Beaufort vicinity, west facade (Jack Boucher, HABS).

L. Rowland has argued, built the first Sheldon Church, or another itinerant bricklayer who, given the expense and scarcity of local brick, can have found only intermittent employment around Beaufort. Questions of attribution are further complicated by alterations and additions. Upper gables are obviously rebuilt, with introduction of cast-iron tie rods and plates following the so-called Charleston earthquake of 1886. The house was abandoned in 1912. By 1939 its roof had collapsed. It is said that fireplace surrounds, floorboards, and other timbers were subsequently stolen. Extensively restored during the 1940s reusing salvaged trim from other historic buildings, the house was extended north in 1950, with a new wing accommodating kitchen and service areas. More living space was later added on the east and a new bedroom installed above.

How far the interior area enclosed by tabby walls retains its original spatial disposition is an open question. At ground level a large heated through hall (west) seems certain, as do two small flanking rooms (east) heated by back-to-back fireplaces arranged on a triangular plan. There is evidence for an enclosed stair (now realigned and partially rebuilt) behind the through hall. Above, the present staircase gives onto a central passage running front to back of the old tabby dwelling, with one large heated room (presumably its main bed chamber) to the east and two smaller rooms (one heated) west, all lighted by dormer windows. Again it is difficult to tell if these configurations represent the original plan, are modern restorations, or reflect long occupation by three generations of the prominent Barnwell family—General John Barnwell, who acquired Retreat before 1783, his son John G. Barnwell (1778–1828); and his granddaughter Sarah B. Barnwell, who deeded the property to her sisters, Anne B. Barnwell and Emily H. Barnwell, in 1840 when it comprised "310 acres and more bounded N. by creek leading from the Battery river, on south by Battery River and on west by lands belonging to Rev. Stephen Elliot Jnr."[23] Antebellum maps show a single slave street at Retreat, several buildings having tabby chimneys if Civil War–era stereoscopic photographs are correctly identified. Recent archaeological surveys failed to locate these features, so there is no reliable evidence about when this settlement was founded or if anything tangible remains of the site's prerevolutionary slave housing.

Two more ruined structures—the Edwards House, Spring Island; and the William Sams House, Dataw (Datha) Island—encapsulate tabby dwellings built during the 1770s or possibly earlier.[24] Like many Sea Island houses of the period (though not all, as Brick House, Edisto Island, demonstrates), these were relatively small in scale, one room deep, and lacking any architectural pretension beyond what could be achieved by simple symmetry. Excluding porches and external chimneys, early houses of the kind rarely exceeded one thousand square feet in area, 650–800 square feet being nearer the norm (see table 2).

The original Edwards House, Spring Island, exemplifies dwellings built by families who began settling outlying portions of Beaufort District following cessation of the Yemassee War in 1728. Granted to John Cochran as early as 1706, the island was the site of an Indian trading post. By 1733 it had passed to his son James Cochran, subsequently descending through the female line to Mary Barksdale, who, in 1773, married John Edwards, a Beaufort merchant already administering the property. Whether Edwards or his father-in-law, George Barksdale, built the ruined house constituting the present enlarged building's nucleus is uncertain, though documents indicate that Barksdale—a

Table 2. Tabby Plantation Houses of the Prerevolutionary Period

Location	Name of Plantation or House	Exterior Dimension
Jekyll Island, Ga.	Major William Horton	41'6" × 18'2"
St. Helena Island, S.C.	Lawrence Fripp	49'10" × 28'10"
Port Royal Island, S.C.	Retreat	36'1" × 28'4"
Port Royal Island, S.C.	Prospect Hill	32' (?) × 22'10"
Spring Island, S.C.	George Edwards, Phase I	37' × 19'9"
Dataw Island, S.C.	William Sams, Phase I	38'4" × 20'3"
Lady's Island, S.C.	Ashdale	37'3" × 18'2"

Loyalist—was living on Spring Island before 1780, when he removed his family to Charleston and safety.

As for the early dwelling, fragmentary tabby walls indicate a rectangular floor plan with two exterior end chimneys now represented by massive tabby bases. Fabricated along with exterior walls, bases were cast solid up to the main living level, raised four or five feet above grade. Wide porches supported on tabby piers once extended along the building's two principal facades (east and west), timber steps giving access via the porches. An attic probably accommodated sleeping areas, but whether the house was gable ended or had a hipped roof is unknown. Not much is understood about the island's economy before the Revolution either. George Barksdale's will dated 1775 mentions "cattle on Spring Island," sheep, horses, and hogs, while John Mark Verdier's account books attest indigo production.

On Dataw (Datha) Island, the original William Sams House—now ruined—closely resembled Spring Island's main dwelling in size, organization, and construction although when first built it had only one external end chimney (east), a second chimney being added (west) during an early nineteenth-century reconstruction phase. James Julius Sams, who lived here when young, said the house "consisted of two rooms, a narrow passage between, two attic rooms above and two cellars below"—a good description of the raised 1½ story central hall plan type almost universally favored by Beaufort area planters for country residences before the Revolution. The same account mentions "a narrow piece above the stair"—which must mean that stairs rising out of the central passage gave onto a landing linking two garret spaces where numerous Sams children slept.[25]

The dwelling's south facade was organized symmetrically about the central entrance into five bays. Archaeological excavation exposed cruciform brick piers for the porch fronting this side of the building. Whether another porch extended along the north face is uncertain, alterations and rebuilding obscuring the evidence. Several courses still in situ indicate that the earliest hearth and chimney stack were fabricated from tabby brick. An undated drawing by Eugenia Sams (1845–1920) illustrates a gable-ended dwelling with gables carried up in tabby, the roof shingled, and dormer windows lighting attic spaces. Relatively narrow timber steps leading up to the south porch from an adjacent yard along with six Tuscan-style porch columns are depicted. Fragments of the latter still exist, these stuccoed-brick features having been installed around 1810 to match new columns that Dr. B. B. Sams introduced when adding two new tabby wings to what his family called the "Middle House."[26]

When and by whom was the original dwelling built? James Julius Sams stated it was old in his grandmother's time. If he was referring to Elizabeth Hext Sams, this could mean construction was completed during the second or third quarter of the eighteenth century by Lewis Reeve (1739–74) or possibly Sarah and Robert Gibbes, who sold Dataw to William Sams (Elizabeth Hext's husband) in 1783. Architectural details are consistent with prerevolutionary traditions, but no archaeological evidence has emerged from the heavily disturbed site to unequivocally substantiate a late colonial date for the first building episode beyond scatters of imported mid-eighteenth-century pottery, scraps of oriental porcelain, and slave-made colonoware.

Disassociated chimney and fallen wall sections of the Lewis Reeve Sams House, located off the north end of Dataw, displays similar tabby construction. The building itself has been drowned by the Coosaw River, currents and storms taking most of the superstructure. Probes show this dwelling was exactly the same length and perhaps the same width as the original William Sams House. Construction is usually attributed to Lewis Reeve Sams (1784–65), brother of Dr. B. B. Sams (1787–1855). Having examined the admittedly deficient structural evidence during low spring tides, when wall fragments and remnants of the two end chimneys were visible, I believe construction by his father or grandfather more plausible.

Incomplete tabby walls of the main house at Ashdale Plantation, Lady's Island, held by the McKees before 1862, further represent unpretentious dwellings of the sort favored by local landowners before significant money was made from Sea Island cotton. Elevated on a bluff near the confluence of Lucy Point Creek and Morgan River, it was rectangular with external end chimneys. Like the first Edwards House, rooms faced east and west, thereby taking advantage of prevailing river breezes and panoramic views extending from Coosaw to Dataw Island. Variant architectural formulae are illustrated by ruins of St. Quenten's main house at what is now Walling Grove, Lady's Island. Measuring 36' × 20' in plan, the dwelling possessed two external end chimneys of tabby and fired brick erected on tabby bases.[27] Exterior walls were timber framed rather than tabby, with timber sills supported by tabby piers. Raised only about 2'6" above grade, "the building while perhaps two stories in height may have been 1½ stories" incorporating a hall or passage giving access to single rooms on each side and stairs linking them with garret or attic spaces.[28] So simple an arrangement is not necessarily indicative of the owner's social or financial status. If the dwelling was erected during the 1770s—as the archaeological record indicates—then its builder cannot be determined. It is known through Trinkley's research that in 1825 the property belonged to Joseph and Sarah Fickling. Both husband and wife were propertied, holding acreage in St. Helena Parish, lots in Beaufort Town, and some eighty-seven enslaved people, which placed them among the higher echelons of planter society.

❧ After the American Revolution ❧

Over the late 1780s and early 1790s rural structures damaged, destroyed, or abandoned in the Revolution were repaired or replaced. Plantation houses neglected during the lawless and uncertain period that followed or now too small for owners spending more time and money on coastal estates were reconfigured, old work being absorbed or integrated into expanded building programs. The forced departure of Loyalists and subsequent sale of their lands along with an influx of speculators and other new settlers also stimulated development, country houses becoming larger, better appointed, and occasionally grander than prerevolutionary predecessors, unprecedented growth of slave holdings requiring establishment of new settlements and extension of preexisting slave rows.

Across the Sea Islands, development was fueled by profits from long-staple cotton, planters abandoning indigo following the new crop's first commercial success in 1790. For Wormslow Plantation near Savannah (a city that grew rich on both rice and cotton), Swanson calculated that "acre for acre a well-managed sea island cotton plantation was twice as profitable as a rice plantation in 1800."[29] Before planting could begin, places where the new crop thrived—whether pine lands, Atlantic forest, or palmetto-dominated scrub—needed clearing followed by installation of drainage ditches (cotton not tolerating saturated soils), levees, passable roads, docks, and landings. Onerous to planters and enslaved alike, the labor involved in preparing cotton fields, fencing and ridging, gathering marsh mud, or cutting Spartina for fertilizer was less intensive—even when reclamation of salt marsh was involved—than the vastly expensive business of making and keeping rice fields in production. William Elliott, among the lowcountry's most experienced and successful cotton planters, observed in 1828: "when I speak of the Agriculture of this Parish (St. Helena), I confine myself almost of necessity, to the production of Sea-Island cotton; for the Parish is composed exclusively of sea islands; and excepting the provisions produced for plantation supply . . . the only staple cultivated, is that to which our insular location and salt exposure give us peculiar aptitude."[30] For owners with sufficient capital, credit, tenacity, and agricultural knowledge, this was an era that, lasting down until the Civil War, saw unprecedented population growth, marked increases in property values, and opportunities to accumulate significant wealth. Of course success was never assured even with the best cotton land. Speaking from his own experience when returns were low, Elliott remarked that planters were then ready "to substitute any other culture which may offer to industry and skill, a competent and an equal remuneration. For if the high rate of valuation, at which the lands and Negroes have been acquired, whether inherited or purchased, be taken into the account, it will be confessed, that few investments of capital have yielded for the last ten years so trifling a return as that, of the Sea-Island planters."[31] Market fluctuations or crop failures could (and often did) wipe out profits, while population growth was largely accounted for by involuntary immigration of African slaves needed in ever-increasing numbers to work fields newly brought into production. On St. Helena Island, one of Beaufort District's most intensively cultivated and valuable cotton growing areas, as Rosengarten estimated there were about 250 whites and two thousand blacks in 1845—"1 white person for every 8 slaves, and 1 slave for every 8 acres of arable land."[32] By 1850 U.S. Census returns tabulated 1,063 white and 7,725 black residents of St. Helena Parish.[33]

Once picked, sorted, whipped, ginned, moted, and baled, Sea Island cotton was mostly shipped to international markets, the great mills of northern England with their new and improved steam machinery having insatiable appetites for the best Sea Island product. Blackburn, Lancashire, for instance, had only one cotton mill in 1816. By 1838 there were forty-four mills employing 10,460 people—men, women, and numerous children—unbridled expansion (and appalling human exploitation) occurring through-out newly industrialized districts of Britain, with Lancashire alone processing three hundred thousand bales of lowcountry cotton in 1820.[34]

Across the Sea Islands tabby construction helped fuel recurrent building booms stimulated by unprecedented demand—so much so that it is rare to find any historic settlement in eastern Beaufort County or coastal north Georgia where the material does not occur. Many surviving domestic, agricultural, and industrial structures are attributable to the period 1790–1820, decades that saw production of long-staple cotton increase more than eightfold, prices reaching an unprecedented seventy-five cents per pound in August 1818. After 1819 falling prices inhibited investment in land and building, until the late 1840s, when cotton moved upward again.

• Planter Houses •

Standing preeminent on its own estate, the main house at the center of plantation activity has become a mythological symbol for residents of the American South, regarded by some as embodying a genteel, even aristocratic way of life, long gone. For others, structures of this kind, irrespective of their hierarchical status, were the focus of what Peter H. Wood called an agricultural gulag, dependent on forced labor performed by enslaved individuals primarily of African origin.[35] Not entirely irreconcilable these divergent views should not blind us to the prosaic fact that Sea Island planter houses were largely functional, facilitating productive businesses tied to the uncertainties of distant markets besides offering—with varying degrees of architectural ambition—shelter for the landowner and his or her family, servants, and agents. Although proprietors respected inherited planning modes and rarely financed extravagant decorative schemes, plantation houses evolved over time as older formal traditions were modified or selectively discarded and new architectural fashions adopted. Domestic building types familiar yet uniquely responsive to local environments emerged, these setting new standards of comfort and adding picturesque or scenographic elements to island landscapes not previously imagined.

Planter houses did not stand alone. Rather they stood amid clusters of dependencies, outbuildings, and subsidiary settlements. "In contrast to European and English farmsteads that often collected different functions under one roof, farmers and planters throughout the South erected numerous small buildings, each with its own specific purpose."[36] Structures near the main house might formally relate to the master's dwelling, with a kitchen and office positioned symmetrically right and left. Gardens modulated the scene, putting distance between owner and workers, drawing in carefully orchestrated views, providing color, scent, and, most important, shade to mitigate suffocating summer heats. Sea Island layouts were never as extensive as those early Ashley River showplaces belonging to the Draytons or Middletons, whose parterres, canals, and lakes perpetuating English Georgian prototypes were already out of date by the time of the first great cotton boom. Farming journals—notably the *Southern Agriculturalist,* which declared itself dedicated to "Improvement"—and proliferation of agricultural societies ensured that a sense of scientific inquiry informed certain owners, spurred initially by the success of innovative farming practice in England and France. William Elliott, William Seabrook, and William Washington—cotton magnates whose opinions carried great weight among Sea Island growers—put the results of their experiments into print, discussing everything from seed selection to manuring, planting, soil chemistry, crop rotation, harvesting, and building besides addressing crucial questions concerning the selection, management, and health of enslaved labor.

◈ Single- and Double-Pile House Plans ◈

There is no reason to believe relatively simple, rectangular, one-room-deep plantation houses resembling those examples built before the Revolution fell out of favor by the end of the eighteenth century. Indeed—without building accounts, daybooks, or inventories—prerevolutionary building and postrevolutionary building of the type cannot always be distinguished. Well-tried planning formulas proved resilient, with living spaces organized about central hallways or passages opening into porches front and back remaining standard long after the plantation era had passed. Generally Federal-era and antebellum planter houses of the Sea Islands were larger than their colonial and provincial predecessors: raised two-story structures with rooms symmetrically disposed in four-over-four, two-over-four, or two-over-two plan configurations remained popular down until the 1860s, when traditional architectural traits were disrupted by war. Most were products of vernacular builders who, with various degrees of skill, integrated fashionable Federal, Greek Revival, and occasionally Gothic elements into domestic building forms with long historic antecedents.

Attrition has been very high. Of thirty-two properties enumerated by Union assessors in 1862 on Lady's Island just across the river from Beaufort Town, only four currently retain traces of early houses, represented by mutilated and incomplete tabby foundations. St. Helena Island was apportioned between fifty-five separate plantations before the Civil War, almost all boasting an owner's or tenant house, various outbuildings, and slave settlements. In 2018 five plantation houses stood occupied, the rest gone or fallen into ruin. Extrapolating from this inadequate sample and from places where survival rates are slightly better (Edisto Island for example) it appears that the majority of Sea Island residences were timber framed, many—how many one cannot say—elevated on high masonry (tabby, tabby-brick, and fired-brick) foundations. Myrtle Bush, one of Port Royal Island's Barnwell plantations, was typical, old photographs picturing a two-story framed main dwelling with hipped roof and end chimneys raised over an elevated tabby basement fronted by an attractive entrance porch, supported on masonry (brick or tabby) arches. Well-proportioned Tuscan columns carried the porch roof, splayed steps (resembling those of the Barnwell Gough House, Beaufort) giving access. Mention of six rooms suggests that, much like an unidentified house from St. Helena Island known from an early image and several other "big" houses, the two back rooms at entrance level had shed roofs, this giving the "two-over-four" plan configuration illustrated by Sarah Fick's valuable analysis of Sea Island cotton plantations.[37]

With an assessed value of $5,752 and comprising 1,438 acres, Coffin Point, located near the northern end of St. Helena, was the largest and most valuable of the island's holdings when surveyed in 1862. Today it is among the few for which any early building records are preserved. Occupied by teachers attached to the Freedman's Bureau, the main house survived Reconstruction, though not unscathed. Despite additions, mutilations, and unsympathetic alteration, it still encapsulates the two-story framed dwelling raised over a high tabby basement commenced by Ebenezer Coffin of Boston and Charleston in 1803, who came by the property through marriage to Mary Mathews in 1793.[38] Measuring about 46'8" × 30'6" overall (excluding porches), the original plan incorporated four basement rooms with four principal living spaces above organized about the usual central passage and staircase. The uppermost floor could have been similar or alternatively incorporated only two front rooms, extensive alteration obscuring what might or might not have been. Giving distant prospects of Edisto Island, a wide "piazza" fronting the pedimented river facade has been rebuilt several times, the present iteration a clumsily cut-down version of the original. Careful proportioning, tall sash windows, finely worked exterior cornices, a tripartite window lighting the upper stair hall (now incongruously converted into an exterior doorway), and lunette piercing the pediment are hallmarks of late Georgian taste—and a wealthy client—here transposed to what, at the beginning of the nineteenth century, was an exceedingly remote subtropical wilderness more easily reached by sea than by land. Transformed first by Ebenezer Coffin, then (after 1818) by his son, Thomas A. Coffin, this place ranked among the most productive, well-managed, and best-regarded lowcountry cotton producers, its "pinched coffin" bale mark recognized as the sign of exceptional quality in Liverpool as well as in Charleston.[39]

Incomplete plantation journals name a "Mr. Wade" contractor responsible for the main house, kitchen, stable, "Negro houses etc.," who along with five carpenters was engaged at "two dollars per diem for himself, one dollar for each workman finding them in provisions." Previously either Coffin or Coffin's agent listed "scantling" required for the job, presupposing that plans were already drawn and material specifications made.[40] Never priced the list is detailed insofar as it goes, specifying timber sizes, lengths, and quantities. Reference to masonry brick for chimneys and story-height tabby walls enclosing the basement is also lacking, there being no indication that Mr. Wade made "boxes" necessary for tabby casting or oversaw bricklayers following their trade. Later entries record Colonel Talbird received payments (for unspecified services or goods), meaning perhaps he, with his expert knowledge of tabby construction, had some hand in the work. Captain Francis Saltus (whose tabby townhouse is described in chapter 5) was also acquainted with the owner, one of the earliest journal entries giving the captain's sailing directions for safely crossing the bar of St. Helena Sound when approaching Coffin's plantation by sea. Fleetwood noted Ebenezer Coffin brought eight shipwrights down from New York in 1816 to build or repair vessels, making it possible that Saltus assisted with this enterprise in addition to any involvement with building activities.[41]

Despite surface erosion tabby walls (approximately fifteen inches thick) enclosing the dwelling's lowest level have carried timber-framed construction above through numerous Atlantic storms and the terrifying wall of water generated by the "Sea Island Hurricane" of 1893. To the west, ruins of an associated slave settlement—possibly the one built by Mr. Wade—stand near Coffin's Creek. Badly damaged it includes several tabby chimney bases arranged in two opposite, not quite parallel rows along the shore. Most bases represent single slave houses. A larger structure (about 30'8" long) stood near the north extremity of one row. This timber-framed building had an external chimney at both ends, indicating occupancy by two or more enslaved families.

Tombee's main house is another fortunate survival, the diary of its last antebellum owner, Thomas B. Chaplin Jr., giving intimate pictures of the day-to-day activities, trials, frequent tribulations, tragedies, uncertainties, and small pleasures of planter life. In 1862 his estate was middling in value and size, comprising 376 acres assessed at $1,504. Located near the southwest end of St. Helena Island off Seaside Road—an old public highway linking plantations strung out along the island's eastern shore from Coffin Point thirteen miles south to Jenkins Plantation overlooking Port Royal Sound[42]—this house was probably built with profits derived from Sea Island cotton by the diarist's father, Thomas B. Chaplin Sr., who formerly occupied a smaller tabby house nearby called Riverside.[43] Tombee's main house is T-shaped, the plan type becoming popular during the late 1700s and early 1800s among Beaufort's elite. Here there are two timber-framed stories with principal rooms organized about the usual central hall, the head of the T fronted south by an apparently reworked double-height porch and provided with brick end chimneys.

Rosengarten linked the building's longevity to its "magnificent foundation," describing "broad tabby piers rising eight feet out of a subterranean slab that prevented the house from settling unevenly."[44] I can vouch that tabby piers supporting upper floors and tabby bases of the two end chimneys are still intact but know nothing of the slab. Assuming it is not concrete introduced during later twentieth-century restoration by James Williams of Savannah, but fabricated of tabby, it would be an unusual, possibly unique construction feature. Not far distanced from the Atlantic, Thomas Chaplin's house rattled and shook in storms, and the roof leaked prodigiously. Nevertheless, despite inadequate maintenance before and after the Civil War, Tombee, like Coffin Point, weathered the 1893 hurricane that carried off substantial proportions of St. Helena Island's population, livestock, and countless cabins and outbuildings besides wrecking or destroying many former planter dwellings.

The largest T-shaped plantation house so far discovered stood at Haig Point near the northern tip of Daufuskie Island. Known primarily from Lepionka's archaeological work (1988), this now wrecked 2½ story tabby dwelling was built by the Reverend Hiram Blodgett either shortly before the death of his wealthy first wife, Sarah, in 1833 or soon thereafter.[45] Each story had an area of about 3,203 square feet excluding porches and

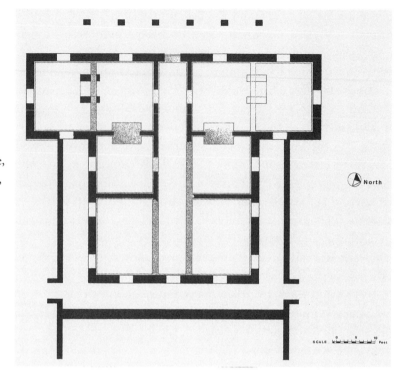

Haig Point House, Daufuskie Island, restored lower floor plan.

North

SCALE   0   5   10   Feet

excavated wall falls indicating exterior building corners were reinforced with fired brick. The elevated basement was subdivided by tabby partition walls into eight separate spaces, organized symmetrically about a relatively narrow central passage. How upper parts of this house were arranged is not clear—we might conjecture spaces were ordered about the usual central stair hall, basement partitions possibly supporting heavy loads generated by large sliding doors linking first-floor rooms above.[46] The building's main block (creating the stem of the T) was surrounded by porches on three sides. Massive tabby foundations suggest that double-height supporting columns of exceptional scale faced out toward Calibogue Sound, an often-treacherous channel separating Daufuskie and Hilton Head Islands.

Insofar as understood, exterior wall openings were relatively small and sparsely distributed. Over seventy-five feet long, the landward (west) elevation was divided into five bays, a single-story back porch raised on brick piers enlivening the otherwise severe facade. Incorporating just three bays, the east (river) front betrays stylistic links with rationalist architectural trends current during the 1820s. Whether any professional designer was involved is not known. Distant influence of the English architect William Jay—who introduced an austere Greek revival style with William Scarbrough's Savannah house in 1819—seems possible. Jay's penchant for monumentality and sparse fenestration (enunciated in an article published by the *Savannah Georgian,* January 22, 1820) were perhaps reflected by Haig Point's overscaled porch and wide expanses of masonry minimally interrupted by voids for windows and doors.[47] Other than scraps of scored stucco, nothing survives of external wall finishes, scavengers having destroyed the interior during the Civil War, fires and demolition crews leaving only carbonized floor timbers, brick hearths, and incomplete foundations now backfilled after excavation.

What Reverend Blodgett was trying to prove or say with his prominently positioned mansion can only be guessed. Visible from Tybee the house was an excellent landmark for steamships or sailboats plying between Savannah and Charleston as well as mariners approaching or leaving Hilton Head Island.

On Hilton Head itself, the tabby-built Stoney-Baynard House, occupying an unusually high prehistoric dune ridge rising above Calibogue Sound's southwestern shore, was, before falling into complete ruin, another nautical landmark. Straddling an ill-understood position somewhere between orthodoxy and innovation, too little is preserved to draw definite conclusions about influences inspiring its design. Almost square (measuring 40'6" north/south × 46'6" east/west excluding porches), the house incorporated one main story raised over an elevated basement, porches apparently enclosing upper and lower building levels along all four sides. In these respects it may have resembled Kelvin Grove, St. Simon's Island, where the first main house (razed after the Civil War) was forty-four feet square, of tabby, incorporating two stories and a basement with twelve-foot-wide "piazzas" on three sides.[48] Stoney-Baynard's principal living spaces were lighted by generously proportioned double-sash windows, with basement fenestration featuring small, rectangular openings. Facades were stuccoed and carefully scored in imitation of regularly coursed stonework.

If correctly interpreted, overall massing finds few extant parallels from Beaufort County. Double-height porches completely surrounding living and storage accommodation is more reminiscent of West Indian models or French colonial prototypes, the galleried houses of Louisiana and, at a more distant remove, Jamaica coming to mind. Best preserved the northwest exterior corner indicates the main roof was hipped rather than gabled; however, questions regarding junctions and details remain obscure. Were roof members carried over porches and supported by colonnades as often done in Louisiana? Alternatively could roofs enclosing porches have been built independently from the roof frame—a more common local treatment? So far neither documentation nor excavation has provided answers for these tantalizing architectural puzzles. Archaeological analysis does reveal the house was erected between 1800 and 1815 either for James Stoney or for John Stoney, two Charleston merchants who were then buying up prime cotton lands across Hilton Head Island as speculative ventures. It is also likely living accommodation was organized in a double-pile configuration about the customary central hall, tumbles of fallen brick and depressions marking paired internal chimneys and cross partitions dividing front from rear spaces. This conjectured arrangement isolates Stoney-Baynard House from many contemporary tabby dwellings where desire for comfort during sultry summer months caused adoption of narrow, linear plans allowing ample cross ventilation. Still the Hilton Head site opened onto Calibogue Sound, wrap-around porches—if present—sheltering and shading the entire building besides giving views out toward the sea.

⊕ Tripartite and Linear Plans ⊕

Preference for linear rather than deep, double-pile plans is exemplified by three tabby dwellings already mentioned, the William Sams House, Dataw Island; the George Edwards House, Spring Island; and Thomas Heyward's White Hall Plantation, near Grahamville. Through enlargement and addition during the early nineteenth century, each of these late eighteenth-century structures became the centerpiece of a new tripartite building scheme, responsive to its own locale and prevailing climatic conditions. Best understood is the William Sams House, where successive building episodes gradually transformed the original tabby dwelling to accord with the family's increasing size, prosperity, and status. Work probably started soon after 1783, when William Sams acquired the old raised 1½ story property. To improve head room, an existing basement was dug out, floor joists supporting the first floor raised three feet or more above their former position, and wall openings adjusted using tabby brick for necessary patching and blocking work. External walls were subsequently made higher by casting new tabby on top of original tabby all around the building, an exercise requiring first removal and then reinstallation of roofing materials. While it improved storage capacity, this episode provided no additional living space, which remained near 650 square feet until about 1812, when Dr. Berners Barnwell Sams—who inherited the house along with half of Dataw Island from his father, William Sams, in 1798—conceived an ingenious, yet simple, scheme. Two almost-identical tabby wings were constructed to flank the original building right and left.

Each new wing (measuring about 38'9" east/west × 20'9" north/south) had one living floor raised over an elevated basement with spaces organized about a centrally positioned chimney stack. Set back from the old house, new construction was linked by an enclosed corridor constructed behind the preexisting dwelling. J. J. Sams accurately described the arrangement, his terminology reflecting how the three building masses were considered distinct units even though they constituted parts of one architectural composition: "the two wings were connected by a large passageway, running back of the middle house, not only connecting the east and west house, but also connecting the middle house. The narrow passage [i.e., central hall] in the middle house opened into this large passage on its side. The two ends of this passage were entered from two doors respectively in the parlors and piazzas of the east and west house. The three houses each had its own piazza."[49]

The sense of economy dictating alteration, rather than demolition, of the original dwelling is further shown by the way B. B. Sams set about fabricating east and west "houses." These were erected together, slight dimensional differences showing use of two separate formwork sets—one set for each wing—an expedient calculated to save time and maximize output of his workforce. East and west "houses" underway, operations commenced on the narrow-roofed area linking them, the high tabby facade enclosing the link's north side, creating a screen wall complete with windows extending between the two new building masses. Ultimately this link became the enlarged building's principal entrance, visitors and family members entering its upper level from the north via two opposed staircases leading up from an open court fronting the old dwelling's north

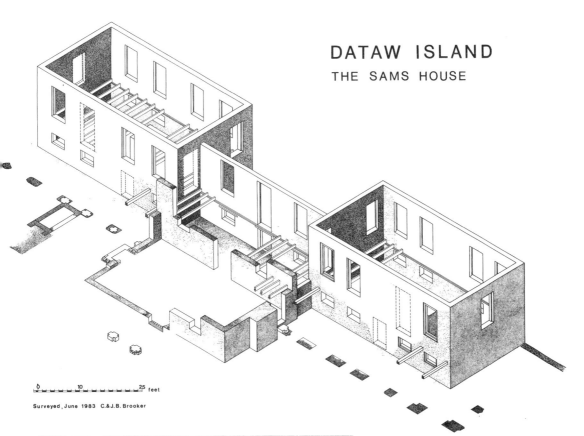

DATAW ISLAND
THE SAMS HOUSE

0   10   25 feet

Surveyed, June 1983 C.&J.B. Brooker

Above: William Sams House, Dataw Island, partly restored isometric of dwelling as enlarged ca. 1812; viewed from southeast.

William Sams House, Dataw Island, north entrance and east wing.

face. To the south, entrance was made via porches, each "house" approached by its own timber staircase. Tuscan columns of stuccoed brick raised over tabby foundation blocks and cruciform brick piers supported all three "piazzas."

East and west "houses" comprised two principal spaces at the upper level separated by timber partitions running north/south about the central chimney. J. J. Sams stated: "the extreme west room was my Father's chamber, the next that a parlor."[50] He implied the east "house" was analogous, remarking, "the first room was called the drawing room," but this leaves us guessing about the second room's function. Living spaces were probably interconnected by doors positioned right or left of chimney stacks and main chambers paneled. Facade organization mirrored the hierarchy established by placing habitable areas above and service or storage functions below. Long elevations featured four small lower windows grouped in pairs. Insofar as known second-floor openings repeated the same rhythm but were much taller and slightly wider. The drawing by Eugenia Sams shows hipped roofs enclosing the two wings with roof spaces lighted by dormer windows. If correctly portrayed the arrangement requires access stairs communicating with the attics, but where such features were positioned is entirely uncertain. Storage rooms below had dirt floors overlain by clean sand resembling those still existing at Hampton, the great, rambling Horry family house overlooking Wambaw Creek, an area of St. James's Parish formerly called French Santee.

Dr. B. B. Sams was almost certainly his own contractor and probably used enslaved laborers when hauling shell, burning lime, and casting tabby, J. J. Sams reporting how "generally he superintended [tabby] work himself knowing how particular it was necessary to be."[51] The clever, idiosyncratic, yet amateurish plan of the enlarged plantation house surely means B. B. Sams was also his own architect, the design appearing less dependent on any formal or bookish model than contemporary or near-contemporary works undertaken by owners of Spring Island and White Hall near Grahamville.

Spring Island's prerevolutionary tabby house was also enlarged near the beginning of the nineteenth century by its then owner, George Edwards (1776–1859). In 1801 Edwards married his cousin Elizabeth Barksdale, who brought valuable real property with her marriage portion, including Ferry Plantation, North Santee, a house on Tradd Street in Charleston, and twenty slaves.[52] Edwards himself was already well provided for, the 1800 U.S. Census listing forty slaves under his name and an additional sixty from the estate of George Barksdale, which he (Edwards) apparently controlled. Some time before 1810, the couple took up residence on Spring Island, where over the next two decades the number of enslaved people increased rapidly, reaching 336 persons by 1830, making Edwards one of Beaufort County's largest slaveowners. Cotton was always the plantation's mainstay, census returns listing 150 bales in 1850, along with quantities of foodstuffs (Indian corn, peas and beans, sweet potatoes) primarily for maintenance of the enslaved population. Destined for eventual sale, cattle (two hundred head); hogs (one hundred); sheep (seventy), butter, and, surprisingly, twenty-eight hundred pounds of rice are also listed, the agricultural schedule valuing Edwards's plantation itself at $50,000 excluding machinery worth $2,300.[53]

Growing wealth funded major improvement of the owner's residence and its surroundings. Two symmetrical, mirror-image, but otherwise identical flanking wings of tabby were added to the old main dwelling. Measuring 22'3" × 25'4" overall, each new wing enclosed one light and airy living area raised over an elevated basement or service floor. Lower (first-floor) exterior walls were cast fifteen inches thick using twenty-four-inch-high timber forms except where interrupted by door and window openings. At second-floor level, the thickness of exterior walls was reduced to twelve inches, formwork height remaining constant. Roofing has left no trace, but hipped solutions would answer local preferences and overall plan dimensions. Inside, upper living spaces were heated by means of a brick-built fireplace and chimney erected independently against the inside face of an outer wall, this arrangement protecting tabby, the building, and its occupants from fire, structural destruction and ultimate ruin coming only after everything was deliberately set alight sometime during or after the Civil War. The usual complement of timber baseboards, chair rails, architraves, and cornices enriched plastered walls. Besides ensuring good ventilation, eight tall double-sash windows gave views extending over rivers, islands, fields, and gardens, allowing the family to witness arrivals and departures by sea, George Edwards himself traveling—it is said—by barge rowed by seventeen or more enslaved men. New rooms and the old house were linked by porches (U-shaped in plan) wrapped along the enlarged building's riverfront, doors symmetrically placed at

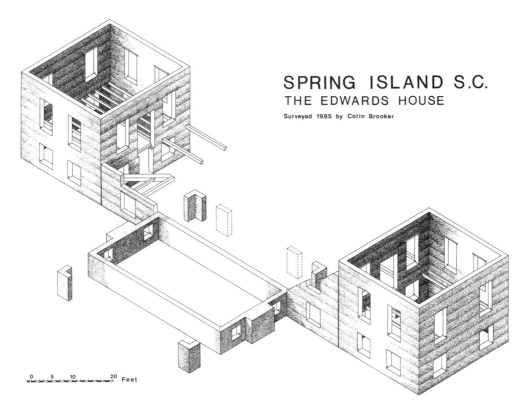

SPRING ISLAND S.C.
THE EDWARDS HOUSE
Surveyed 1985 by Colin Brooker

0 5 10   20 Feet

George Edwards House, Spring Island, restored isometric of dwelling
as enlarged ca. 1812; view from southwest.

George Edwards House, Spring Island. South facade of main house, north wing
and view of north flanker to right (HABS, Jack Boucher, 2003).

upper and lower levels giving interior access. From the land side, porches were masked
by tabby screen walls extending between new and existing construction, these essentially
freestanding elements not being bonded into adjacent tabby. Besides providing linkage,
porches and screen walls performed aesthetic functions, unifying the tripartite massing
produced by architectural additions. Convenience was sacrificed, movement between
different parts of the house requiring passage along covered but otherwise open exterior
walkways—no real hardship during warm weather, inconvenient when rain was blowing
in from the river, well-nigh impassable in cyclones or summer storms.[54]

◦ Environmental and Planning Determinants ◦

The enlarged Edwards House echoed new fashions permeating lowcountry plantation
architecture over later decades of the eighteenth century characterized by loose, fragmented, or linear plans that seem vaguely Palladian although no house or villa illustrated
by the *Quatro Libri* offers more than distant parallels. Instead the group occupies territory standing between "polite" and vernacular, adapting "bookish" architectural models
to local climatic and social needs. Describing two similar assemblages from the Santee
Delta, Stoney observed, "in the 1790s El Dorado and Harrietta, with their elaborated
wings, mark attempts to give with some architectural distinction more and better spaces
for windows and the cross ventilation so necessary for comfort in the Low Country."[55]

Built overlooking the South Santee River by Rebecca Bruton Motte, Eldorado (now ruined) incorporated three principal blocks arranged with an open court on one side. Alice Ravenel Huger Smith's drawings show that roof frames enclosing each block were hipped, tall end chimneys of the central block giving the overall composition an agreeably broken silhouette. George Edwards may have remembered this house when enlarging the old Spring Island dwelling since his wife's Santee plantation, then called the Ferry (now Crow Hill), was located only a mile from Eldorado. If so, changes were made. Rebecca Motte's model was refashioned through processes typifying folk building in peripheral geographic areas—Edwards disassembling the design into separate parts then reassembling them into an original, yet not altogether unfamiliar building form.[56] He also followed local precedent, choosing tabby construction common about Beaufort District, rather than Eldorado's timber framing supported on brick arches, having doubtless found that unlimited quantities of shell might be quarried from prehistoric middens within several hundred yards of his building site. Exterior tabby surfaces of the enlarged Edwards House were stuccoed, old surfaces perhaps restuccoed and scored to simulate high-quality stone construction beyond the means of even the wealthiest Sea Island planters. Results must have been convincing after these finishes were burnished and newly tinted with lime or color washes. Fragmented massing created another illusion, making the house seem grand even though actual living spaces were few in number and relatively modest in area for an elite residence of the period.

Aesthetic and climatic factors played complimentary roles. Despite the practical disadvantage of long, external circulation paths, the tripartite planning formula, whether executed in tabby, timber, or brick, offered new levels of domestic comfort for affluent families residing intermittently on coastal estates during hot, humid weather, large sash windows and distribution of living spaces into nearly discrete building masses allowing entry of every sea or river breeze. La Rochefoucauld-Liancourt observed of late eighteenth-century Charleston, "everything peculiar to the buildings of this place is formed to moderate the excessive heats, the windows are open, the doors pass through both sides of the houses. Every endeavor is used to refresh the apartments within with fresh air. Large galleries are formed to shelter the upper part of the sea house from the force of the sun's rays. . . . Persons vie with one another, not who shall have the finest, but who the coolest house."[57] Tripartite and linearly organized houses with their long porches, screen walls, and other linking devices produced an architecture characterized by vigorous articulation and picturesque massing. An amateur designer could make schemes of this kind appear deceptively imposing, especially when seen from the great waterways dominating the lowcountry. Sailing the Broad River, William Elliot gave an idealized picture of the scene "fringed with tall green wood, descending to the water's edge, and varied here and there by clusters of white houses grouped in the prospect like fleets of sail at anchor, . . . the occasional glimpse of a cool vista leading the eye up to some cheerful retreat of hospitable wealth."[58]

Linear planning shortened structural spans thereby minimizing loads and saving money. It also allowed experimentation and occasional eccentricity of expression. The anchor-shaped Bellvue (Charles Floyd House), Saltilla River, Georgia, is an oddity, boasting an exceptionally large semicircular bay at one end enclosing a "great circular drawing room" at first-floor level and the owner's bedroom above.[59] Upper construction was of wood, the basement story, standing about eight feet high, of tabby. Excepting certain Jeffersonian echoes, Thomas Spalding's own Sapelo Island house (built in tabby 1810–12), now a mere ghost absorbed into twentieth-century construction, defies characterization, its central flat-roofed block organized about several semicircular bays with an entrance porch *in antis.* To the right and left, open colonnades linked main living areas to small flanking wings housing the kitchen and office. Spalding himself said: "my house at Sapelo is one story, 4 feet from the Ground & Sixteen in the Ceiling, 20 feet in all. It is 90 feet by 65 feet in depth, besides the Wings. The roof is of Tar and Sand upon Sheathing paper three-fold resting on 2 layers of Boards."[60] The latter detail typified the owner's pragmatic and experimental approach whether concerned with building or agricultural ventures. Modernist architects have learned flat roofs rarely succeed on the Sea Islands. Spalding's was no exception, the tar melting during summer months, consequent rain water penetration causing the building's ruin and abandonment by 1851.[61]

• The Plantation Owner's Personal Domain •

George Edwards is credited with installation of an unusually evocative landscape design centered about his enlarged Spring Island residence. Although modern golf course construction has distorted early nineteenth-century design concepts, what exists indicates that full advantage was taken of the location. Along the Chechessee River paralleling the shore, two small tabby flankers created an architectural composition centered symmetrically about the main house, extending 244' from north to south. Relatively intact below eaves level, both flankers are about fifteen feet square, contain one undivided space raised three or four feet above grade entered by a single doorway, and probably featured a pyramidal roof. The northern flanker was perhaps a store for high-quality foodstuffs—hams, rice, game, and preserves. The southern one is more puzzling. It incorporates a service space sunk below ground reached by a full-height doorway approached via some sort of trench blocked and concealed by earthen fills during antebellum times. The single room above had large glazed windows and was plastered, shelves or cupboards lining interior walls. Excavation yielded elegant Chinese porcelains, specially commissioned for Edwards and shipped from Canton. Was this an office then or personal retreat? Answers are complicated by the subterranean lower space, which cannot be without functional significance. Could this be a wine cellar or some repurposed industrial structure? Without comprehensive investigation of the entire garden layout using ground-penetrating radar and new analytic technologies, we shall never know.

John Frederick Holahan—one of several marauding Union soldiers who landed near the main house in 1862—found that "the immediate grounds were enclosed by a fence of Osage orange, trimmed as rectangular as a stone wall. . . . Flowers grew everywhere in profusion and everything about us was calculated to delight the eye and overpower the senses with beauty and fragrance."[62] There were smaller timber-framed outbuildings nearby—the owner's kitchen (not yet located), a cotton house, and stables.

George Edwards House, Spring
Island, slave tenement, west facade,
main house in background (right).
(HABS, Jack Boucher, 2003).

Often absent Edwards entrusted management of his plantation to an overseer or other trusted individual, but where this person and his family lived is not known for sure—near Spring Island's north settlement seems likely, where ruins of a storage barn, several unidentified tabby structures, and landing stage were developed to accommodate the owner's comings and goings, loading and unloading of supplies, and shipping of cotton off island. Between this area and the main house, Edwards installed consciously naturalistic or "picturesque" elements including an artificial lake and curved or semicircular slave row built back from its far bank.

Describing the richly furnished, already-despoiled main house, where paintings had been cut from their frames and carpets carried off, Holahan (who admitted stealing books, besides an admirable reading stand) mentioned three "magnificent avenues leading away at least half a mile."[63] These projected an axis westward into surrounding landscapes, providing cool, shaded passage to and from the island's interior besides contrasting with flower gardens around the main residence and cultivated fields extending far inland. Still framing, as always intended, the main house, the memorable live-oak allée underplanted with drifts of daffodil and *Gladiolus byzantinus* to its west, is the chief surviving remnant of these early landscape features. Off to one side within sight but not close enough to impinge on his personal space, Edwards built a tenement probably occupied by favored or indispensable domestic slaves. This tabby building incorporated four separate dwelling units arranged on two stories. Measuring 36'3" × 20'2" overall, exterior walls mostly stand despite a fire that consumed stairs, roof framing, floors, and internal timber partitions. The lower level was divided into two single-room apartments separated by a narrow central hall and stairs leading up to two similar apartments on the floor above. Each lower apartment had its own entrance opening south toward the allée, excavation and faint impressions revealing that two tabby-brick chimneys stood against the opposite (north) elevation's inside face, each serving one upper and one lower room.[64] The north facade was left blank except for a single window lighting the upper stair, no apartment having any direct view of the artificial lake behind. Individual units encompassed about 262.5 square feet, making them slightly smaller than Beaufort District's "best" single slave houses, but Spring Island's tenement units were lighted by large glazed windows—an

White Hall Plantation, Grahamville vicinity, flankers (Library of Congress).

almost-unheard-of luxury among plantation slave dwellings—protected by exterior timber shutters. Fabricated from local materials (tabby and tabby brick) the tenement's appearance, organization, and appointments are more reminiscent of urban models than rural ones, personal servants who traveled with their owners to and from Charleston perhaps living here intermittently.

Not far away Heyward's White Hall Plantation exhibits variant planning arrangements. Here two double-height flankers were erected northeast and northwest of the enlarged main building. Reliable informants have said that the east structure was a kitchen, with kitchen workers probably occupying the space above. Badly ruined the west flanker resembled Spring Island's tenement building. It was of similar overall size (39'6" × 19'3") and divided into four single-room apartments. Windows and chimney positions differed, old photographs showing the White Hall structure had a two-story timber-framed porch running along its entrance side, an external staircase accessing upper apartments.[65]

Associated landscape installations were extensive. The most interesting survival is a low, curved tabby wall defining what was probably an elaborate semicircular garden positioned overlooking Euhaw Creek south of the main building. Today an avenue of live oaks leads away northward. Largely replanted during the later twentieth century, several older trees hint that the present avenue follows an earlier allée. Altogether the layout of house, flanking wings, symmetrically disposed dependencies, and formally arranged landscape elements have strong Palladian overtones more reminiscent of mid-eighteenth-century than early nineteenth-century taste.

On Dataw Island a large yard (measuring about 122' north/south × 233' east/west) enclosed by timber-post-and-rail fences erected on continuous twelve-inch-wide tabby foundations fronted the main Sams House on its south side. Development about this residence was more functional than decorative, more opportunistic than formal, and compared, with White Hall, far less dependent on Palladian prototypes. Seven or more small tabby structures opened into the yard. To the west stood the principal kitchen, distinguished by its oversized tabby fireplace and tabby-brick chimney of shouldered form. Opposite, strung linear fashion along the east side, stood three single-story dwellings (now reduced to their foundations) each with an end chimney. The northernmost house measured about 23'8" × 12', the two others smaller (about 18' × 12') but otherwise similar, with thin (eight-to-nine-inch-wide) tabby walls enclosing a space divided—if divided at all—by timber partitions. These dwellings were almost certainly erected for house slaves or domestic workers whose duties demanded constant attendance on the owner and his family. Considering that one was larger than the rest, some kind of social hierarchy existed, the northernmost slave house perhaps occupied by the housekeeper, who left quantities of needles and buttons behind. Cast-off goods from the main house discovered during excavation—imported feather-edge plates, transfer-printed bowls and dishes—demonstrate that individuals living here received privileges unknown to field hands accommodated in far less substantial quarters elsewhere about the plantation, all of which have now disappeared. Positioned near the yard's southeast corner, a relatively narrow framed building erected on tabby footings perhaps stored goods coming in or ready to be shipped off island. An enigmatic tabby structure occupied the yard's southwest corner, this having been cut down and leveled while the site was still owned by the Sams family. Its outsize chimney base and scatters of mid- to late eighteenth-century ceramics may indicate it was an early kitchen, predating the present kitchen standing a few yards to the north.

Altogether the Sams installation is characterized by its functional layout. Although symmetry governed organization and massing of the main residence, the yard was not centered on the "Middle House" but extended further east than west. Within the enclosure independent structures were kept small and distributed according to utility rather than any preconceived notions of balance or formality. Access was strictly controlled. There are indications for one gate on the yard's south side entered from pathways or tracks leading down toward a landing where small ships or flats sailing from St. Helena Sound via Jenkins Creek tied up. Otherwise visitors passed through the main house or its peripheral dependencies before gaining entry. This ensured security for "family" members, who probably feared for their own safety, heavily outnumbered as they were by the island's enslaved population. Limited access further underscored the owner's unassailable position at the apex of day-to-day plantation life. Presumably only domestic slaves were given leave by the master or mistress to enter the main house, all other hands dealing with an intermediary overseer, manager, or butler. An early nineteenth-century painting (ca. 1820?) of Rosehill-on-the-Combahee, Colleton County (possibly built before 1800 by John Gibbes), provides analogies, illustrating an enclosure defined by post-and-rail fences fronting this lost, timber-framed plantation house, which, like the Dataw building, was tripartite in plan. Similarly Rosehill's yard incorporated the owner's kitchen and numerous smaller dependencies.[66]

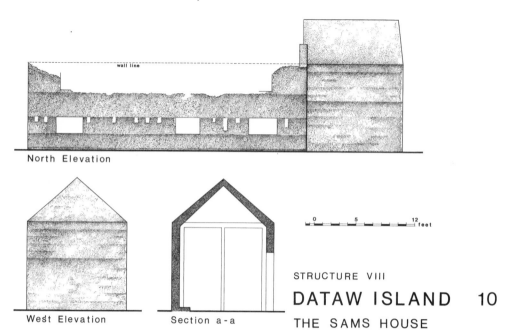

North Elevation

West Elevation   Section a-a

0   5   12 feet

STRUCTURE VIII

DATAW ISLAND 10
THE SAMS HOUSE

William Sams House, Dataw Island. Elevations and
plan of dairy with section of cold store.

The dairy standing almost opposite the main Sams building's north (land) entrance
shows the owner was mindful of his status in the larger world of planter society. Most
family guests would have come this way and seen not only the dairy but also service
structures (pigeon house, barn, stables, and overseer's house) all neatly laid out in linear
fashion right and left. The dairy was the product of more than one construction phase.
The earliest, a narrowly rectangular tabby structure called "the blade house" (for storage
of corn blades used as animal fodder), was at some undetermined time reconfigured, the
altered building incorporating two stories with the lower floor sunk several feet below
ground. Milk was probably cooled to raise cream and churned butter at the lower level,
having first been scalded in the room, or rooms, above. To the west Sams added a new
component, constructing exterior walls and—remarkably—its roof of tabby. About 10'3"
square, the interior as preserved is undivided. There are no windows, only one wide doorway opening south. Vertical grooves suggest framing for insulation, perhaps sawdust or
Spanish moss. Lack of light and ventilation must indicate storage functions, crudely fabricated floor channels (for collecting runoff), insulation, and a wide door (for deliveries)
typifying icehouses, which became popular after 1800 among lowcountry planters able
to afford the luxury of ice brought by ship from New England. On Dataw the building
probably doubled as cold store for butter and other perishable dairy products destined
either for sale or for the owner's own table. Most important, ice was needed to keep milk
and buttering tables at proper temperatures.

Regarding social and aesthetic notions, Robinson noted that English dairies during
the Georgian period "were polite buildings, the province of the lady of the house and her

daughters . . . [having] taken on some of the connotations of goodness and simplicity, popularized by contemporary philosophers."[67] Fox-Genovese confirmed that wives and daughters of wealthy southern plantation owners, who rarely, if ever, set foot inside kitchens, smokehouses, or other ancillary buildings, were not above supervising "buttering" themselves, the perceived gentility of operations, not to mention the delicious example of Marie Antoinette's *laiterie* at Rambouillet, causing French, English, and American landowners to give dairies prominent positions on their estates.[68] For the Sams family, the carefully located Dataw dairy and icehouse could not be missed by inquisitive or envious visitors, though B. B. Sams, true to his plain-dealing ways, did not emulate John Townsend of Bleak Hall, Edisto Island, by dressing up his structures in Gothick or other fanciful detail.

The larger landscape made or inherited by B. B. Sams was described by his son: "in front of the house were two sycamore trees. . . . Between the house and pond was a large grove of poplar trees. . . . On the other side directly to the north, was the pear orchard. East of it the old plum orchard . . . northwest of house was the orange orchard. There were pear, fig, apple and orange trees everywhere."[69] Allowing for exaggerations and the author's bias, which excluded all mention of slavery or the enslaved status of his boyhood companions, a sylvan and idealized scene emerges, the pond and poplar grove, orchards, and lack of symmetrically disposed outbuildings near the main residence signifying a layout conceived on picturesque rather than strictly formal principles. Practicality and plainness characterize architectural development around the house, qualities enhanced by use of tabby and tabby brick, which enabled erection of substantial and relatively permanent outbuildings without excessive expenditure.

Consciously or otherwise the owner was putting into practice ideas advocated by British authors, who from about 1750 onward propagated notions of agricultural "improvement." Nathaniel Kent's *Hints to Gentlemen of Landed Property* (London, 1775) with numerous later editions) recommended that landowners use "skill and frugality" when erecting farm buildings, admonishing them "not to build anything but what will be really useful. To build upon a small compact scale, and as much as possible upon squares or parallelograms, not in angles or notches. To build at all times substantially, and with good materials."[70]

We can be reasonably certain those local planters who, along with northern farmers, realized that successful entry into regional and world markets depended on achieving efficient production through "improved stock, improved buildings, improved implements, improved orchards, gardens, mowing, pastures, improved everything" would have recognized the innovative, picturesque, and improving qualities of areas around the main house.[71] Dr. B. B. Sams had brought together a web of ideas woven by the Enlightenment. But improvement and social reform went together—once let loose across the lowcounty, these dual concepts would ultimately dispossess an elite planter class and destroy its economy, sweeping away houses, estates, ancillary structures, and the entire structure of forced servitude on which almost all that they valued ultimately rested.

8

Slave Dwellings,
Settlements, and the Quest
for Rural Improvement

Quarters, as groups of slave dwellings were colloquially called, varied in relative proximity to the main plantation residence. Domestics might be housed within sight or earshot, while ordinary hands were housed nearer the fields they worked, machines they operated, or services they performed. Although scattered huts or collective barracks were the norm before the American Revolution, no early examples are known from Old Beaufort District. After 1810 larger settlements usually comprised regimented rows of single or double dwellings ordered along a "street," the street sometimes forming an avenue or allée leading toward the owner's own domain. On Edisto, Seabrook's place boasted a triple row of slave houses located near the main residence. Additionally numerous double slave dwellings were situated along the shore of Steamboat Creek, these following the waterway's natural course rather than any rigorous geometric pattern. Occasionally groups of slave houses were assembled more artfully to give the impression of picturesque villages—a clear sign that new attitudes about enslaved labor had entered the lowcountry.[1] Confronting abolitionist sentiment and the economic, if not humane, necessity of keeping their slaves healthy, certain planters emulated English and French reformers who from the late 1700s onward argued that improved housing for agricultural workers would boost productivity of landed estates and reveal the hand of progressive or paternalistic owners. Thomas Spalding was a leading southeastern exponent of these views, linking tabby with notions of "improvement" formulated by European writers, extolling the material's permanence and comparing it to the pisé de terre favored for inexpensive rural housing by fashionable and reform-minded landowners on both sides of the Atlantic.

Although statistics are missing, it is evident from architectural surveys that tabby-walled slave houses were always in the minority, examples now occurring on a mere handful of plantation sites scattered from the Combahee River to Georgia's "Golden Isles" and Northeast Florida. Yet being qualitatively near the top end of the broader spectrum of local slave housing, these dwellings have higher survival rates in coastal zones than comparable wattle, log, and framed structures or those miserable cabins patched together with whatever came to hand pictured by mid-nineteenth-century photographers. This does not mean that under poor management or absentee ownership better-than-average slave living conditions did not deteriorate or that improvement was always sustained. Visiting her husband's neglected plantations on St. Simons Island in 1839, Frances Kemble found the once-proud main house unfit for occupation and its attendant settlements filthy, unsanitary, and squalid to a degree that demeaned occupants and owner alike. Productivity had tumbled, and an enslaved man called Friday had died like some worn-out "beast of burden," unattended except for flies on the dirt floor of the dark, dilapidated plantation hospital, as the powerless writer watched.[2] This anecdote and similar ones illustrate the fundamental dichotomy between progressive ideas espoused by enlightened owners and the savagery inherent in plantation slavery that neither wishful thinking nor building improvement could ultimately hide.

⚬ Pisé, Tabby, and Notions of Improvement ⚬

Before proceeding with descriptions of slave houses fabricated from tabby it is worth considering the convoluted intellectual process by which the material assumed connotations going beyond the strictly practical among individuals inspired or alarmed by radical notions of rural reform. Discussion best starts with the British commentator Nathaniel Kent (1737–1810), whose views about farmyard planning have already been cited. "Estates are of no economic value without hands to work them," Kent observed, arguing that improving living conditions of agricultural workers was in the landowner's own interest. "The laborer," he wrote, "is one of the most valuable members of society," adding "there is no object so highly deserving" as providing for the laborer's welfare.[3] His *Hints to Gentlemen of Landed Property* (London, 1775) offered practical solutions regarding erection of better dwellings illustrated by various model designs. After examining workers' housing across the English countryside, John Wood enunciated similar views in 1792, writing—much as Olmstead traveling through the Deep South later wrote about slave housing—"the greatest part of the cottages that fell within my observation, I found to be shattered, dirty, inconvenient, miserable hovels, scarcely affording a shelter for beasts of the forest; much less are they proper habitations for the human species."[4] More advice about building design aimed at landlords followed, his stipulations that estate cottages should be "dry and healthy," economically built yet strong, and provided with small garden plots subsequently gaining currency among West Indian proprietors as well as planters across the southeastern United States.[5]

For "improving" British and American landowners who baulked at replacing decrepit structures that many agricultural laborers inhabited near the turn of the eighteenth century—and long after in most instances—a French commentator, previously mentioned in Chapter 3, François Cointeraux, who variously styled himself "master mason, agriculturalist and architect," offered economical solutions. Cointeraux's commentary praised pisé's cheapness, simplicity, and durability, which made it eminently suitable for country building. After a brief history, fabrication methods were painstakingly described. The reader was told how he or she might select suitable earths, make sturdy forms, insert windows, and finish completed walls, practical information that helped this publication gain international circulation and widespread admiration.[6] Thomas Jefferson corresponded with its author, the English architect Henry Holland, whose clients included the Prince of Wales and some of Britain's greatest Whig landowners, becoming so engaged by the work that he prepared an edited translation published by the British Board of Agriculture in 1797 under the title "Pisé, or The Art of Building Strong and Durable Walls, to the Height of Several Stories, with Nothing but Earth, or the Most Common Materials."[7] Holland observed, "there is every reason for introducing this method of building into all parts of the kingdom; whether we consider the honor of the nation as concerned in the neatness of its villages, the great saving of wood which it will occasion, and the consequent security from fire, or the health of inhabitants, to which it will greatly contribute, as such houses are never . . . [exposed] to extremes of heat or cold. . . . It saves both time and labor in building, and the houses may be inhabited almost immediately after they are finished."[8]

The same volume included sections devoted to country cottages of types later romanticized by the Victorians. Holland contributed the leading article, embellished with architectural drawings showing an attractive, if small, two-story duplex dwelling.[9] His text recommended that walls of houses for farm laborers "be built of the earth or topsoil, compressed in moulds, and shall be at least 20 inches thick. The roof should be covered with thatch on rafters of young fir, or poles of unsawn timber. Such a building well executed will last half a century." Always obligatory cost estimates followed, earth or mud walling (twenty inches thick) being calculated at three pence per foot run, foundations (22 inches wide) of brick, rough stone, or flint at sixpence per linear foot. Lasting, cheap, and fashionable—an irresistible combination for his landed patrons—"the sort of walling proposed," Holland emphasized, "admits of an outside colouring to imitate stone which added to the soft and elegant simplicity that always attend good thatching would make cottages one of the greatest ornaments of the country."[10]

Holland's client the fifth Duke of Bedford, England's preeminently wealthy and innovative farmer, must have agreed, commissioning several pisé cottages for his celebrated estate at Woburn, Bedfordshire, closely resembling the architect's 1797 design. Intrigued with "composite" walling materials, Robert Salmon, the duke's agent, built his own house of pisé and developed fabrication methods involving the addition of lime to rammed-earth mixes. His formula "comprised three parts fine loam (not sand or clay) and one part lime slacked with water rammed into molds in three inches layers," finished on the exterior with lime plaster and an ochre-colored wash.[11] To improve adhesion each course (round) was sealed with a thin layer of grout composed of one-fifth part lime to four-fifths parts earth.[12] In 1808 models of an improved "mould" earned Salmon twenty guineas prize money from London's Society of Arts.

By then the topics of pisé and improved domestic building for agricultural workers, along with aesthetic notions about the "purity" of rustic and primitive structures, had been inextricably linked by architectural theorists. John Plaw's *Ferme ornée* (London, 1795) was probably the first English book to recommend pisé, an advance advertisement referencing new methods of "building walls for cottages, etc." by Cointeraux.[13] Drawing heavily on the same French authority in 1802, William Barber (who was no stranger to poverty judging by his imprisonment for debt in Dublin) published an illustrated treatise

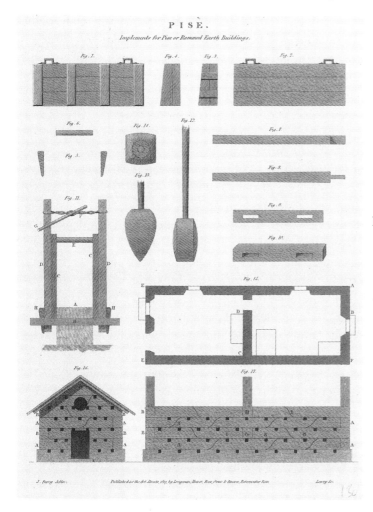

Pisé, plate from Abraham Rees, *Cyclopaedia or Universal Dictionary of Arts, Sciences and Literature,* London, 1817.

encumbered with the title *Farm buildings; or rural economy. . . . Containing designs for cottages, farmhouses, farm yards etc. etc. . . . Also a description of the mode of building in pisé. As adopted in several parts of France for many ages; which would be attended with great advantage if practiced in this country, particularly in cottages and farmyards.* In 1819 Abraham Rees's *Cyclopaedia or Universal Dictionary of Arts, Sciences and Literature* treated pisé "at very great length and with very great ability," nine densely set pages laying out almost everything previously written about the material by French and English authors. Issued separately from the text, a single plate (dated 1817) illustrated necessary implements, formwork, and a rudimentary pisé dwelling copied largely from Henry Holland's earlier works. Although lacking originality, the *Cyclopaedia* became a compendium for generations of pisé enthusiasts scattered from Europe to Australia, Africa, and the Americas.[14]

As late as 1825, John Hall, secretary of the [British] Society for Improving the Condition of the Laboring Classes, was promoting pisé for workers' cottages, which he thought conducive to the physical health, cleanliness, "delicacy," and "industrious morality" of residents.[15] Yet despite—or perhaps because of—aristocratic patronage, testimonials from landowners, publication, and reissuing of articles describing construction methods improved or otherwise, Britain's taste for pisé was already over, the material never entering into the mainstream of vernacular building. Today very few late eighteenth- or early nineteenth-century essays survive, even at Woburn, where most pisé structures were demolished or patched with brick rather than being rebuilt with formed earth after they fell into disrepair.[16]

Outmoded or not pisé was still touted as an economical building material on the Atlantic's opposite side. The *American Farmer* reprinted Henry Holland's translation of Cointeraux in 1821, ignoring the fact that publications more or less dependent on it were already circulating, the fifth edition of the *Complete Farmer: A General Dictionary of Husbandry* (1807) including designs for pisé cottages.[17] Mills Lane cited S. W. Johnson's *Rural Economy* (New Brunswick, N.J., 1806), observing how this book, dedicated to President Jefferson—again reproducing Cointeraux's plates along with the author's own practical observations—inspired construction—or rather reconstruction—of the Borough House, Stateburg, South Carolina, a two-story framed structure of mid-eighteenth-century date transformed by Dr. William Wallace Anderson, M.D. (1789–1864), who added single-story gabled wings right and left of the original timber-framed building in 1821. Cast in pisé using dark-red clay obtained near the site, the wings created yet another variant of the tripartite planning formula for plantation houses already described. Simultaneously the original structure was updated with an exceedingly handsome two-tiered portico, supported on finely detailed Roman Ionic columns.[18] Similar columns front Dr. Anderson's temple-fronted office, located some small distance away from the main house next to the old King's Highway, where he treated enslaved patients. This attractive single-roomed building, which brought dignity and grace to the cruel practice of premodern medicine, measures 25'2" × 16'3" excluding its entrance porch. Exterior pisé walls eighteen inches thick cast on stone foundation plinths are stuccoed and color washed. Other pisé structures erected near the main residence include a two-room schoolhouse with peripteral colonnade, kitchen, weaving shed, dairy, and well house. Dr. Anderson was graduated from the University of Pennsylvania in Philadelphia, where he possibly met S. W. Johnson or bought a copy of the latter's book. This book still remains at Borough House, along with correspondence from General John Cocke, who was improving one of his Virginia estates using similar formed-earth techniques during the early 1820s.

Cointeraux's treatise circulated along the Eastern Seaboard for decades. In 1830 Thomas Spalding, thinking subscribers might find the work "advantageous," sent the *Southern Agriculturalist* what he called "a small book in French, upon the construction of Terre or Pisé"—almost certainly the original 1790 Paris edition.[19] Despite promises the journal's editor made no immediate move to reprint Cointeraux's text, this not appearing until January 1837, when Bartholomew Carroll sent in his own translation. Stirred by newly enacted ordinances, another of Carroll's contributions published during the following year addressed the "distressed inhabitants of Charleston" now obliged to build using incombustible materials following fires that devastated houses and public buildings alike. Though the image of the Holy City's grandees, merchants, or tradesmen using dredged harbor mud when erecting structures for their personal occupation is implausible, Carroll advocated use of the latter material or rammed clay instead of expensive brick for walls, describing methods of making pisé borrowed—without acknowledgment—from Cointeraux's much plagiarized publication. Assuming the article is not a parody (echoes of earlier publications indicating it might be), Carroll's most useful remarks concern the pace of pisé construction: "my pisadore (or rammer) and one attendant, carried up sixteen inches of the wall sixteen inches thick in four days, all round a house 48 feet long, and 24 feet wide." What was presumably the same house contained six large rooms, and three wide passages "built in nine months by two men only (who hired out at $10 per month before employed on the work) assisted by a black country carpenter." Less convincingly it was argued that if dead shells were rammed into mould's of the kind he described, the result would be a "fine wall equal to Tabby, well known to many of us."[20] Spalding's aforementioned article "On the Mode of Constructing Tabby Buildings and the Propriety of Improving Our Plantations in a Permanent Manner," printed by the same journal, was not referenced. Admittedly this piece had limited appeal for residents of coastal South Carolina, who practiced their own building traditions. Experienced Beaufort-area planters such as Dr. B. B. Sams (a known subscriber and occasional contributor to the *Southern Agriculturalist*) found little useful information about tabby manufacture nor much use for fanciful designs distributed by English and French book dealers depicting workers' cottages in various picturesque guises—rustic, primitive, Gothic, Italianate, even oriental. Nevertheless Spalding's article was not without novelty, interest, or utility. By extrapolating from fashionable literature describing pisé and emphasizing how it resembled tabby, Spalding furnished his readers conceptual frameworks linking formed materials they made with ancient antecedents and, what is more important, modern philosophical ideas of agricultural improvement, the latter point emphasized by his article's title.

These arguments and similar ideas put forward by contemporary proponents of Cointeraux's methods were especially attractive for owners who sympathized with abolitionist sentiment. General John Harwell Cocke, a friend and kinsman of Thomas Jefferson, erected several pisé slave cabins at Upper Bremo, Fluvanna County, Virginia. Paraphrasing Henry Holland he said they provided "the warmest shelter in winter and the coolest in the summer of any buildings their size I ever knew."[21] Their location near Cocke's own residence, coupled with the preponderance of timber-framed slave housing elsewhere on his holdings, suggests the pisé dwellings were erected to accommodate small numbers of relatively privileged individuals—domestic slaves perhaps. Further south, along the Atlantic coast, tabby filled pisé's role, offering inexpensive means of improving slave dwellings and the health, if not industriousness, of their occupants. After visiting Hamilton Plantation, St. Simon's Island, in 1830, the *Southern Agriculturalist*'s editor wrote, "We were much pleased with the construction and arrangement of the Negro houses, they are built in parallel rows, facing one another and extending some distance forming a wide avenue or street. . . . In the rear of the houses are the small gardens and hen houses of the occupants. The old buildings are of wood, but all of those recently erected, are of tabby, which adds much to the neatness of appearance, and the comfort of the inhabitants. They are constructed by the plantation hands at leisure time."[22]

Earlier Edward Swarbreck, who owned Chocolate (Le Chatelet) Plantation on Sapelo Island, said he used tabby for his double row of slave dwellings because "it makes my Negroes more comfortable, and I desire to leave my estate as valuable to those who may inherit it." There were other practical reasons for building this way, as he probably learned from Thomas Spalding, who owned part interest in the property: "the walls are of tabby, which in a little while becomes like stone, requiring no repair: this causes considerable saving to the Negroes, for it is generally expected that they will make the repairs as they become requisite, unless they are so to much extent, and then the plantation mechanics are employed."[23] Now roofless and substantially ruined, at least nine of these dwellings (typically measuring 14' × 20' overall) arranged as a double row survive along with incomplete remains of the two-story tabby-built main house, barns, and other outbuildings.

Praise for neat rows of slave houses, regularly arranged, reflected another facet of contemporary plantation life. For planters slave dwellings arranged linearly along streets or more rarely on grid patterns allowed orderly growth when holdings increased while facilitating cleaning and refuse collection. Always near the forefront of planters' minds, supervision, discipline, and control were enhanced. Avenues of tidy workers' dwellings, if properly maintained, also projected affluence, management skills, and enlightened leanings, or so certain owners hoped. For the enslaved, single or double houses helped mitigate the evils of overcrowding and strengthened family ties, with produce from cultivation of adjacent garden plots supplementing diets and generating income. This came at the expense of regimentation of course and any opportunity for self-expression, cultural or otherwise, by the enslaved.

Regimentation is strikingly illustrated by Hubbard and Mix's stereo view of Perryclear Point, showing eight identical slave cabins closely packed linear fashion along an unmade track, one of two larger structures at the row's extremity perhaps being the cotton house mentioned by northern missionaries who thought the settlement better and cleaner than most. Despite the plantation's scenic position bordering the Beaufort River near Port Royal Island's northeastern extremity, we see little except an open field, fences, and scattered pine trees offering hardly any shade. This stark arrangement is matched by the small size of individual dwellings and their mean openings, comprising a door flanked by one unglazed window on entrance facades. Still shouldered brick or tabby-brick end chimneys appear substantial, walls consist of vertical boards and battens, floors are raised somewhat above ground, and roofs are shingled, everything looking well maintained even though the former owner, Eliza Perryclear—who married the plantation's overseer, Captain John Hamilton, C.S.A.—had decamped several years earlier. Now abutted by Beaufort's U.S. Marine Corps Air Station, Perryclear Point's antebellum infrastructure is apparently lost.

❧ Combahee River Plantations ❧

A survey dated 1834 of Hobonny on the Combahee River, then owned by Henry A. Middleton II (1770–1846), preserves the plan of Beaufort District's leading rice plantation. At this period there were two separate slave settlements here (both considerably larger than Perryclear's), located some small distance from the main (probably manager's) house, which occupied an enclosure situated near the nexus of a complex series of canals and trunks regulating the flow of water into rice fields extending north toward the river. To the west the smaller settlement comprised twenty-four structures arranged in two almost-parallel rows, the "street" bordered by rice fields on both sides, trending toward Hobonny Creek, a small tributary of the Combahee.

Situated in what was called pasture land the bigger eastern settlement comprised thirty-six units also arranged in two parallel rows. Citing James Tuten, Linder observes: "at the far end of each street was a tabby structure. One was a single house and the other a duplex." This circumstance having led Tuten to conjecture "that a driver was living in each settlement, with the members of his respective gang… the location of the [tabby] houses at the far end of the settlements, with the big house in between, gives the impression that there was an attempt to police the slaves's avtivity."

Photographs published in 1935 by Todd and Huston show the duplex was then intact. Long facades each had two doorways flanked by small windows. Gable ends were carried up in tabby and featured windows at ground an upper window lighting sleeping or storage areas. The roof appears shingled with a central chimney serving back-to-back hearths.[24]

Two tabby slave houses survive above ground at Hobonny today. Both are ruined, making it difficult to tell if these are the same dwellings mentioned during the 1930s, though it seems likely, an informant saying he lived in both extant buildings as a child, moving from a single house (which often flooded) in the former western settlement to the other in the east as late as the 1960s. If so, the double dwelling is now missing its original roof, tabby gables have collapsed, and the chimney looks rebuilt; the eastern unit has a concrete floor, newish window frames, and a small north-facing addition. Still enough fabric survives to establish that the earliest structural foot print measured 44'1" east/west × 20'1" north/south. Mirroring lost gables a tabby party wall extending up into the roof created two discrete living units. This partition was not bonded into north and south exterior walls. Rather it was an infill, cast separately after the stack and side walls stood complete. Thus divided each unit was entered back and front by centrally positioned doorways. Window openings on north and south facades were 2'6" square, an unusually elongated opening (4'9" wide × 2'7" high) piercing the lower end (i.e., gable) wall. Smaller window frames were of swamp cypress (measuring 3½" × 2¼" in section) and give no sign of being glazed before the twentieth century. Rather, solid, side-hung wooden shutters offered minimal protection against the elements. Direct evidence for internal planning is also missing. Fortunately chance survival of framing for the lower gable window shows that this unusually proportioned feature was divided equally into two parts by a central post. This probably means that interior space was organized into three rooms, one large through room at the chimney end of the unit and two smaller ones at the opposite gable end—an arrangement still in operation when half of the double dwelling was occupied by an informant in the 1960s—small end rooms then functioning as sleeping areas. Even with doors and window shutters open, ventilation can never have been adequate especially when cooking was in progress or the weather sultry.

Nor was there much light. Worse, lack of glass meant that in winter the enslaved inhabitants "had to choose getting fresh air and being cold or smothering and keeping warm, whereas in summer they could not prevent massive invasions of insects," the latter especially dangerous given the high incidence of malaria and other insect-vectored disease in swampy areas of Beaufort District.[25] The duplex itself was solidly built (though not as solidly as some) with nine-inch-wide tabby side walls standing on robust tabby foundations. Gable walls were cast twelve inches thick at the lower level, this dimension reducing to nine inches above ceiling height. Evidence for flooring is missing, and nothing is known about the presumed loft or its access. Living space was slightly better than average in area, incorporating approximately 362 square feet per unit excluding the loft.

Spatial allocation was greater for occupants of the west settlement's surviving tabby dwelling—a single house facing north and south with gables carried up in tabby. Equipped with an internal brick (now rebuilt) chimney, it measures 27'6" × 18'2" overall. Including the chimney and hearth, this gave an enclosed space of about 429 square feet at ground level plus an equal, if less usable, area in the attic assuming it was floored. Facade treatment followed usual conventions with central doorways front and back flanked right and left by a small (2'7" × 2'7") window opening. Two windows pierced the west gable, the higher lighting the assumed loft, the lower lighting two subsidiary rooms if interior planning followed the scheme described for the east settlement's double house.

Still well-preserved, the large (4'8" wide × 2'6" high) lower gable window opening is an oddity, impressions in tabby to its right and left suggesting it was equipped with either a pair of internal side-hung shutters that folded back against the wall or some kind of sliding-shutter arrangement. Construction details are close enough throughout to indicate this building and the duplex are contemporary, most likely built by the same crew working under the same master mason. One feature unique to the single building is a set of impressions left by timber wall plates set in place at the top of the two long facades to receive roof framing, which has now disappeared. Each plate measured 7" deep × 5½" wide in section, ran along the entire length of its supporting wall, and was anchored in place by tabby cast to form the end gables. Flooring, rafters, and all carpentry work are missing—presumably looted after the building was abandoned during the late 1950s or 1960s.

Tuten's suggestion that both tabby structures were allocated to workers who enjoyed an elevated status in Hobonny's enslaved population is probably correct. But, I doubt if this was the whole story since remnants of another tabby structure measuring 28' east/west × 19' north/south with sixteen inch wide walls standing at or near the east settlement's west end along with fragments of tabby walling emerging as undergrowth is cleared from the west settlement suggesting that more tabby structures remain to be found that are distinct from groups of brick walled double slave houses known from Hobonny and two adjacent former Midlleton holdings- Nieuport and Old Combahee Plantation. Remnants of another tabby structure measuring 28' east/west × 19' north/south with sixteen-inch-wide walls standing at or near the west end of the east settlement along with fragments of tabby walling emerging as underbrush cleared from the west settlement suggest more tabby structures exist that are distinct from groups of brick-

Hobonny Plantation, west settlement, single slave house viewed from south.

walled double slave houses known from Hobonny and two adjacent former Middleton holdings—Nieuport and Old Combahee Plantation. At Hobonny one house of the latter type in the east settlement is under restoration, and another reduced almost down to its foundations.

Situated near the main Combahee River crossing, Nieuport had a row comprising six or possibly eight of these gable-ended double slave houses (typically measuring 42'6" × 19'8" overall), only four of which survive (three ruined). Exterior walls and party walls are of fired brick raised just high enough above ground on tabby foundations to resist rising damp. Organized about a central chimney, individual units followed an established formula with entrance doors front and back flanked by windows right and left, an interior partition of vertical boards dividing off living and sleeping areas. Principal facades of fired brick are laid up in a variant of Flemish bond called "rat-trap" or Chinese bond in nineteenth-century England—where used almost exclusively for low-status rural cottages and farm buildings—this pattern involving laying bricks on edge rather than on their bed.[26] An unusual treatment for Beaufort County, it reduced the quantity of brick required by about 30 percent compared with similar walls built using much stronger English bond, saved mortar, and reduced labor. There were snags, the bonding technique producing cavities at the center of walls—a weakness causing eventual structural disassociation followed by collapse at Nieuport and Hobonny. Still similar brick cross walls going up as high as the roof are surprisingly lavish features, designed to protect the spread of fire between adjacent apartments, a frequent occurrence among slave settlements. The tenant's or overseer's house, as it is called, upstream at Old Combahee Plantation has lost its brick superstructure, tabby foundations much like those at Nieuport outlining an enclosed space measuring about 48'8" × 20' overall, depressions and brick scatters suggesting similar internal organization about the lost central chimney. Resemblance between dwellings from the two sites is made closer by subfloor tabby sleepers bisecting each building along its long axis. Halving the span of floor joists, this feature is shared with near-identical slave houses at Hobonny. Taken together structural similarities displayed by these brick-walled slave houses are close enough to indicate all three of the Middleton Combahee River holdings saw "amelioration" of slave living conditions at much the same time.[27] Because this estate was divided after Henry A. Middleton died in 1846, he might have been the driving force behind the program, slave populations at the time of his death numbering 289 at Hobonny, 224 at Nieuport, and sixty-three at Old Combahee Plantation. Williams Middleton, who managed his late father's estate on behalf of his siblings after the division, is an alternative candidate if construction took place between 1850 and 1860 as concluded by the College of William and Mary Field School in 2010.[28]

In 1850 these three plantations were "valued at seventy-seven thousand dollars" and produced an estimated income of more than $95,000, Hobonny alone producing 1,507 pounds of rice.[29] Three years later productivity plummeted, Middleton's son John Izzard writing: "I am quite mortified at the ill results that have so often attended our efforts to make a crop at Hobonny. I know that the season was a very bad one, but not so bad to

Nieuport, Combahee River crossing vicinity, double slave house,
brick construction on tabby foundation.

justify a half crop."[30] Worse—much worse—was to come when the old order collapsed in the face of Union incursions. On September 25, 1865, Williams Middleton wrote: "I have just heard that the Hobonny steam engine has been carried off and also that a man went to Old Combahee to take down the engine there and removed it." In February 1866 he said, "upon my arrival there what should I find? At Old Combahee one single negro house left standing & nothing else. . . . Banks broken, fences burnt, & fields grown up."[31]

After the Civil War, a few slave dwellings were patched up to provide minimal shelter for tenants—but most disappeared leaving little but broken foundations behind. With them went reminders of the brutal, unremitting toil that all aspects of rice cultivation entailed under forced, often coercive management, plantation settlements such as Hobonny's—however picturesque to the casual observer—remaining little more than crowded, insanitary labor camps no matter how much any owner—luxuriously ensconced at Middleton Place on the Ashley—tried to ameliorate living conditions of the enslaved.

◦ Tabby in Slave Settlements of the Sea Islands ◦

Sea Island settlements exhibiting patterns of reuse, extension, or addition indicate that enslaved populations were seldom static, numbers fluctuating with their owner's fortunes. High prices for plantation crops combined with profitable management practices fueled expansion of slave holdings. Conversely market declines, low yields, neglect of fields, extravagance, debt, the death of a master, absentee ownership, foreclosure, or—all too often—protracted lawsuits over inheritance could initiate marked diminution in slave numbers. These ebbs and flows, coupled with differing levels of investment over time, are reflected by the composition and physical appearance of settlements incorporating structures or groups of structures resulting from different building episodes. Similarly individual dwellings varied in size and quality. Whatever their shortcomings brick slave houses raised on tabby foundations and tabby-walled dwellings at Nieuport and Hobonny were among the best, care having been taken and the master's money spent to prevent rising damp and provide a measure of insulation against excessive summer heat and winter chills. Other slave houses incorporated milled timber, but this was an expensive item that entailed either shipping from urban centers or purchasing costly machinery. Most structures were more cheaply made using materials that came readily to hand—logs, wattle and tabby (lime-mortar) daub, or wood scavenged from abandoned buildings. For islands cut off from distribution centers, transportation of manufactured products by boat or raft raised never-ending practical problems, problems that made tabby cast on the spot an ideal solution when building foundations, floors, walls, pathways, and occasional roof slabs if the planter had sufficiently experienced labor.

This material might profitably—at least for owners if not enslaved workers—be combined with even less expensive ones, General Macomb remarking in 1827: "*tapia* . . . cottages or Negro quarters of one story, a hearth of clay or brick and a back of the same materials with simply a chimney of sticks and mud, or clay, mixed with course straw or pine leaves formed like an inverted funnel will answer well."[32]

Blurred and indifferently composed, Hubbard and Mix's Civil War–era stereo photograph "Street and Negro Quarters at Battery Plantation" substantiates the coexistence of more than one construction mode on a property owned by Miss Mary B. Elliot (daughter of William Elliot), which, except for tabby foundations of an unidentified outbuilding, was swallowed up and destroyed by later development of Port Royal Town. In this image we see three tabby slave houses standing in line.[33] One is a duplex with two doors opening onto the "street," each door flanked by a small window. The gable of an adjacent building (probably another duplex) is timber framed and clad with horizontal timber boarding. Opposite several timber-framed single houses are visible. Compared with the strict architectural discipline imposed at Perryclear Point, settlement planning appears haphazard, with uniformity of materials and building size lacking and setback dimensions from the street varying from house to house.

At Riverside, near the south end of St. Helena Island, stricter planning parameters prevailed. Roofless and overgrown two double slave houses still stand. Each had tabby walls, end chimneys, and an unusual central hallway shared by the occupants. Accommodation was skimpy by European standards, two families occupying a structure measuring 26' × 18'10" overall excluding chimneys. Nevertheless the situation was advantageous, a survey dated 1862 showing these dwellings were part of a larger settlement comprising two parallel rows built linear fashion along an unusually high bluff overlooking the Broad River.[34] This settlement's west (river) side incorporated the two structures described plus two others now destroyed. To the west stood eight additional dwellings (all lost), which backed onto fenced gardens extending along the settlement's land side.

Floyd described a variant type of double slave house from Couper's Point, St. Simons Island. His field notes (dated 1934) record a tabby structure measuring 30' × 15' overall, the chimney most likely centered between the paired units.[35] Walls were ten-inches thick cast using eleven-inch-high forms as recommended by Spalding. To the north two entrance doors mirrored windows punctuating the south facade. Small, round openings protected by shutters piercing gables lighted garrets probably reached by way of ladders. Informants gave Floyd conflicting dates, one saying John Couper erected it around 1812, another reporting her grandfather James Hamilton Couper was responsible for construction sometime around 1845.[36]

Slave houses fully enclosed by tabby walls are now unusual. By contrast, tabby chimney bases are relatively common above-ground features associated with Sea Islands slave settlements, having proved far more durable than mud, wattle and daub, or timber chimneys known from several excavated plantation sites.[37] At Fish Haul, General T. S. Drayton's Hilton Head Island property, six single dwellings distanced about eighty feet apart were added during the 1830s or 1840s to an existing slave street leading toward the main house. Civil War–era photographs show that each gable-roofed house was timber framed, with an end chimney fabricated from tabby and tabby brick, individual dwellings measuring about 16' × 20'.[38] Ground sills were elevated above grade presumably on timber piers and wall framing enclosed by horizontal timber boards. Central doorways front and back with flanking windows probably indicate interior division into separate living and sleeping spaces defined by one or more cross partitions. By 1862 Drayton's settlement

Above: Drayton Plantation, Hilton Head Island, view of slave settlement during Union occupation in the Civil War (Library of Congress).

Drayton's plantation, Hilton Head Island, slave settlement, tabby chimney base, side view.

numbered at least eighteen units of variable size and appearance, the six houses described arranged in a single row. Today only chimney bases of the latter remain in situ, carefully preserved by Beaufort County authorities, their tabby-brick stacks having disassociated and fallen apart.

An evocative Civil War–period stereo card by Hubbard and Mix of what might be T. J. Fripp's place on St. Helena Island pictures a group of enslaved or recently emancipated adults and children standing beside two timber-framed, gable-ended slave houses.

Each building featured centrally positioned doorways and an end chimney.[39] Chimney bases were of tabby standing about six feet high cast in three successively higher pours. Construction above comprised regularly coursed tabby brick. Corbelled the lower ten courses produced shoulders that made the transition between base and stack. Today similar chimneys are seldom, if ever, preserved complete, Hubbard and Mix's image being the best-known contemporary record from the Carolina Sea Islands of this distinctive building detail. Although looking the worse for wear, it is obvious that horizontal timber boarding enclosing these buildings had been whitewashed some time before hostilities began.

Slave dwellings at Battery Plantation and Retreat, Port Royal Island, were whitewashed too—an annual activity while slavery lasted. Writing to the *Southern Agriculturalist* in September 1836, an anonymous planter explained, "being situated where my Negroes procure many oysters, I make them save the shell, which they place in one pile, of which I burn lime enough each year, to whitewash my Negro houses, both outside and inside. This not only gives a neat appearance to the houses, but preserves the boards of the same and destroys all vermin which might infest them."[40] At Pierce Butler's Hampton Plantation on St. Simons Island, Frances Kemble complained that "great heaps of oyster shells are allowed to be piled up anywhere and everywhere, forming the most unsightly obstructions in every direction." Lost for words she continued: "they are used in the composition [i.e., material] of which. . . . [slave] huts are built, and which is a sort of combination of mud and broken oyster shells which forms an agglomeration of a kind very solid and durable for building purposes."[41] Matthew Brady's image (dated 1862) of an unidentified St. Helena Island plantation documents shell heaps resembling those that so irritated Frances Kemble, shell saved for whitewashing the timber-framed slave houses filling this photograph's background.

Owners informed by "picturesque" notions occasionally broke the visual monotony of linear slave rows. William Washington's Combahee River settlement was asymmetrically clustered. Other owners organized their slave houses in arcs or semicircles, these moves reflecting trends current across the English countryside, where conscientious landlords "tried to return people to a Rouseauesque state of nature—a state they have never left in the first place." Tidy peasant houses carefully sited would theoretically ensure "sobriety, goodness, and perhaps productivity too."[42] Milton Abbas, Dorset, typifies the trend and the lordly self-interest of its patron. Built to rehouse displaced estate workers whose decrepit cottages spoiled a local landowner's view, this model village comprises forty well-spaced double ("semidetached" in English parlance) dwellings built of cob and thatched. Each unit has one room "front and back" upstairs and down, "a midget communal vestibule" giving access from the street. Along with an inn and workhouse, houses are arranged in two parallel yet sinuous lines sloping down toward an artificial lake, a design attributed to the gifted landscape architect Lancelot "Capability" Brown, who received £105 for his plan in 1773.[43]

Considerably less ambitious George Edwards erected something reminiscent, U.S. Coastal Survey charts depicting eight structures arranged along an arc overlooking an artificial lake near his Spring Island residence. Regrettably whatever remained of the

Milton Abbas, Dorset, England, view of village street,
model village, ca. 1773 (courtesy R. C. Brooker).

settlement was buried unexplored under fill during construction of the present golf course. The lake too has been reconfigured—episodes indicative of the scant respect given historic landscapes by local and state preservation authorities. Kingsley Plantation, Fort George Island, Duval County, Florida, is far less disturbed, its slave "village" comprising twenty-four individual tabby-walled dwellings arranged as a semicircle bisected by an avenue leading toward the owner's residence.[44] Zephaniah Kingsley Jr., who bought the plantation in 1817, probably executed the scheme, perhaps having learned about tabby through his Loyalist father, Zephaniah Kingsley Sr., who owned Beaufort's DeSaussure House during the Revolutionary War (see chapter 5).

Less rigorous the north settlement at Haig Point, Daufuskie Island, presents analogies.[45] Construction was probably initiated by the Reverend Herman Blodgett after his marriage to Sarah, widow of the former owner, John David Mongin, in 1825. The self-indulgent son of an influential and wealthy Savannah landowner, John David (died 1823) had become "incapable of society from drinking brandy and consequent stupidity and ignorance." Haig Point's slave population (numbering ninety-three individuals in 1820) suffered from the younger Mongin's incapacity, their "physical, intellectual, and moral" state becoming "abject" beyond description, said the Reverend Jeremiah Evarts, who spent three days on Daufuskie in 1822.[46] Everything was slack and listless at Haig Point, or so Evarts's companion Mr. Eddy thought. Without supervision, the enslaved—who had as much land as needed for their own use—sold whatever they could "for liquor." Whether, once possessed of Sarah's property, Blodgett—whom the Mongin's thought an adventurer—changed conditions for the better is uncertain though four single slave houses attributable to his ownership period were larger than average and more substantial than most. Following the Cooper River's gently curved shoreline, they were clearly visible to passing ships and situated well away from standing fresh water, a prime source—though planters were only dimly aware—of mosquito-borne disease causing fevers, agues, disability, and premature death besides providing habitat for cottonmouth "moccasins," dangerous reptiles common all over the island.

Ruins of three now-roofless dwellings with incompletely preserved tabby external walls still exist. Rectangular with an external end chimney of tabby and tabby brick, each measures about 24' × 16' overall (excluding the chimney). Construction required three separate pours above ground, the two lower ones each raising walls 2'4", the uppermost pour adding another two feet around the building. Exterior walls were about nine inches thick, irregular pour lines indicating the tabby mix was unusually liquid when cast. Facade treatment followed local conventions, with doorways centered on the long walls flanked right and left by small unglazed windows fitted with wooden shutters letting in little light and less air. Inside, floors were tabby, cast as thin slurries over sand, and walls were plastered with shell lime. Roofing has long gone, photographs from the 1930s or 1940s showing early coverings—probably timber shingles—replaced by corrugated metal sheets. Triangular-shaped upper section of gable ends were originally timber framed and boarded rather than cast in tabby. Since the same formwork set was used for all three structures we can assume building operations—cutting timber, making forms, gathering and burning shell, casting, and finishing—were all organized by the owner or his agent, no opportunity for self-expression by the enslaved being countenanced.

Nothing is known about how these buildings were furnished; neither it is certain how exactly the occupants utilized interior space. Notably absent is evidence for internal partitions. Nevertheless one primary division separating sleeping and living areas probably existed, most likely consisting of vertical timber boards (possibly tongued and grooved) simply nailed to a rafter or ceiling joist, as I have seen in Antillean slave houses. Neither is there any evidence preserved for sleeping or storage lofts, which might have alleviated cramped living conditions and provided minimal privacy for parents with children. Heating and cooking arrangements are better understood, the fireplace standing at one end of each structure serving both purposes. Each hearth was of fired brick laid over layers of oyster shell, the fireplace opening relatively wide (4'9") and high (3'8"), spanned by a 6½" deep × 9" wide timber lintel, allowing use of iron pots suspended on cranes, small ovens, long-handled pans, and all other paraphernalia customarily utilized by enslaved women when cooking for their own families.[47] Over time fires heating food or keeping dwellings warm would gradually burn through a tabby firebox. There is no indication at Haig Point for any serious conflagration the kind of which regularly destroyed chimneys or consumed settlement houses constructed of flimsier materials.

The north settlement site is rarely without an ocean breeze. This means slave houses situated along the shore line might be ventilated and cooled during spring and summer

Haig Point, Daufuskie Island, slave house, detail of end facade showing
tabby brick at base of lost chimney stack (HABS, Jack Boucher, 2003).

by merely opening front and back doors if sand flies and mosquitoes were not too trou-
blesome. Over winter months, when winds off the Atlantic became cruel, tabby had the
advantage of blocking drafts far more efficiently than poorly insulated timber framing or
badly chinked log construction. A well-drained site ensured that tabby walls stayed relatively dry as long as stucco exterior coatings and roofing materials were kept in good order. Situated near Haig Point's ambitious main residence, the north settlement probably
first accommodated domestics or other favored slaves, an inference supported by Kingsley Plantation, where the largest, presumably most desirable tabby slave dwellings—
occupied by drivers according to Vlach—were almost identically sized, each measuring
25' × 18'7 exclusive of the end chimney.[48]

At undetermined times before 1838, the Haig Point settlement was enlarged twice,
first west and then east, the final arrangement incorporating twelve or more dwellings
and one communal structure—a kitchen perhaps or "sick house"—now represented by
an oversized tabby chimney.[49] No additional tabby-walled houses were built during the
second and third building phases, tabby being used solely for chimney bases. Little is
known about the size or character of these later dwellings except that they were barely
elevated, if elevated at all, above grade.

Information is even less certain for Haig Point's south settlement, where excavation
in 1989 revealed eight slave dwellings arranged in a single row running almost exactly

east/west, with individual dwellings distanced about seventy feet apart. Each featured an exterior chimney at its east end represented by a tabby base, all but two subsequently being destroyed by resort development.[50] Floors were of lime mortar poured three to four inches thick over compacted top soil. Plan dimensions proved impossible to recover, drip lines suggesting these houses were about fourteen feet wide and gable ended. The fact that floors were not elevated above ground coupled with an absence of nails, brads, or spikes opens the possibility that walls were log built.[51]

Tabby used in conjunction with log, a mode more reminiscent of frontier than coastal building, is not unexpected considering the physical and social isolation of Beaufort's outer islands. If this is actually the case, cost may have dictated the choice of log rather than framed construction since the south settlement was erected during the late 1840s or early 1850s, when the plantation's productivity was in decline. Desire for speed and inadequate supervision probably played contributory roles, chimney bases displaying hasty and slipshod workmanship that produced ill-compacted tabby of poor quality. Altogether construction seems at odds with dwellings erected during earlier expansion phases, indicating "improvement" could not be sustained in the face of diminishing agricultural returns or, alternatively, that a different owner was responsible for Haig Point's late antebellum building.

The second explanation is probable seeing that William "Squire" Pope bought the property in 1850, Blodgett having moved his operations to Causton's Bluff, about one mile east of Savannah, a rice plantation where his Daufuskie slaves, then numbering ninety-five individuals, were almost certainly relocated. Frequently enveloped by "miasmas," an overseer complained, this place was far less salubrious than Daufuskie. But here the slaves stayed until 1852, when they were sold for $32,400, along with what remained after several land divisions of Blodgett's holdings, to Robert Habersham, already a highly successful rice planter who owned property suitable for cotton cultivation along the Broad River (Beaufort County).[52] At Haig Point we might speculate that Pope, who lived elsewhere (primarily Hilton Head Island) gave an agent responsibility for renewed construction work instead of supervising it himself, an action not necessarily conducive to maintaining high standards.

Whatever the true case, Haig Point's slave settlements encapsulate cycles of prosperity and hard times, of improvement and retrenchment, these conditions, together with ultimate removal of Blodgett's "people," underscoring the obvious, inescapable fact that enslaved families were always victims of circumstance.

Workplaces and Gardens

Processing and Storage Facilities

Plantation raised cash crops—principally rice and indigo before the Revolution; rice, cotton, and sugar during the antebellum period—required different types of processing facilities. Yet no building for related activity survives intact anywhere in Beaufort County. Likewise mechanized technologies are scarcely represented, early machines developed to cut costs, minimize labor, and increase agricultural productivity having disappeared or been scrapped, stolen, or destroyed by the elements.[1] We must therefore deal with mere skeletons when considering the most valuable pieces of architectural infrastructure punctuating the plantation landscape—indigo works, rice mills, cotton houses, gin sheds, and sugar manufactories.

• Indigo Production •

Ubiquitous before the American Revolution and largely displaced thereafter, indigo production has left few physical traces across the Sea Islands. One notable exception occurs at Burlington, bordering the Broad River and Laurel Bay Plantation, Port Royal Island, where an apparently unique series of tabby vats remain almost whole except for their pumps, taps, stopcocks, and drain linings. Resembling a windowless and roofless military installation rather than an agricultural building, the structure comprises a large rectangular block of tabby defined by twenty-inch-thick exterior walls rising about six feet above grade. The interior is divided into four cells, each with walls and flooring of cast tabby. Observing that the two northern cells are larger (12' × 13') than the two southern ones (10' × 12'), Brockington and Associates have considered that the feature represents two separate vat sets.[2] But wall drains linking northern and southern components raise the possibility that everything worked together as one unit.

    If this is actually the case, the arrangement recalls the West Indian installation illustrated by Jean-Baptiste Du Tertre's *Histoire Generale des Antilles Habitées pour les François* (1667), which, rather than Burlington's square disposition, shows four masonry vats stepped linear fashion against sloping ground to facilitate drainage.[3] Here stems and leaves of the indigo plant were first macerated with water in what English-speaking planters called a "steeper" for twelve or fifteen hours, which released the dye, liquids then being transferred into the second vat (battery), where lime was added and the mixture thoroughly beaten. After resting eight or ten hours this was strained and drained. Drained water was let into a third vat, leaving behind the indigo to slowly precipitate. The mixture having dried, resulting cake was cut into squares. Du Tertre's fourth unit being labeled "reservoir" might explain Burlington's fourth cell, fresh water perhaps being pumped into the unit from an adjacent brick-lined well.[4]

Carroll's *Historical Collections of South Carolina* (1836) informed his readers that for every twenty acres under indigo local growers recommended a steeper vat measuring 16' square × 2½' deep and battery measuring 12' × 8' × 4½'; that indigo cake was usually bagged, pressed to remove any remaining water, and then cured in a log-built drying house, where it was carefully turned three or four times each day.[5] Burlington's installation overlooks an open, relatively deep tidal creek, suggesting harvested indigo plants were brought for processing by boat or flat. Whether one or two sets of vats are present, their overall size indicates an extensive operation rivaling Orange Grove on St. Helena Island, where the owner, John Evans Jr., had two sets of indigo vats, thirty-eight slaves, and indigo "on hand" worth two thousand pounds currency when he died in 1775.[6]

Macerating indigo created an overpowering and sickening stench attracting swarms of flies. Consequently planters positioned habitations well away from processing activities. Rumors of tabby house foundations are persistent at Burlington, but if foundations exist, archaeological surveys have not discovered them. Assuming construction during the later eighteenth century for the vats and supposed house, their builder was possibly Benjamin Guerard (1739–89), who married Sally Middleton, daughter of Thomas Middleton, who owned neighboring Laurel Bay, another plantation heavily involved with indigo production.[7] In 1799 Guerard's holding was raided by British privateers, who made off with forty-seven slaves. Whether any buildings were burned, looted, or damaged during this episode has not been determined.

◦ Rice Production ◦

Federal retaliation during the Civil War caused widespread destruction among rice-growing areas of Old Beaufort and Colleton Districts. Planter houses "known to belong to notorious rebels," mills, slave dwellings, everything combustible went up in flames. "Sluices were opened, plantations flooded and broad ponds and lakes made where, a few hours before, luxuriant crops of rice and corn were putting forth their leaves."[8] Over seven hundred enslaved people were carried away by gunboat to the safety and freedom of Union-occupied Beaufort, agricultural traditions and skills that had helped make their masters rich traveling with them. After the war lack of capital and shortage of experienced labor essential for keeping operational the network of dykes, canals, causeways, gates, and trunks that successful rice cultivation entailed meant the industry never fully recovered its former prominence. Today so much is overgrown and the ruin of buildings so complete that the role tabby construction played in processing across the ACE Basin during the period when rice dominated the scene can only be guessed.

Located off River Road—a highway running along the Combahee River's south side, Old Combahee Plantation's brick-built rice mill was probably typical. The only tabby

Combahee River basin, Beaufort County, S.C., abandoned rice fields,
canal with tabby abutment for flood gate.

now visible are several largish blocks with embedded metal bolts for the heavy mechanical equipment robbed in 1865.[9] Flood gates regulating water levels of an adjacent canal linking inland rice fields fed and drained by Izard's Creek hint that the material found wider usage than currently apparent. The gate itself (now lost) was of diamond style with pairs of inner and outer timber leaves closing and opening with the tides.[10] Parallel tabby walls (about twelve inches thick and seventeen feet long) run along opposite sides of the canal's bank. Both walls retain impressions of a ten-inch-wide central post designed to carry the heavy timber gates. Tabby was carefully made, with outer faces stuccoed and the top protected from weathering by rolled cappings of Roman cement. Like most gates or locks of similar character, it had a hard floor installed three feet or so below median water level but whether of tabby or timber has not been determined. An almost identical installation exists a few miles downstream at Hobonny Plantation. Both diamond gates were installed by the same owner—Henry Middleton II or his son Williams Middleton— during the 1840s or 1850s. Nothing similar is reported from old rice plantations bordering the Savannah River, where construction often featured brick raised on timber pilings, Pennyworth's mill built by Nicholas Cruger or James Hamilton providing an example (see below).[11]

◦ Cotton ◦

For coastal cotton growers few things were more important than careful processing of their principal crop, quality of product determining the very existence of most Sea Island plantations. Following picking, cotton was assorted, whipped, ginned, moted, baled, and stored before loading onto droghers ("drogers")—small freight boats calling along water-ways linking Beaufort, Charleston, and Savannah from where the staple was transhipped to northern or international markets.[12] Along with everlasting questions of current market conditions, contemporary letters frequently mention cotton houses, gin sheds, and storage barns. Paradoxically Beaufort County preserves very few of these structures.[13] On St. Helena Island antebellum processing facilities built on what were among the most productive agricultural holdings in the antebellum South have mostly gone. Neighboring islands exhibit similar losses. Now reduced to enigmatic foundation features, the few fragments of early nondomestic buildings still extant often defy identification in the absence of estate maps or contemporary accounts.

Many processing facilities—exactly how many one cannot say—were burned in November 1861 by owners who would rather destroy their crops and outbuildings than let them fall into Union hands.[14] William Elliott wrote that he burned his cotton houses along with ninety-seven thousand pounds of Sea Island cotton "in the seed" along with one hundred or more bales of cotton at his three plantations on Port Royal Island (the Grove, Shell Point, and Ellis) by direction of General Lee, Stephen Elliott of the Beaufort Volunteer Artillery overseeing this desperate operation. John Townsend of Edisto Island told John Ashworth—a cotton manufacturer with mills in Bolton, Lancashire, who had visited him in happier times—how cotton and provision crops, "the product of the year's labor" raised at his plantation called Sea Cloud, "were consigned to the flames." The full account is graphic, recording conflicting emotions felt by elite planters forced by duty—whether misplaced or not—to destroy the very substance of their privileged life:

> the last sight I had of my "Sea Cloud" plantation . . . was that of blazing buildings containing cotton, grains, and fodder which I had no time to remove, and therefore fired with my own hands, rather than they should come into the hands of the Enemy. I felt it to be a stern necessity, but a very sad one to me;—and I turned my back upon the burning ruins, & have not since revisited the plantation, although the Enemy have for many months abandoned Edisto Island. Since that time, my cotton houses, my gin house with all its machinery, together with other outbuilding, at my Bugby Plantation on Wadamalaw Island, have been burnt down with all they contained; and during the early part of the last year [1863] . . . when the enemy were for a short time in large force, in the North Edisto River on Edisto Island, they burnt down my steam mill at my Bleak Hall plantation, with all the cotton machinery it contained, together with the whole group of buildings which clustered around it: my saw mill, corn-mill, sugar mill and cotton buildings.[15]

Altogether Townsend estimated the combined loss for crops, machinery, and buildings at Sea Cloud and neighboring Bleak Hall (now Botany Bay) slightly over $41,000. At Bugby losses were higher, the total including six "Negroes" (five of whom "ran away to the enemy" and one man who was hanged) valued at $18,200, bringing the total to something over $60,000. While crops and slaves figured largely in claims submitted by Sea Island planters seeking compensation from Confederate authorities, capital investment

"Preparing Cotton for the Gin on Smith's Plantation, Port Royal Island,"
ca. 1862 (Timothy O'Sullivan, Library of Congress).

required for the infrastructure of a well-run cotton plantation was significant. A. R. Chisolm of Coosaw Island (Beaufort District) valued his two-story steam-powered gin house at $2,500 and two other cotton houses at $1,200.[16] Townsend's claims were more modest, including one cotton house (location not specified) valued at $400 and the Bugby gin house with its machinery at $1,100. These valuations and dozens of similar ones do not reveal the character of the buildings themselves nor how they were ordered in the larger plantation landscape. Survival rates are so low we must rely on contemporary accounts by planters, survey maps, and photographs of lost structures to supplement the often-incomplete evidence that archaeological excavation supplies. It is obvious that workplaces were often tied to transportation networks existing on or near the plantation. For Sea Island planters, waterways were the only feasible route to markets either domestic or foreign, a circumstance requiring investment in docks, piers, flats, and boats, even shipping lines by the most affluent. Loading and offloading activities became points of convergence for tracks and paths linking settlements and work areas alike, with processing facilities and rows of slave dwellings frequently being found in close proximity. A photograph taken in 1862 by Timothy O'Sullivan of Old Fort (Smith's plantation) located just outside Beaufort Town overlooking Fort Frederick and the Beaufort River illustrates

Retreat Plantation, stereo image (1860s) of settlement showing cotton house or gin and slave dwellings (Library of Congress).

a timber-framed, 1½ story, gabled structure raised on brick piers identified as the gin house. Entrance into the building was from one end, another opening above allowing goods or materials to be hoisted into the upper floor/roof space. Windows were apparently unglazed but protected by top-hung shutters. Women sitting in the foreground are picking through mounds of cotton; on one side there is a high-wheeled wagon that had transported freshly harvested product to the site. Related photographs show the row of timber-framed buildings in the background were double slave houses, each having a brick chimney and loft over living spaces. At least one slave house was supported in part by an irregular chunk of tabby probably salvaged from Fort Frederick, the ruined building giving this property its name.[17] Portraits of occupants survive, the photographer having posed five generations of what was perhaps the same enslaved family in front of their dwelling.[18]

Comparable spatial arrangements are illustrated by an anonymous stereoscopic view dating from the early 1860 of Retreat Plantation, Port Royal Island. Another raised two-story timber-framed structure similar—though probably smaller—than Smith's supposed gin stood in line with one of two facing rows of slave dwellings. Although Retreat's main tabby-built house still stands and its outbuildings are shown by antebellum plats preserved among papers submitted by owners attempting to recover their property after the Civil War, all service structures have now disappeared, archaeological surveys failing to find any trace either above or below ground.

The layout of service areas was not static, a circumstance illustrated by Cotton Hope Plantation, Hilton Head Island. Formerly called Skull Creek Plantation, this property belonged to Thomas H. Barksdale, a close relative of George Edwards of Spring Island until acquired by "Squire" Pope soon after 1832. Excavation by Chicora Foundation in 1990 established the site's changing aspect over the first half of the nineteenth century. Slave houses were removed from the edge of Skull Creek "to the interior of the tract" and subsequent use made of the waterside area "for the processing or, more likely, the storage of cotton." The old overseer's house was also demolished and replaced with a new timber-framed building better orientated to "catch the breezes coming off Skull Creek."[19] Altogether the site presents a dynamic picture of building and rebuilding in response to changing market conditions and restructuring of plantation activities.

❦ Cotton Processing ❦

The various stages of cotton production from sorting, grading, cleaning, and packing to warehousing for eventual shipping were invariably accommodated in separate buildings. The precedent had been set across the Bahamas, late eighteenth-century Bahamian wills and inventories making frequent mention of cotton houses, gin houses, and barns, installations soon to become familiar across the Sea Islands along with processing technologies developed by exiled Loyalists, transmitted through family members and business associates living in the lowcountry. A Bahamian planter's residence commonly doubled as both cotton and dwelling house. To ensure maximum protection against pirates and the elements cotton was stored at the lower level (usually in a substantial stone-walled

space) and dried on piazzas accessed from living rooms above. Whether Sea Island planters ever stored cotton in elevated basements of their own dwellings has not been verified.
It is clear that, like their West Indian cousins, most realized the necessity for scrupulous
cleanliness during all stages of production, outbuildings so used having glazed windows
and plastered finishes. As technologies improved, building types were modernized, but
Sea Island planters, always mindful of product quality, were very cautious about and even
suspicious of change.

Cotton processing was a lengthy and intensive business demanding close supervision
by the planter or his agent and constant assistance—skilled and unskilled—from a predominantly enslaved workforce. First freshly collected cotton had to be dried. William
Elliott (1838) noted that if damp or wet, fiber was spread "on a scaffold, and exposed to
the sun, but if gathered in dry weather, on the floor of the house, to suffer whatever moisture it has imbibed to escape, before it is stowed away in bulk."[20] Robert Ralton (1824)
recommended that "scaffolds be unconnected with but contiguous to the cotton house in
case of rain when the cotton might easily [be] thrown in." Once dried, cotton was carefully picked over to remove leaves "then passed once though a patent whipper," which
removed any foreign materials that remained.

Cotton houses were collection points and temporary storage facilities for fiber prior
to ginning, their design varying with the size of projected harvests and relative affluence
of owners. In 1844 Governor Whitemarsh Seabrook's cotton house on Edisto Island was
divided into six separate spaces lined on the inside with planed boards and provided with
glazed windows. John Townsend said his cotton house at Bleak Hall measured 20' × 40'
and was two stories high, "sealed and glazed"—a description consistent with the substantial two-story structure depicted in a painting by Kallie Sosnowski (c.1860) slightly west
of his principal residence.[21] Less exalted planters made do with simple timber-framed
buildings. Ashland's cotton house on Lady's Island (measuring 35' × 20') was called "a
rough shed" when adapted by northern missionaries for a school in 1866.[22] Whatever
their character cotton houses were lively centers of activity when freshly picked cotton
was brought from the fields in baskets or on carts, unloading usually demanding some
kind of cover, a platform, and hoists.

Today Beaufort County's best candidate for an antebellum cotton house is located
several hundred feet northeast of the William Sams House, Dataw Island. Continuous
tabby foundations (thirteen inches wide) outline the perimeter of an otherwise demolished building measuring 38'3" × 38'10" overall. A similar foundation bisects the plan,
but if this had spatial significance—supporting framed partitioning or merely halved the
length of floor joists—cannot be determined. Foundations were strong enough to carry
two stories assuming the superstructure was timber framed. Excavation yielded quantities of disassociated plaster and broken glass indicating that enclosed spaces were well
finished and had glazed windows, but, significantly, no evidence for domestic occupation
emerged. Size alone indicates the outbuilding's importance, an undated map naming a
nearby track "Cotton House Road" offering evidence for its identification. Unfortunately
no inventory of the estate survives; neither does the memoir of J. J. Sams, who lived
on the plantation as a child, provide confirmation for the structure's use. Nevertheless

processing cotton (Dataw's chief cash crop) near enough to the main house to be observed by the master, but not close enough to intrude on the "family's" privacy, is likely. Two adjacent features deserve notice. These are flat, unenclosed areas of uncertain extent, carefully surfaced with tabby (i.e., shell) mortar extending north and south of the present ruin. Civil War–period photographs depicting enslaved women sitting outside the gin at Smith's Old Fort Plantation described above provide an analogy, Dataw's hard tabby surfaces perhaps constituting an exterior sorting area of similar kind. Alternatively, or perhaps in addition, these carefully made features might be drying places, where cotton was spread, turned, and allowed to dry.

In 1857 a correspondent of *Du Bow's Review* remarked: "cotton gains no good by keeping in bulk, but on the contrary it deteriorates, losing a great deal of its soft silk feeling and natural vegetable oil, which qualities are much desired by consumers," adding that planters should gin their cotton: "as soon after harvesting as possible." Ginning was the mechanical process separating seed from cotton lint, an operation housed in buildings usually designed to accommodate moting and packing activities as well. Before introduction of McCarthy's improved gin during the mid-1850s most Sea Island planters used foot gins, old-fashioned machines consisting of wooden or iron rollers turned by foot-driven treadles, the single slave operating each machine feeding cotton through the rollers. Thomas Spalding believed these models were first introduced locally by Alexander Bisset of St. Simons Island in 1778, who "took the idea from communications from the Bahama islands."[23] Based on *chuka* gins—primitive Indian devices used for centuries, foot gins had developed before the 1760s, Denis Diderot illustrating an example from Martinique or Santo Domingo in his great *Encyclopedia* (Paris, 1762).

Later experience proved that, unlike Eve's gin, which could be animal, wind, or tide powered, and Eli Whitney's better known device (patented 1794), foot gins did not damage delicate long-staple fibers. Visiting Beaufort in 1796, La Rochefoucauld-Liancourt saw foot gins owned by General John Barnwell, Robert Barnwell, or Major Wright, all called pioneer cultivators counted among Port Royal Island's largest producers. The distinguished visitor also viewed a large horse-driven barrel gin equipped with mahogany rollers. Although this cumbersome, rather dangerous machine significantly enhanced productivity, quality suffered, and cotton so processed brought an inferior price.[24]

On Edisto John Townsend had improved, yet still-simple, foot gins "3½ feet high, 2 feet long and 1 [foot] wide with an iron fly wheel . . . upon each side, working a pair wooden rollers made of hard oak," and might have twenty or thirty in use when his crop was brought in. The "task" for operatives was "20–30 lbs per day according to quality."[25] Soon after Christmas 1838, Frances Kemble—who found the Carolina Sea Islands "dismal and wearisome to the eye"—visited William Seabrook's Edisto plantation, renowned for cotton of the highest quality. She observed: "a short distance from the landing we came to what is termed a ginning house. . . . It appeared to be open to inspection and we walked through it." The building was divided into "about eight or ten stalls on either side, in which a man was employed at a machine, worked like a turner's or knife grinders wheel, by the foot, which as fast as he fed it with cotton, parted the snowy flakes from the little black cause, and gave them forth soft, silky, lean and fit to be woven into the finest muslin."[26] Thomas Chaplin's diary tells us ginning operations continued regardless of cold weather—all too common an occurrence even on St. Helena Island and Edisto at harvest time (usually late December and early January)—when icy winds might blow in from the Atlantic, snow sometimes adding to the misery of ill-clad operatives working among unheated buildings or open sheds.

Cleaned lint was next taken to the mote table, where women, half frozen or not, removed any stained fibers or broken pieces of seed taking care not to remove too much good product in the process. Elliot said this operation often (but not always) utilized "frames of wire, or latticed wood, at a rate varying from fifteen pounds to thirty pounds to the hand."[27] Subsequently "sewing the end of a bag over a hoop, and suspending it through a hole in the floor," an operative standing over the bag would commence packing using an iron or wooden pestle, this time-consuming job maintaining full and proper weight only if closely supervised. Most Sea Island planters eschewed mechanical presses, preferring hand packing, which, like foot gins, preserved long cotton fibers intact, and preferred old-fashioned round bales containing "320–400 lbs merchantable cotton."[28] *Leslie's Newspaper* (April 17, 1869) gives a variant account about packing cotton on the Sea Islands:

> once ready for this, doors are opened in the floors of the cotton-room, underneath which are suspended large bags, composed of flax twine manufactured in Kentucky, or of gunny cloth, of foreign importation. The cotton is thrown into these bags, and tamped into considerable solidity by the feet of the workmen. When the bag is full, the open end is sewed up, the planter's name is branded on it[s] surface, and it is hauled away to the landing.[29]

In 1852 it was reported: "Governor Seabrook spent $5,000 in experiments and others equally as much to get a substitute [for foot gins] but have been compelled to go back to the primitive machine described."[30] Although arousing hostility at first, McCartney's so-called Florida gin processed long-staple cotton better than any foot gin. European spinners considered the product superior, paying between 3 and 10 cents more per pound for cotton ginned by the new machine. In 1853 Robert Alston reported McCartney's invention was: "attracting much attention, and the planters are putting them up as fast as they can. A gin costing one hundred dollars, propelled by a good horse or mule, or still better by steam will clean from 150 lbs to 200lbs per day."[31]

Steam power was a significant departure. Only forward-thinking landowners capable of producing long-staple cotton of the best quality were ready to make the considerable investment in money and labor this technology required. Among Beaufort District's proprietors these included A. R. Chisolm (whose two-story "steam gin house" stood at Coosaw); William Henry Trescot, who listed a cotton house and steam engine in "perfect order" on Barnwell's Island; and the Coffins of St. Helena Island, heirs of Mary Coffin itemizing—along with thirty-two "gins" (presumably foot gins)—a steam engine worth $250 probably installed at Coffin Point. All were erased from the plantation landscape after Union occupation in 1862–63. No certain antebellum gin houses are known from Beaufort District, nor have any ruined or subsurface examples been fully explored. Conversely Sunnyside on Edisto Island (formerly Colleton now Charleston County) preserves that great rarity among Sea Island cotton plantations, an assemblage of service structures including a two-story barn with root cellar erected on tabby foundations, several small timber-framed structures probably designed to accommodate house servants, and ruins belonging to an exceptionally large cotton gin, all clustered around the owner's dwelling, which stands on its own miniature island overlooking Store Creek.

Old photographs (probably early 1900s) show that before its destruction, the gin incorporated two timber-framed stories raised over an elevated tabby basement. Walls of the latter remain more or less intact and belong to two separate building phases. Phase I tabby defined a structural footprint measuring about 40'4" east/west × 20' north/south. Phase II extended the original building eastward by about 20'6" and included insertion of a massive rectangular base cast solid in tabby, which still retains metal bolts cast into its top surface designed to anchor machinery, indicating the extension was built to house steam equipment. Standing more than four feet high, early tabby walls are battered in profile measuring about twenty-two inches thick at ground level and sixteen inches wide at their highest point, the top surface preserving impressions of 9" × 9" timber sills supporting the building's framed superstructure. Two relatively wide doors positioned opposite one another on the two long sides gave access, and two smaller openings provided light. Phase II tabby is generally less robust, measuring twelve to fourteen inches in width and standing about two feet high. There is one wide opening at the extensions far end. The tabby base is higher and measures about 6'6" × 5'8" in plan. Structural details are unusual, there being no evidence for timber cross ties holding formwork in place during the casting process. In this respect the Sunnyside gin resembles a barn erected several miles away at Bleak Hall, apparently a pre–Civil War structure probably built for John F. Townsend, the same crew perhaps working on both buildings. Although day-books survive, Sunnyside's chronology is uncertain. The present mansion, a raised two-story framed structure with wrap-around porches, mansard roof, and cupola, was built circa 1870–80 by Townsend Mikell, who was born at Bleak Hall. It is thought attendant outbuildings, including the gin, could be earlier, but how much earlier is unclear. Ginning operations at Sunnyside continued well into the 1900s.

Very few planters returned to Beaufort's islands after the Civil War, those who did, such as Thomas Chaplin at Tombee, facing difficulties with labor (or lack thereof), credit, and mounting debt, which soon put them out of business. Federal agents and missionaries who took over holdings abandoned by their former owners achieved disappointing or—at best—mixed returns. Nevertheless breakup of old estates into small units and government property auctions in 1864 allowed newly emancipated slaves to farm on their own account, their efforts along with those of northern entrepreneurs, who bought up the best cotton lands at bargain prices, producing about half the late antebellum quotient by 1880.[32] The same period saw establishment of George Ricker's Beaufort steam mill and gin, which then had six gins operating (replaced by four Platt's patent roller machines before 1899), this enterprise along with G. M. Pollitzer's nearby "Cotton Warehouse and Gin" bringing new, more efficient technologies to local growers.

Once ginned, moted, and packed, cotton began its long journey by sea or waterway toward eventual milling. Owners might transport bales using their own flats, barges, or schooners. Many relied on commercial riverboats or small coasting vessels, which picked up goods from individual plantations when signaled. Before widespread advent of steamships this could be a frustrating, tedious, and uncertain business. Describing Captain Daniel Bythewood's schooner *Delancey* plying between Beaufort and Charleston, the planter and poet William John Grayson said waiting for fair winds could delay passage for days or even weeks.[33]

How steam power helped revolutionize this archaic transportation system can be judged by Frances Kemble's account of her voyage from Charleston to Savannah in 1838 on the *William Seabrook,* an old, far-from-comfortable steamboat belonging to Seabrook's Edisto Island Ferry Company. Deemed unfit for ocean travel, it made the trip (calling at Edisto and Hilton Head) via the "inland passage" in just over two days.[34]

Mindful of possible shipping delays, inclement weather, and the vagaries of distant markets, prudent planters provided themselves with solidly built barns where their precious cotton might be safely stored until conditions improved. Grayson's own St. Helena Island plantation called Frogmore has Beaufort County's best-preserved example. Relatively undifferentiated this tabby structure was essentially a "concrete shell" protecting the valuable fiber from fire and storm, as Vlach characterized another similar building (built by James Couper) at Hamilton Plantation, a neighbor of Butler's plantation on St. Simons Island.[35]

Measuring 48' × 25'3" overall, Frogmore's barn is single story with an attic above. Cast uniformly in two-foot-high vertical increments, twelve-inch-thick exterior tabby walls stand over thirteen feet high, the quality of their construction ranking not much below contemporary domestic work. Wide doorways (probably enlarged) back and front of the building give entry to an interior space uninterrupted except for a row of timber columns supporting the attic floor above. No provision other than the doorways was made for light or ventilation, three small windows cut through the north facade representing an alteration phase (late nineteenth or early twentieth century?), which introduced the timber columns. Wall sockets show that original flooring (comprising 5" × 5" joists set 2'1" or 2'6" on center) was raised two feet or more above present grade to ensure stored items stayed dry when storm-driven tides flooded the site, an unusual but not unknown occurrence. Today the barn's main body is enclosed by a timber roof of gabled form, the roof space floored and lighted by an east-facing dormer. Perhaps replaced after the Great Sea Island Hurricane of 1893, roofing was extensively repaired in 2001 only to be damaged again by Hurricane Matthew in 2016. During the later antebellum period stabling for horses or mules was introduced along the building's west side, tabby cross walls defining stalls for animals and covered storage for carts or wagons.

Construction details indicate initial construction occurred between 1800 and 1825, when not only Frogmore but the whole of St. Helena Island saw unprecedented agricultural development based on profits derived from long-staple cotton.

In 1816 Frogmore produced 133 bales. In 1818 the enslaved population numbered 139, the main settlement comprising an overseer's dwelling, barn, provisioning house, cotton house, ginning shed, and "18 Negro houses."[36] By 1853 Thomas Aston Coffin (1795–1863), who owned Coffin Point on the same island, had obtained the holding. Losses sustained

Frogmore, St. Helena Island, barn, exterior from northeast
(Jack Boucher, HABS, 2003).

Frogmore, St. Helena Island, barn, interior looking north
(Colin Brooker, HABS, 2003).

during the Civil War give an idea of the scale and value of his enterprise, the list (excluding livestock) including 28,334 pounds of cotton valued at sixty cents per pound ($17,000); eighteen bales of cotton, three hundred pounds each ($8,3510); one steam engine ($400); forty-five foot gins ($540 total); four cotton whippers at $50 each; twenty-five ox carts ($1,250); four mule carts ($100); four boats ($600); and four flats ($400). Buildings are not mentioned except for twelve "Negro" double houses (valued at $100 each). These were occupied by some—if not all—the 193 slaves collectively valued at $135,100 who worked 1,365 acres valued at $40,950 by the estate's trustees rather than the derisory figure of $2,240 set by the U.S. Direct Tax Commission in 1862.[37]

Spring Island was an even larger operation. In 1830 its enslaved population numbered 336 persons—an astonishing figure that made the proprietor, George Edwards, one of Beaufort County's largest slave owners. Edwards was not immune to financial setbacks, selling eighty-two slaves in 1837 and fifty-five more in 1847. However, cotton remained Spring Island's mainstay. In 1850, 150 bales were produced along with quantities of foodstuffs, the plantation itself then being valued at $50,000 excluding machinery worth $2,300.[38] Exactly where this machinery—including one McCarthy gin (valued at $50), one Shaffer gin (valued at $50), and one lot of foot gins ($25)—was housed is not known. Frederic Holahan, a Union soldier looting Spring Island in January 1862, said that "dwelling houses for overseers and larger buildings for the storage of cotton were [located] at intervals along the shore where landing were made."[39]

One of these storage buildings is almost certainly represented by tabby foundations located just north of the main plantation house. Features excavated in 2016 define the footprint of a timber-framed structure measuring 32'6" × 19' in plan now completely lost above ground. The foundation system—consisting of rectangular intermediate piers and L-shaped corner supports—was strengthened during an alteration phase presumably to accommodate exceptionally heavy floor loads, a circumstance consistent with storage of cotton bales. Equally telling are timber piles driven into the adjacent marsh. These attest docking facilities or a pier extending into channels leading out into the Chechessee River, Broad River Sound, and Atlantic Ocean beyond. Despite exposure to storms and hurricanes, these channels have shifted little, if at all, since first mapped at the end of the eighteenth century. Never ideal except at high tide, it was the only possible access near the main house for traffic coming or going by water, a fact explaining why other relatively large-scale tabby structures now reduced to broken, shapeless tabby blocks were clustered along adjacent shorelines. An associated slave row, perhaps housing boatmen who ferried goods to and from waiting vessels, was torn down in the 1940s. The few surviving remnants were subsequently bulldozed into oblivion.

• Sugar •

Paradoxically Beaufort County's most extensive agricultural processing complex where tabby construction predominates is a sugar works, the ruined installation and its settlement reflecting short-lived enthusiasm for cane developed in response to tariffs imposed following the War of 1812. Overlooking the northeast shore of Callawassie Island, three

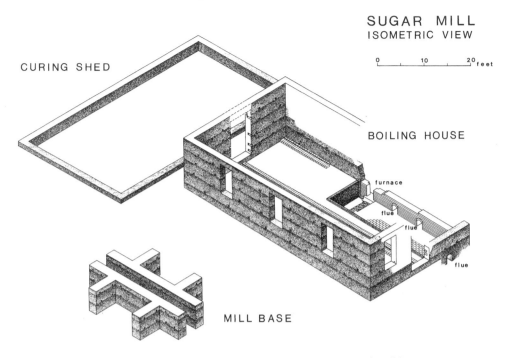

SUGAR MILL
ISOMETRIC VIEW

CURING SHED

BOILING HOUSE

furnace

flue

flue

flue

MILL BASE

Hamilton sugar works, Callawassie Island, isometric of mill base,
boiling house, and curing shed after excavation.

separate processing structures are represented, a mill base, boiling house, and curing
shed, the last two components making a "T" when viewed in plan. The latter arrangement, not otherwise known from the state (but common in Jamaica, where William
Beckford, perhaps the island's richest sugar magnate thought it most "commodious"),[40]
closely resembles Thomas Spalding's Sapelo Island sugar house described and illustrated
by his "Observations on the Method of Planting and Cultivating the Sugar Cane in
Georgia and South Carolina," published by the Agricultural Society of South Carolina in
1816.[41] Like Spalding's scheme the Callawassie works relied on animal rather than steam
or tidal energy to power the mill proper. Rather than housing milling operations inside
an octagonal structure built of tabby boasting two distinct levels as on Sapelo, a relatively primitive arrangement was employed on Callawassie, nothing emerging from Dr.
L. Lepionka's excavation (1982–83) to suggest the mill was ever enclosed, palmetto fronds
or some equally flimsy covering incapable of leaving any trace sheltering crushing equipment, if not the enslaved operatives, from sun and rain.

Well preserved an elevated base supporting milling machinery comprises two parallel tabby walls, standing about five feet above present grade. Over twenty-one feet
long, distanced 2'6" apart, each wall is buttressed externally by two tabby spurs running
at right angles. Remarkably the form created reproduces the footprint of braced timber frames commonly fabricated by plantation carpenters across the Antilles to mount
cane-crushing machines incorporating three vertical iron or oak rollers driven by oxen,
horses, or mules harnessed to a sweep arm. Cane was hand fed between these rollers—an

Animal-driven vertical cane mill, Diderot and d'Alembert, *Encyclopédie*, Paris, 1762.

exceedingly dangerous operation that caused grisly accidents—and crushed to express the juice, with waste (bagasse) being collected for fuel when enough accumulated

Exiting the mill cane juice passed by means of a chute, gutter, or similar device into the boiling house. Calawassie's was a freestanding, single-story tabby structure measuring 45' × 25' overall. Here the juice was first clarified in a tank (now lost) by addition of lime, then ladled into four or five graduated metal kettles set linear style along the building's south side over a common closed flue or "Jamaica train" as it was called, principally built of brick with the requisite furnace positioned at one end and chimney at the other.[42] Nearly eight feet wide, the train bed was sunk approximately 2'5" below general floor level. Tabby walls of uncertain height defined the boiling area on its two interior sides (north and east). The exterior (south) side was built of fired brick, and the exterior west end of tabby. An arch piercing the building's south facade accessed the brick-lined furnace (measuring about 3'6" × 4' in plan). Combustible materials (bagasse and wood) were doubtless supported on a metal fire grate now lost. A second, elliptically arched opening, positioned immediately below the furnace feed, allowed for cleaning the brick-lined ashpit, which still contained ash and carbonized pieces of pine when excavated. Both furnace and ashpit were furnished with metal access doors—these too have gone, robbed or sold for scrap.

The boiling train's bed of fired brick laid over clay fills remains. Most associated features are either fragmented or lost. Missing elements include the boiling pans (kettles) and their masonry supports. However small flue openings on the train's exterior side indicate use of four copper or cast-iron boiling pans arranged in line. The largest (*grande*) would have been positioned near the clarifier. Juice was ladled successively from there into smaller pans until it reached the smallest, known as the *teache* or *strike,* positioned directly over the furnace. Heat passed beneath each kettle in turn, flue openings along the exterior (reached by way of a shallow firing trench) allowing temperature regulation. By closing or opening these vents with wooden or possibly iron dampers, individual pans could (in theory) be made to boil more or less rapidly. Another small opening piercing the building's lower west facade probably communicated with an exterior freestanding chimney (now lost)—the boiling train's principal flue.[43] Spalding's arrangement differed, his plan showing a chimney located outside the building, not at one end as presumed on Callawassie, but rather against his long "copper wall." After boiling, concentrated juice was transferred into shallow wooden coolers where the sugar slowly crystallized. A narrow brick path running along the Callawassie building's center (found almost intact when excavated) indicates coolers were positioned opposite the boiling train in the same general arrangement illustrated by Spalding. The long tabby exterior wall on this side of the structure has largely collapsed. Fortunately it fell outward, splitting horizontally but retaining enough coherence to allow theoretical reconstruction of the original fenestration pattern, which incorporated three large window openings, each measuring approximately 2"-9" wide × 6'2" high. Judging by quantities of glass found during excavation, these windows were glazed. Whether smaller windows pierced the opposite (south) facade above the boiling pans cannot be said. Old illustrations of Caribbean sugar houses suggest this was probably the case, but the wall in question is so extensively robbed as to preclude anything but speculation about details. Nevertheless the size of known window openings indicates the boiling house was well ventilated, although nothing tangible survives of the roof, making it impossible to know if a "lattice work Cupola" like Spalding's was ever installed along its length.[44]

Prints of West Indian boiling houses provide other possible roofing solutions. William Clark's *Ten Views in the Island of Antigua* (London, 1823) gives an interior view of the boiling house on Delap's estate showing skimming operations underway and sugar being ladled into long wooden spouts that conveyed "liquid sugar" into a row of rectangular coolers. Through clouds of steam, small louvered windows can be seen above the boilers. Larger windows occur all along the opposite wall immediately over the coolers. The pitched roof supported by parallel king-post trusses has barely visible "apertures" above the boiling train to allow the escape of "vapours." James Hakewill's *Picturesque Tour of the Island of Jamaica* (London, 1825) illustrates louvered clerestories running along the entire length of boiling houses at Williamsfield Estate; St. Thomas in the Vale; Trinity Estate; St. Mary's; and Montpelliar.[45] Large quantities of nails discovered during investigation suggest Callawassie's boiling house had something similar if Spalding's "cupola" was not emulated.

With only a narrow passage separating them, the curing shed on Callawassie was positioned east of the boiling house. Here casks containing the now-granulated sugar were drained, the process taking two or three weeks to purge the sugar of its molasses—a valuable by-product that might be sold or distilled into rum. Apparently timber-framed Callawassie's curing shed is reduced down to its tabby foundations. About 1'3" wide these provide more evidence of dependence on Spalding's model. Besides producing a T-shaped footprint with respect to the boiling house, it was of about the same size (measuring about 45' × 24'10") as the example (measuring 45' long × 22' wide) illustrated by Spalding's drawing of 1816. Spalding gave no information about the construction of his facility except to say that curing houses should be lighted with very few windows to keep in heat and prevent premature coagulation of sugar, adding that "stoves also would facilitate the sugar discharging its molasses." Inside he described how the bottom of the house comprised "two inclined planes, of two feet descent." These discharged "the molasses into a gutter in the middle," which emptied into "a close cistern containing two thousand gallons."[46] The cistern, he said, "may be made of cypress plank, rammed at the bottom and sides with clay." If anything similar ever existed on Callawassie, it has left no trace.

Located within yards of the island's northeast shoreline and easily approached from Chechessee Creek, which is tidal—though not salt enough to discourage alligators—the mill's location would have allowed delivery of cane by bateau or flat. Callawassie is low and often floods, conditions severely hindering movement of carts loaded with cut cane during November and December, when harvesting occurred. Alternatively, or perhaps in addition, the site's open situation allowed river breezes to cool the boiling house and disperse clouds of water vapor it produced.

South and southeast of the sugar works, Lepionka exposed remains of slave houses, several outbuildings including one or more barns, and a middling- or possibly high-status timber-framed dwelling with two end chimney bases made of tabby. All these structures (most neither published nor properly recorded) along with the small wattle-and-tabby-daub cabin described in chapter 4 were destroyed by residential development during the late 1980s.

Probably erected in 1816 or soon before, the sugar works proper (mill, boiling house, and curing shed) were likely built by Major James Hamilton Jr. (1786–1857), an attribution made on the basis of a notice published by the *Southern Patriot and Commercial Advertiser* (Charleston, S.C.) in January 1817 (reprinted by the *City Gazette and Daily Advertiser*, January 24, 1817; and *New England Palladium and Commercial Advertiser*, February 7, 1817). It reads as follows:

> We understand from unquestionable authority that SUGAR of an excellent quality was made at the plantation of Maj. James Hamilton Jun. on Callawassie Island, St. Luke's Parish, on the 6th of Jan from canes which had been cut and stacked since the 12th of November last. It is remarkable that their exposure to a severe frost on the night of the 11th, did not prevent a perfect granulation of the juice, notwithstanding some few of the exposed ends of the canes were partially accidulated. We learn that the product in quantity per acre is sufficient (when the last most unfavorable season is considered) to warrant and encourage a continuance of its cultivation on a more extensive scale.[47]

James Hamilton had acquired Callawassie by marriage in 1813 to Elizabeth Heyward, who was awarded the island (along with nearby Rose Island and other plantations) when still a minor in 1806 following settlement of long-drawn-out family disputes over the estate of her great-grandfather Daniel Heyward (died 1777).[48] It is likely the newlywed couple lived near Callawassie's western extremity, an airy peninsula (now called Tabby Point) overlooking Colleton River and Chechessee Creek affording wide views toward other Heyward properties occupied by large Yemassee settlements during early colonial times. The Hamiltons' Callawassie residence was forty feet square, broken and incomplete tabby exterior walls (eighteen inches thick) showing the lower story was divided into four rooms organized about a central hall wide enough (measuring 7'8" across) for stairs. Of the upper story or stories, nothing definite is known.[49]

Elizabeth Heyward's stepfather, Nicholas Cruger Jr. (1779–1826), was from the then Danish island of St. Croix (Lesser Antilles), where his family had long established mercantile interests revolving around wholesale shipment of sugar, rum, and molasses to the northeastern United States (principally New York) in exchange for lumber, livestock, and foodstuffs.[50] The birth of Elizabeth's half brother Henry Nicholas Cruger on St. Croix in 1800 suggests she herself lived there during childhood. Subsequently Nicholas Cruger Jr. migrated to South Carolina, purchasing (ca. 1815) two Back River properties with the potential for rice cultivation situated just north of Savannah, called Pennyworth Island and Rice Hope respectively. At Pennyworth Cruger installed a rice mill partly overhanging the river, an installation enlarged and reequipped with the latest steam-driven machinery after James Hamilton purchased this holding along with Rice Hope following Cruger's bankruptcy in 1824. Vast amounts of borrowed (some said misappropriated) capital were poured into these plantations to good purpose at first, but with ruinous effect on Hamilton's reputation in the end. The scale of his activity is still attested by heavy timber cribbing supporting the Pennyworth rice mill's brick foundations, segments of an associated wharf, and extensive bulkheads along the river.[51]

Close family and business relationships documented between Hamilton and his father-in-law make it likely that Nicholas Cruger or operatives he employed helped design the Callawassie sugar works. This would explain certain practical refinements, including the small exterior vents along its boiling train—features found in contemporary West Indian boiling houses, including several known from the Virgin Islands, including Cardon Bay, St. Croix; Estate Annaberg, St. John; and the well-preserved sugar factory at Estate Reef Bay on the latter island.[52] The intriguing possibility that Hamilton corresponded with Thomas Spalding also exists. Writing from Sapelo in July 1816, Spalding said he enclosed a letter (now lost) written "to Maj. Hamilton in South Carolina," giving detailed information about the construction of tabby roofs.[53] Whether Spalding's "Maj. Hamilton" and the proprietor of Callawassie were one and the same individual remains unproven. It is certain that Hamilton soon lost interest in an island too isolated from urban centers for his restless ambition. Moreover he probably realized Spalding's sugar-making technology was outdated, inefficient, and unlikely to yield returns commensurate with those promised by rice, conclusions later confirmed by experiments with steam power at Rice Hope and Pennyworth. Callawassie was purchased by John A. Cuthbert in 1819,

Hamilton then removing to Charleston, where he resumed an interrupted legal career and began a controversial political one, becoming a U.S. representative in 1822 and governor of South Carolina in 1830.

Whether Cuthbert (lieutenant governor of South Carolina, 1816–18) or his son James, who inherited the island in 1826, cultivated sugar commercially is not known. It is probable that cane was abandoned here before 1829, Edward Barnwell, who had erected "a small mill driven by one horse near Cossa river," making no mention of Callawassie in his survey conducted to determine if sugar was a viable crop for coastal South Carolina.[54] Neither did he find any local grower approaching John Hamilton Couper's productivity, the latter's Altamaha River plantation called Hopeton with 202½ acres under cane yielding 32,350 pounds of sugar and 5,349 gallons of molasses in 1829. The following year Couper planted 303 acres of cane, annual yields reaching an astonishing 107,811 pounds of raw sugar and 21,895 gallons of molasses.[55] These figures reflect how steam power revolutionized production. A "fine fourteen-horsepower steam engine . . . built by Bolton and Watt in England" was installed by Couper in 1830 to drive both his sugar mill and a rice-pounding mill.[56] Having abandoned technologies reliant on animal power, Couper ignored local architectural conventions, milling, boiling (using a double Jamaica train), and curing operations associated with his sugar manufactory all taking place under the roof of one very large tabby building. Measuring 240' × 38' the design was inspired by sugar houses in Louisiana, where linear plans, including the first one erected at Ashland/La Belle Helene, Ascension Parish (measuring 146'6" × 45'3" before extension), were common.[57] Edward Barnwell realized that despite Spalding's stubborn resistance to steam, competition from Cuba, the British West Indies, and Louisiana demanded better,

Plate No. V

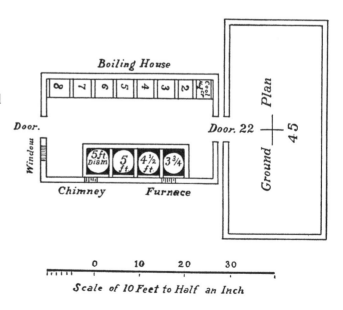

Thomas Spalding, plan of boiling house and curing shed from "Observations on the Sugar-Cane in Georgia and South Carolina," Charleston, Agricultural Society of South Carolina, 1816.

Hamilton sugar mill, Callawassie Island, S.C.,
flue openings along exterior side of Jamaica train.

more efficient processing if sugar producers of coastal South Carolina and Georgia were to stay in business. He remarked:

> It appears to be the opinion of most, if not all, the gentlemen I conversed with upon the subject, that when cultivated as a large crop, the power of steam would be required for this essential purpose. Tide mills are uncertain, and animals are liable to various accidents, are expensive, and often too weak. For, when the crop is ripe, every exertion must be made to expedite its manufacture, and no delay for want of water or animals must be experienced.[58]

It also became evident that cane's best varieties were cold sensitive, doing better when planted further south, James Hamilton perhaps learning this after one or two seasons. Nevertheless South Carolina continued commercial production down until the Civil War, producing seventy thousand pounds of sugar in 1860, when U.S. Census returns record that Beaufort County contributed twenty thousand pounds. But the state's production never approach that of Georgia (329,000 pounds in 1840 and 846,000 pounds in 1850), much less Florida, where, after a slow start across Flagler and Volusia Counties, the sugar industry shifted southeast and northeast, planters producing 2.7 million pounds in 1850. Robert Gamble Jr. of Manatee County (Tampa vicinity) anticipated yields of two thousand to three thousand pounds of sugar per acre in 1851, his property producing 230,000 pounds during the previous year.[59]

Desperate for cash as federal troops raided along the Combahee, the Middleton family agent at Hobonny and Old Combahee planted about twenty acres of cane in March 1864. "I ought to make a quantity of syrup," B. T. Sellers wrote, "but as I have no experience in it, I cannot make an estimate."[60] Cane was still cultivated very near Callawassie in 1911, Harford C. Eve recording: "I went over to Spring Island and [with] old Mr. Pinckney—brought back 25 gals fine cane syrup that I bought at Spring Island @40c per gal."[61] Photographs of St. Helena Island taken around the same time or slightly later illustrate primitive horse-driven mills and single kettle boilers, a few farmers, then as now, producing cane and syrup for roadside sale and home consumption.[62]

◦ Garden and Landscape Features ◦

Garden buildings are among the most transient of structures. Bleak Hall (Botany Bay), Edisto Island, retains two. One is a small (12'2" × 12'4"), windowless storage building entered from the north, with thirteen-inch-thick tabby walls and bell-cast shingled roof (restored). The other is an altogether delightful Gothic-revival-style gable-roofed icehouse. Timber-framed it stands on tabby foundations, concealing beneath its floor a square subterranean well enclosed by two parallel tabby walls (each nine to ten inches wide) on all four sides separated by narrow cavities once filled with sawdust, charcoal, or possibly Spanish moss for insulation. Additional insulation was provided by enclosing interior space above the well within another timber-framed shell, similar multiwalled techniques having proved successful in Cuba, alternative designs with deep brick or masonry shafts seen in Europe and New England being impractical across the lowcountry. Shipped from Pennsylvania or Massachusetts by the entrepreneur Frederic Tudor or his competitors who began supplying Charleston during the early 1800s, ice was delivered at one end of the Edisto building through tall double doors. When filled the well was concealed and further insulated by flooring hinged (if the current ingenious arrangement matches the original) to provide access when lifted. Part of an extensive garden layout, now lost, both buildings were erected by John Townsend in the 1840s. The icehouse is just visible in Kallie Sosnowski's painting of Bleak Hall (ca. 1860), its tall, distinctive finials emerging above a bank of trees slightly south of Townsend's mansion.[63]

Once famous for its horticulture, Retreat, near the southern extremity of St. Simon's Island, retains near-shapeless remains of the tabby "hothouse" built by Anne Page King, who inherited the plantation (formerly owned by Thomas Spalding) about 1827. Photographs taken during the 1880s and 1930s, when it was far better preserved, show a rectangular structure with impressions of a sloping shed roof (presumably glazed), the back wall and end elevations cast in twelve-inch-high increments.[64]

Beaufort County's most extensive and enigmatic landscape feature incorporating tabby partially encloses Woodward Plantation, Port Royal Island, an antebellum holding bounded west by the Broad River north of Laurel Bay. Before the Civil War, the property was owned by Robert Woodward Barnwell (1801–92), who probably inherited it soon after 1814. The stylish timber-framed main house, perhaps erected by Barnwell's father, Senator Robert Gibbes Barnwell (1763–1814), is documented only from photographs having

Bleak Hall (Botany Bay), Edisto Island, icehouse.

first been looted by the crew of a Union gunboat looking for firewood and then burned. Principal landscape features still extant comprise two long, low tabby wall sections (both now heavily overgrown and damaged), the first (about 679' long) running almost north/south parallel to the river and the second (now about 329' in length) running almost east/west. Neither feature is complete, modern highway work destroying an almost certainly right-angled junction between them. About fifteen inches wide, tabby walls were the product of a single two-foot-high pour cast directly over topsoil. No evidence exists for anything higher, suggesting that walls stood not much more than eighteen inches above grade. Investigation by the author disclosed a ditch approximately ten feet wide paralleling the north/south segment along its outer (east) face. Interpretation presents difficulties. The division implied by these features hardly constituted more than a token barrier between separate activities or properties. If the tabby supported a fence similar to that enclosing the yard of the William Sams House, Dataw Island, no evidence (such as distinctive post holes) is now preserved. Moreover Woodward's layout would be considerably larger. Indeed, it is hard to imagine an enclosure of the dimension indicated serving primarily domestic functions unless the concept of "yard" is expanded to include some kind of controlled landscape setting. In this context the combination of broad ditch and low wall is informative, recalling one of the most popular landscape devices of the eighteenth century, variously called a sunken fence, ah-ah, or ha-ha.

A. J. Dezallier D'Argenville's account translated and published by John James (*Theory and Practice of Gardening*, London, 1712) describes the device and explains its purpose:

> at present we frequently make Through-views call'd Ah, ah, which are Openings in the Walls, [surrounding the estate] without grills, to the very level of the walks, with a large and deep Ditch at the foot of them, lined on both sides to sustain the earth, and prevent the getting over, which surprizes the eye upon coming near it, and makes one cry Ah! Ah! From whence it takes its name. The sort of opening is, on some occasions, to be preferred, for that it does not shut up the prospect.[65]

In South Carolina a ha-ha was installed at Drayton Hall (Charleston County), Charles Drayton writing: "Nov. 9 [1789] was begun the ha ha across the garden," a note dated eight years later recording how work was then needed "to repair the garden fences of the S. E. ha ha."[66] Ditches still extant show that Drayton's ha-ha maintained an open prospect between the main house and the Ashley River. In Georgia Thomas Spalding was an enthusiastic advocate of similar landscape elements, his article concerning tabby (published December 1830) suggesting to his planter readers:

> instead, then, of wasting labour on fences, let us surround our dwellings with sunken ditches, executed in such manner as to be unobtrusive and unobserved, until approached very near. . . . The dwelling side is made perpendicular, and if there is stone abundant it might for more security be faced with stone, if not, it may, at least, be sodded with grass; and at first be protected by a single rail. Upon the other side it should be sloped, and the earth taken out of the fosse, so raised and dressed up, as when looked at from the house, or from a distance, the sink is not observed, and the eye passes smoothly over, as the glacis of a fortified post.

Noting that "enclosures executed in this manner, would be by much the most beautiful and permanent that could be formed," Spalding added:

> the perpendicular side of this trench should be six feet, which would prevent any animal from attempting to pass up or down . . . and after the earth had become solid and well covered, particularly with Bermuda grass, perhaps no other

protection would be necessary; at any event, a single rail upon the surface would be protection enough, without being conspicuous, or observed at a distance.[67]

It is impossible to know whether or not features at Woodward were analogous. But if for an inconspicuous "single rail" a low tabby wall is substituted, the resemblance between Spalding's model and Woodward's scheme is striking, so much so that we might speculate that the simple device of a sunken ditch surmounted by low tabby walls was designed to keep cattle and other livestock out of a large garden or grassed area overlooking the Broad River besides providing fashionably "picturesque" settings for the main house. These features would have given the owner—Robert Barnwell or his father—unobstructed views eastward besides making the residence visible from the public or semipublic highway bordering the plantation along its inland side. What little is recorded about Woodward's layout just before the Civil War confirms the plantation then had an extensive flower garden. Additionally photographs of its driveway (dated 1863 or 1864) show a winding, sinuous feature 1½ miles long, the main house and surrounding landscapes revealed not at once but gradually from different perspectives in true "picturesque" fashion.

10

Chapels and Cemeteries

Excluding fortifications only one public structure survives from prerevolutionary times on Beaufort's Sea Islands. Called the Chapel of Ease or White Church, this now-roofless ruin with eighteen-inch-thick tabby exterior walls standing fully fourteen feet high is located near the junction of Land's End and Club Bridge Roads, St. Helena Island. It was established around 1748 by Anglican settlers for whom the long journey by road, boat, and ferry into Beaufort Town could be thwarted by storms, dangerous tides, impassable highways, runaway animals, or exactions of ferrymen. Construction is not documented by vestry minutes of St. Helena Parish (to which the property still belongs), suggesting local planters paid for the building out of their own funds. Progress was slow, John Fripp leaving $500 for the still-unfinished project in 1781.

Before enlargement around 1850, the chapel exhibited the standard Anglican colonial church plan as defined by Lindner, "with entrances on the north, south and west sides," a plan repeated with minor variations among several small religious structures ranking among the most attractive components of South Carolina's architectural heritage, including St. James, Goose Creek (1708); Pompion Hill (1763); and Pon Pon Chapel. On St. Helena Island, the original chapel's two long walls (north and south) each incorporated three bays comprising a central arched doorway flanked by relatively tall windows with arched heads right and left. All arches were executed in fired brick. The brick was not rubbed or cut (usual among prominent mid-eighteenth-century buildings) to achieve the tapered shape necessary for tight construction. Rather, they were set in generous amounts of mortar shaped to achieve the desired semicircular form.[1] The same pattern of arched wall openings was repeated by the gabled west end, a porch supported by columns of fired radial brick with applied-stucco profiles fronting the principal entranceway and its two flanking windows. A brick-lined oculus (bird's-eye window) pierced the tabby gable above, with beam sockets on inner building faces suggesting some kind of narrow interior gallery.

Old photographs show that following precedent set by other chapels—Strawberry Hill; St. John's Parish, Berkeley (1725); St. James, Goose Creek—roof framing was of the jerkin-head variety. All structural timbers are lost. Cornices, baseboards, and chair rails are known from "ghost" impressions left on tabby surfaces. If, as I imagine, box pews existed resembling those at the beautifully preserved planter church of St. James, Santee, they have completely gone. Largely buried by debris, scraps of an early tile floor still survive in situ.

Chapel of Ease, St. Helena Island, north facade before ruined by fire in 1865
(Beaufort County Library, Special Collections).

Chapel of Ease, St. Helena Island, west entrance facade.

Chapel of Ease, St. Helena Island, arched window head
showing voussoirs applied in shell lime.

During the second quarter of the nineteenth century, the St. Helena building was
extended. Before operations began the original east facade was torn down. An additional bay enlarging the congregational space eastward and narrower chancel were then
erected in tabby, with stucco and plaster concealing junctions between the two building
episodes. New tabby cast in two-foot-high vertical increments almost exactly matched
earlier work. New window openings replicated earlier examples too in size, shape, and
detail. Semicircular window heads were decorated with built-up stucco voussoirs, which
either copied existing stucco or formed part of new decorative programs introduced to
upgrade and enrich the chapel, now patronized by several of South Carolina's wealthiest
families. Rusticated elements and corner pilasters (also applied in stucco) typify an inherent architectural conservatism common among rural religious structures built or primarily supported by the planter elite, details recalling models first published by the English
architect James Gibbs in 1725 and much copied thereafter.[2] The most remarkable feature
of the entire rebuilding enterprise was an arched opening executed in tabby separating
the main congregational space from the new chancel. Having an unprecedented clear
span of about fourteen feet, the arch still stands though adjacent chancel walls have partially collapsed.

Looking for valuables during the Civil War, Union soldiers broke into an Egyptian-
style brownstone mausoleum erected (ca. 1852) in the adjacent cemetery by W. T. White
of Charleston for Edgar Fripp and his wife.[3] Doubtless the chapel itself was also looted,
final destruction coming in 1865, when the *Palmetto Post* reported, "the tabby church

(Episcopal) in the center of St. Helena Island was consumed by fire on Monday (February 22). Some Negroes were burning new ground and the fire got away and destroyed the venerable old building."[4]

No other tabby-built public place of worship is known, but an analogous although smaller and simpler structure, erected by the Sams Family for its own use, exists in an advanced state of ruin on Dataw Island. James Julius Sams described the setting as remembered when he was a boy:

> West of the orange orchard was our family burying ground. It was shaded all over by the spread of the largest live oak I ever saw. The tree grew in the middle of the graveyard, and threw its limbs out in all directions, even taking under its cover the wall which encircled the yard. On the east of the oak between it and the orange orchard, was a chapel, which was so placed as to form part of the wall.[5]

The tree still lives, and the "yard" remains almost intact, tabby walls standing about five feet high enclosing a space measuring 80' × 70' overall. Marble headstones shipped from Charleston commemorate William Sams (who died on Dataw in 1798) and his wife, children, and descendants through three or four generations, the stones recording demographic information including alliances with prominent local planter clans. There is evidence for intrusive, possibly post–Civil War, burials outside the cemetery enclosure but no indication that anyone other than family members were buried here during plantation times, custom dictating the enslaved, for instance, be interred in their own cemetery or cemeteries located elsewhere on the island. Markers (if any) for the enslaved were of perishable materials that have not survived.

The Sams chapel was rectilinear, the main congregational area (measuring about 20'2" × 30'4") opening into a small sanctuary. The principal (west) entrance had windows right and left. Long facades (north and south) followed patterns established by the Chapel of Ease, each featuring a central door flanked by two windows with double-hung sashes. Drawings indicate the roof was gabled, separate roof frames enclosing the main body of the building and its sanctuary. An interior view by Paul Brodie made during the early 1900s (possibly copied from an earlier drawing now lost) shows that the pulpit was placed centrally near the east end and separated from the congregants by an elaborate balustrade.[6] Nine-inch-wide exterior walls were cast in 19½" high vertical increments. Slovenly execution coupled with an unusually thin wall system provided little resistance to storms. In the 1950s everything collapsed as Hurricane Gracie raged across the island, leaving only broken fragments to delineate the chapel's footprint.

James Julius Sams said his entire family worshiped in the chapel on special occasions—most especially Christmas—and implied the building was complete when he visited his mother's grave in 1831.[7] Dataw's slaves received Christian instruction from their owner, Dr. B. B. Sams. Missionaries evangelized among the island's settlements, the Reverend George W. Moore (who had founded his Pon Pon Mission in 1833) reporting, "we preached on a week day, the Negroes coming out of the fields to assemble at the appointed time in a large cotton house."[8] Whether slaves other than house servants attended the Dataw chapel along with family members is uncertain, the building's size precluding active participation in worship by more than a small fraction of the plantation's total population.

Timber-framed churches are documented from two Port Royal Island plantations—Old Fort and Woodward. On Dataw itself, Reverend Moore stated, L. R. Sams (who shared half of the island with his brother B. B. Sams) "erected a very comfortable house of worship" primarily we must assume, for use of his enslaved community, this structure disappearing along with a schoolhouse and slave dwellings built near the owner's tabby residence, now taken by the sea.[9]

Tabby enclosures surrounding family burial plots added an air of permanence and dignity to lowcountry cemeteries, both private and public. Two exist at the Chapel of Ease, the best preserved with headstones dated during the 1820s commemorating several Fripp family members.[10] Another more extensive Fripp burial ground located within sight of the Harbor River off Harbor Breeze Road, St. Helena Island, is the only surviving relic of Mulberry Hill Plantation, which incorporated 530 acres valued at $2,120 in 1862.[11] Enclosed by fifteen-inch-wide tabby walls (cast using forms made up from three horizontal timber boards of equal width to give a total form height of twenty-four inches), this now-abandoned cemetery resembles an elongated octagon in plan made by cutting off all four corners of a basically rectangular configuration. The result is curious. Headstones are aligned not with the long sides of the enclosure but with its much shorter cut-off corners. The earliest visible headstone is of a John Fripp who died aged forty in 1797.

Luxuriant poison ivy, fallen trees, and dense undergrowth precluded full measurement of the site when surveyed in 1999. It was ascertained the enclosure is not aligned with principal compass points, long walls running approximately thirty degrees west of north. Why is unclear.

Tabby surrounding Piedmont cemetery, located off Olde Church Road, St. Helena Island (where the earliest visible marker is dedicated to William Fripp, died 1794), defines an area about forty feet square.[12] Orange Grove Plantation's Perry family burial ground (measuring approximately 30' × 50' overall) is separated from its surroundings by less substantial walls of the same material rising only eighteen inches above present grade.[13] Low tabby-walled enclosures are also seen at the parish church of St. Helena, Beaufort, where a number were installed side by side by prominent planter families—including the Barnwells, Stuarts, Elliotts, Smiths, and Cuthberts—who preferred a degree of separation in death no matter how closely they may have been related by birth or marriage.

Occasionally tabby was used for tombs and grave markers, the same churchyard possessing an unusual monument erected at the cost of nearly £17 (borne by the parish) above the burial of Reverend William E. Graham (d. October 1800), who married into the plantation elite.[14] It takes the shape of a box tomb ornamented with corner and intermediate pilasters, all (except the inscribed marble ledger) cast solid in tabby. Cheaper than a stone or brick affair, tabby must have required complex formwork, executed perhaps by a local joiner. The rather mean effect was probably mitigated by burnished stucco finishes or marbling.

A unique funerary association comes from Daufuskie Island, where investigation in 1985 of the Mary Dunn cemetery bordering Mongin Creek near the island's south shore

Tabby box tomb with marble ledger of Reverend William E. Graham (d. 1800),
north side prior to stabilization, St. Helena Church, Beaufort, S.C.

briefly exposed four grave markers—one inside and three outside the cemetery's present
boundaries—each comprising a large rectangular tabby slab poured horizontally just above
the burial. None bore any inscription, unrelated epitaphs suggesting that the cemetery
was established during the 1790s by the Dunn and Martingale families, who held plantations on the island.

Apparently no longer visible, a burial ground at Great Ropers, in Prince William's
Parish, remains significant for association with the Kelsall-Bellinger estate, comprising
916 acres situated on a "navigable creek, leading into True Blue River near Coosaw River"
about five miles from Port Royal Ferry.[15] William Bellinger Kelsall's will, dated August
1791, stipulated the property was to be sold for the benefit of his brother's children except
for "a certain part known to be the Family burying place." It was Kelsall's explicit "will
and request" that his wife, Mary Elizabeth DeSaussure Kelsall, his brother-in-law Daniel
DeSaussure (whose Beaufort house and stores are described in chapter 6), and the latter's
protégée, John Mark Verdier (along with several other executors) "cause to be built a wall
of bricks or tabby round said burial place."

There was nothing unusual about the provision, plantation cemeteries often staying
in family ownership despite sale or alienation of surrounding lands. Nevertheless the
document cited has unique aspects, Josiah Tattnall, J.P., having drawn it for signature on
the Bahamian island of Great Exuma as Kelsall lay dying of a long and severe illness.[16]

Orange Grove, St. Helena Island, Perry family burial ground
(HABS, Jack Boucher, 2003).

Lukewarm about the revolutionary cause, though never formally charged with treason, William Kelsall had left Beaufort in 1788, migrating with his wife, enslaved property, and portable possessions to the Hermitage, Little Exuma, a plantation probably established by his brother, Roger Kelsall, previously banished from Georgia because of outspoken Loyalist sympathies.

An intermittent resident of Kelvin Grove, St. Simon's Island, Lydia Austin Parrish spent many years exploring historic sites of the Sea Islands and became well acquainted with southeastern tabby. During the 1940s or 1950s she visited Rolletown, Great Exuma, finding there an old cemetery and several tombs. The only inscription then visible commemorated Ann, wife of Alexander McKay, who died age at age twenty-six in 1792.[17] According to Parish the burial and several others were surrounded by a tabby enclosure, but photographs and measurements are missing from her collected notes.[18] The identity of this Alexander McKay is uncertain, Parrish suggesting—with reservation—he was a Loyalist settler with roots in North Carolina and Charleston, South Carolina. If her assumption is correct, weight would be added to arguments considered in following epilogue, that among many southern refugees who fled the American Revolution, William Kelsall was not the only individual who carried into exile knowledge of building methods familiar along the lower Atlantic coastal plain, the McKays (who soon disappeared from Exuma) also remembering old ways of construction when honoring their dead.

Epilogue

A Legacy of the Loyalists?—
Tabby in the Bahamas

A Visit to Grand Bahama

In October 2008 I was asked by Dr. James Miller on behalf of the Bahamas National Museum and Freeport Authority to examine heritage resources at Old Freetown, a small settlement located on Grand Bahama's south coast. This site was abandoned about 1969 following acquisition of the property by the Freeport Authority for developmental purposes. Residents were then relocated several miles away, the new settlement taking the name of the old one. Subsequently most original buildings were demolished, only two single-story houses escaping complete destruction. When first seen by Dr. Miller and myself, these structures were overgrown and hemmed in by dense tropical vegetation. The larger was better preserved though it had lost most of its front wall, roof, and flooring. Nevertheless about three-quarters of the exterior still stood, several wall openings were nearly intact, and ghost impressions attested timber door and window frames. Examination quickly established construction was of what local informants call "tabby," whereas the other structure retained pieces of wattle-and-lime-mortar-daub walling in situ.

Standing within an enclosure of dry stacked rock located about five hundred feet from the seashore, the tabby house (assuming it still exists after devastation caused by Hurricane Dorian in 2019) measures 20' north/south × 23'9" east/west. Principal facades (north and south) each have a central doorway flanked right and left by smallish (2'10" × 2'8") windows. Another doorway cuts through the building's east side, the west side having two more window openings matching those of north and south facades. No evidence remains for interior partitions, exterior door distribution and the fenestration pattern suggesting some kind of central-hall arrangement with perhaps two rooms east and one larger room west. Foundations further indicate a timber-framed porch looked northward out to sea across an unmade path linking hamlets scattered along the sparsely populated coastline.

Exterior walls (eleven inches thick) rest on wider (fourteen-to-fifteen-inch) stone foundation plinths, this producing an interior ledge that received timber floor joists (now lost). Wall construction above consists of form-cast materials containing quantities of broken limestone rubble bound by copious amounts of lime mortar, finished with lime

231

Old Freetown, Grand Bahama, tabby house, detail of northwest corner showing impression of vertical timber post.

mortar inside and out. Exterior corners preserve impressions of 4" × 4" vertical timber posts, similar post impressions occurring around door and window openings. Local informants say these posts were set into the stone plinth to fix formwork in place, the forms being filled with the rubble-mortar mix round by round. Rafters have gone, old photographs and oral testimony indicating hipped roof frames were covered with timber shingles or, alternatively, palm-leaf thatch—both common Bahamian treatments.

During the course of investigation I met two former residents of Old Freetown, Mr. and Mrs. Jenius Cooper, both then octogenarians. Mr. Cooper, who as a young man built a very similar house at Old Freetown, told me most settlement houses were fabricated of "tabby," and that in his day form boards nailed to vertical timber corner posts were made of plywood. Lime was produced by piling limestone rock onto logs and burning everything together, conch shells when similarly burned producing finer lime used for floors. Windows of houses were not glazed but closed with solid timber shutters. Mr. Cooper preferred timber shingles when roofing his own house—using palm thatch only for temporary covering.

Old Freetown's custom dictated that before young men could marry, a new house either of tabby or of wattle and daub must be completed, sons building their own houses near their parents' dwelling. Mrs. J. Cooper (née Baillou) said her parents' house was built of wattle and daub, having a central hall with one room right and left. Partitions dividing off the hall were less high than adjacent spaces, thereby improving ventilation.[1] Long gone the Baillou House resembled the abandoned settlement's second surviving structure. Only foundations and scraps of walling survive. These indicate it measured 24'5" north/south (front to back) × 24'7" east/west. A relatively wide porch extended along the south side, reached by stone steps more or less centered about the plan. Corresponding steps once existed along the north side, confirming that, like that of the lost Baillou House, its plan was organized about a middle hall. Foundation plinths of limestone rubble about two feet wide supported walls of interwoven wood laths fixed to five-inch-square timber uprights let into the inside face of the plinth then daubed inside and out with lime mortar.

This technique closely follows construction employed for slave houses built around 1815 near the William Walker Great House, Crab Cay, Exuma. Although substantially decayed these structures were certainly elevated on rubble plinths and enclosed by wattle-and-daub walls, the daub being a limestone-derived mortar, pieces now covering the ground beneath old lignum vitae trees and bromeliads. Most likely occupied by domestic slaves and kitchen workers, dwellings were smaller than Freetown's remaining examples, one house measuring 14' north/south × 13'7½" east/west, another 13'8" east/west × about 17'6" overall.

Easily damaged by storms, wattle-and-daub construction was especially vulnerable when standing near the sea. Following a devastating summer hurricane (August 1899), the exasperated acting governor reported from Nassau that "[on] Andros Island alone 311 huts of laboring population have been leveled to the ground. . . . These are nothing but wattle and daub huts with palmetto leaves and might be re-erected or replaced by similar erections with ease in a very short time if the attitude of the population was such as to second the efforts made by the Government on their behalf."[2]

Cast tabby was the preferred building medium for Old Freetown's residents. Despite its relative impermanence, wattle and daub carried no social stigma, some men anxious to marry finding the material easier, cheaper, and, what is most important, faster to build. Neither of the Coopers could be sure when Freetown's ruined houses were erected; they certainly did not remember anyone occupying them, which probably means construction took place either during the early twentieth century or in the 1890s. Royal Air Force photographs show that both dwellings were deserted by 1942. Nothing definite could be learned about the origins of Old Freetown either—whether it was a freedman village built like many Bahamian hamlets by former slaves soon after emancipation or was established by government for liberated Africans—that is individuals seized by the Royal Navy from slave ships just before or soon after 1830. Lifetime residents of Freetown's old settlement, the Coopers had only vague traditions concerning their forebears, possibly enslaved but not necessarily so, coming by boat in one group, perhaps from Abaco, perhaps not. As for tabby, although Grand Bahama retains immense stands of Caribbean pine and pockets of tropical hardwoods including mahogany, Mr. Cooper (a sawmill worker at one time) considered formed material to be cheaper and more convenient than any other. After all, unlimited quantities of rock were available almost everywhere for the taking, and great piles of discarded conch shell littered the foreshore near all coastal settlements when most island residents were desperately short of income. This opinion echoes comments made in 1950 by the commissioner of Eleuthera, who, with condescension not untypical of colonial administrators, wrote: "the coloured and labouring class on Harbour Island [Eleuthera] build their houses of 'Tabby'—rocks of uneven sizes and shapes poured into forms and then plastered—this is much cheaper than block construction."[3]

Old Freetown's tabby is not the same as tabby found along coasts of South Carolina and Georgia. Leaving aside recent use of plywood, it is not fundamentally different either if one remembers that neither General Alexander Macomb nor Thomas Spalding considered shell, oyster or otherwise, an essential ingredient of tabby mixes: "stone broken up by the sledge hammer, if more easily procured, would answer equally well," Spalding said.[4] Although conversation with Mr. Cooper was limited by my inability to fully comprehend his dialect and his trouble with an English accent, we shared the same language when talking about tabby, knowledge I gained through conservation work complementing his learned through direct experience handed down through generations of family members living across the northern Bahamas. Moreover Old Freetown's ruins aid interpretation of structures exhibiting similar construction techniques distributed from one end of the Bahamian archipelago to the other. Many are relatively modern, cottages built during the 1920s far south on Acklins Island having floor plans and walling methods indistinguishable from those already described. Others are older, Albert Town, Long Cay (Fortune Island)—once an important stopping-off place for ships carrying Jamaica mails via the Crooked Island Passage—preserving numerous form-cast houses (now roofless) built around 1860–1900 that are essentially the same.

Near the Forrest, Great Exuma, Richmond Hill's small dwellings—some doubtless slave houses—made of what passes for tabby hereabouts share almost identical construction details. Abandoned during the 1970s or 1980s, when—afraid of contagion and wearied of recurrent cyclones, poverty, and negligible economic opportunity—its inhabitants fled leaving most of their belongings behind, this settlement grew up around a former slave village. Similar patterns of growth and desertion were repeated over and over again across the so-called Out-Islands (now Family Islands). Almost all plantation-era slave dwellings stayed occupied after emancipation, with newer houses built much the same way as earlier ones.

At Millar's plantation located near the southeast end of Eleuthera, tabby construction is exemplified by small structures grouped at the foot of a hill on which the probable great house (a cut-stone building of exceptionally large proportions) stands. The earliest of these subsidiary buildings—now ruined—is located near the present plantation entrance and might easily be mistaken for Thomas Spalding's work. It is a single-story dwelling, measuring 14'4" × 16'8" overall with an entrance back and front plus two side windows. Thin (eight-inch-thick) exterior walls retain clear evidence for their construction, exhibiting pour lines left by formwork measuring approximately twelve inches in

Millar's plantation,
Eleuthera, the Bahamas,
detail of tabby wall of slave
or tenant dwelling.

height. Rich in lime the mortar employed bound small, roughly shaped stone blocks and pieces of unworked rock. Originally exterior surfaces of Millar's small dwellings and probably the great house itself were stuccoed and lime washed, British naval surveyors reporting in 1836 that "the white houses" of Mr. Millar's plantation "can be seen a great distance from the eastward."[5] I suspect the tabby building described was a slave house erected soon before emancipation (1834) or, possibly, a tenant house erected shortly thereafter. Either way the neighboring settlement called Bannerman Town, established on a grid plan in the late 1840s or early 1850s for emancipated slaves, confirms that no major break occurred in vernacular building traditions as former plantation life dissolved, residents—many of whom had close associations with Millar's and its former masters—continuing to utilize tabby for domestic building purposes probably down until the end of the nineteenth century even though spatial standards improved and occupation patterns changed.

Typically each "new" house in the settlement stood in its own stone-walled compound along with an outdoor oven, fire pit, and storage structure. Today structural decay is far advanced, the houses smothered by regenerating tropical forest, their inhabitants,

Bannerman Town, Eleuthera, the Bahamas, detail of ruined tabby dwelling showing lignum vitae posts framing window opening.

driven by famine or lack of economic opportunity, having left for Nassau; the Florida Keys; and Liberty City, Miami, fifty or more years ago. Yet what is left of the built environment remains informative, spalled stucco wall surfaces revealing construction techniques not easily seen in better preserved buildings. Of particular note is a system that strengthened tabby walls and ensured that top plates supporting the roof frame were securely anchored.

Prior to commencement of casting, this involved placing into position an approximately eight-foot-high lignum vitae (*Guaiacum sanctum*) post (measuring 4" × 5" in section) right and left of every planned wall opening. Once all posts were set in place, the foundation plinth was cast around them and allowed to set. Subsequently casting work on the thinner upper wall proceeded until it reached a height slightly less high than the timber posts. Timber top plates were then set in place all around the building and pegged to each vertical post. Casting resumed, with the final, uppermost pour (around nine inches high) tying everything together. Work then began on hoisting rafters and all other elements of the roof frame into position. Finally exterior stucco concealed lower and uppermost portions of the posts from view, these placed about an inch or two in from the finished building face. Whether corner posts were set along with those framing window and door openings is not confirmed. But the technique described must have once been common across most islands of the southern Bahamas judging by the number of telltale post impressions seen in abandoned buildings.

Found among pre-emancipation-period settlements as well as more recent ones, questions arise concerning when and by whom local "tabby" was introduced into the Bahamas—questions not easily answered because architectural and archaeological investigation of the nation's historic sites has scarcely begun. There are also semantic difficulties, Bahamians currently using the term "tabby" when describing two different, if related, building modes. The first, most primitive kind involves unshaped or minimally worked stone simply stacked and bound in mortar without benefit of shuttering. The second mode, involving lime mortar and stone aggregates incrementally cast into frames or molds, fits more comfortably into southeastern building traditions.[6] Without invasive investigation it is hard to distinguish between stacked and formed work especially as unshaped limestone is an almost-universal building material. Writing from New Providence in 1740 soon before he left for Charleston, South Carolina, the military engineer Peter Henry Bruce noted: "all the stone on this and adjacent islands is of so soft a nature, when raised from the quarries, that we could cut and shape them into any form with very little labor."[7] Bruce had probably not observed during his brief sojourn that local stone is soon weathered and eroded by tropical rains and winds blowing in from the ocean. Consequently most ostensibly stone structures were either lime washed or stuccoed, these finishes protecting underlying fabrics and concealing any qualitative deficiencies, Daniel McKinnen (1802) observing how the practice added "consistency and beauty to the exterior."[8]

Kelsall's Hermitage Plantation, Little Exuma, is an example. Standing high on cliffs overlooking the sea, the old great house is abandoned now and falling into ruin. This single-story dwelling (measuring 20'3" × 34'11" overall), with exterior masonry walls and steeply hipped roof, probably retains its late eighteenth- or early nineteenth-century shape. Concrete porch construction along the land side, reconfigured interior spaces, modern cisterns, and a roofed passageway linking the main house to its freestanding kitchen all indicate substantial restoration. The kitchen—narrowly rectangular with the usual end chimney—has undergone fewer structural changes. Where exterior finishes have fallen away it can be seen that underlying walls are built of rubble and roughly trimmed stone laid up with little attempt at coursing using large quantities of lime mortar. Considering how much mortar was employed and the massiveness of construction we might assume timber "molds" aided fabrication although it is impossible to be sure this was the method followed without extensive removal of surface coatings, which help bind loose underlying fabrics together, besides adding pleasing consistency to the building group.

Uncertainty also surrounds construction dates. The kitchen and portions of the original main house were probably erected before 1821 when Mary Elizabeth DeSaussure Kelsall, widow of William Bellinger Kelsall (formerly of Beaufort, S.C.) decamped with

The Hermitage, Little Exuma, the Bahamas, great house, ocean-side facade.

the last of the family's slaves for Cuba, where she died in 1823.[9] Construction by her husband, his brother Roger Kelsall or, alternatively, the latter's son John, who died young in 1805, are all possible scenarios, but without better documentation any specific familial attribution is speculative. Tabby or not the chosen wall material fits the environment. Having weathered hurricanes, storms, and hard times severe enough to depopulate most southern islands, Hermitage Great House is the only historic plantation residence standing in the Exumas.

The difficulty of assigning dates and patrons is further demonstrated by Adderley's plantation, located one hundred miles southeast of Millar's, near the desolate north end of Columbus's Fernandina, aptly renamed Long Island before the end of the eighteenth century. Established about 1788 by two Nassau merchants and shipowners with Bermudan family connections—Abraham Adderley (died ca. 1811) and his son-in-law Abraham Eve—Adderley's is one of the few if not the only local agricultural holding that remained operational from its inception down until the early twentieth century. A long occupational history has left an interesting and diverse suite of building modes.[10] Principal structures are ordered linear fashion along an elevated, rather narrow and abrupt limestone ridge running roughly east/west. Retaining walls installed against the steep south side effectively isolate the central settlement area from rocky outcrops and old dunes sloping down toward the sea. Raised on stone plinths, the ruined great house is a single story measuring 22'6" × 36'4" overall, with an exterior skin (1'5" thick) built and partly rebuilt using mixes of lime mortar and broken rock cast into timber forms fixed to vertical posts

The Hermitage, Little Exuma, the Bahamas, kitchen, view from land side.

during the construction phase and stuccoed once dry. Inside ghost impressions indicate a seemingly conventional middle-hallway plan with two rooms positioned right and left. Far less conventional (but not unusual for elite Bahamian planter houses) are multiple points of access, seven doors opening directly to the exterior—an arrangement necessitating steep entrance steps south and west. Roofing has gone except for timber nibs cast into uppermost wall levels, which anchored top plates supporting rafters of what was almost certainly the hipped roof frame.

The main house is flanked east by a freestanding kitchen built of dressed stone. To the west, partly cut into the hillside, ruins remain of what, I imagine, is an exceptionally large stone-built cotton house incorporating two full stories, the lower one designed for storage. Slave dwellings (measuring 12'9" × 16'2" and 15" × 14'2" overall) arranged about twenty-three feet apart located away from the main house are of unshaped rock, dry stacked, and finished inside with lime mortar troweled smooth. Several small work or service buildings grouped near the supposed cotton house display telltale details of Bahamian tabby construction—impressions of vertical posts around wall openings and where exterior stucco has spalled, impressions left by timber forms measuring about fourteen inches in height.

Construction sequences are not securely established. Displaying stonework of the highest quality, the likely cotton house (which could have doubled as the main plantation residence when first built) was probably erected around 1790, cotton production on Long Island then reaching its zenith. I believe formed walls of the present great house are from the 1830s—a time when Abraham Adderley's son Nehemiah was buying up lands abandoned by neighboring planters forced out by negligible returns and labor markets thrown into chaos by emancipation. Good-quality masonry suggests the kitchen was built when craft skills of high order were still available—perhaps around 1800. Rudimentary slave houses must have been erected before 1830. Small tabby outbuildings probably belong to the same period or an undocumented post-emancipation expansion phase.

Identical with what Governor Rawson called "one of the most flourishing estates at the north end" of Long Island, Adderley's was devastated by hurricane-force winds in late September 1866, "when all the building[s] of the proprietor were thrown down, except the kitchen in which 47 persons had to take refuge."[11] The estate was patched up soon thereafter, another hurricane completed the settlement's ruin in 1926.

◦ Loyalists in the Bahamas ◦

Chronological and topological issues aside, the broad distribution of formed materials containing quantities of lime coupled with their frequent appearance among plantation settlements indicates association—if not direct linkage—with expansion initiated during the Loyalist period, when planters—mostly from South Carolina, Georgia, and Florida—banished or alienated by the United States moved from island to island seeking better, more productive, less exhausted soils, taking their slaves or sending out overseers to supervise forced labor among distant, desperately lonely places near the edge of the Atlantic World where they themselves preferred not to go. Fueled by Liverpool's entrepreneurial talent and near-insatiable demand for long-staple cotton fiber, movement into previously uninhabited or very thinly populated areas was highly intensive and short lived. Satellite imagery of Long Island and Crooked Island, two of the most productive cotton producers before 1800, reveal an intricate network of stone walls covering the landscape, walls that repeat line for line property grants made to Loyalist settlers soon after 1785. Exploitation was predicated on cadres of slaves shipped out of southeastern coastal areas, some of whom, like their masters, must have brought knowledge of tabby building methods with them. Major George Millar, who settled on Long Island, was one of these masters, a stray reference recording not only that he came from Charleston but that he brought at least one enslaved woman captured in Africa with him named Dianna. She was later taken by Millar's son Robert to Eleuthera, where she died in 1858.[12]

Clifton, an estate near the western end of New Providence, demonstrates how complete the architectural transfer could be, William Wylly, late of Savannah, then attorney general of the Bahamas, adopting planning principles familiar among contemporary plantations of the southeastern coastal plain soon after 1809.[13] His own house (stone built by John Wood, another Loyalist émigré from Chatham County, Georgia, who abandoned cotton cultivation on New Providence for Long Island in 1789) was enclosed within a rectangular yard and ranges of outbuildings including office, kitchen, stables, carriage house, and accommodation for servants. To the north the enclosure opened toward the ocean; west, into a track accessing provision grounds lined both sides by rows of single slave dwellings and their associated gardens, cooking areas, and animal pens.

Clifton Plantation, New Providence, the Bahamas,
interior of ruined slave dwelling.

Wylly was conscious, if not wholly supportive, of abolitionist sentiment, publishing his own paternalistic (though by no means selfless) *Regulations for the Governance of Slaves at Clifton and Tusculum in New Providence,* printed by Royal Gazette (Nassau) in 1815. This laid down rules of conduct for workers and supervisors besides making provision for their spiritual and physical welfare. Article 13 of the *Regulations* conveys the document's general tone, reading: "people in the sick list are not [to] remain at their houses, except by express permission; but on being taken sick, are forthwith to repair to the Hospital, where proper attention will be paid to them; while in the Hospital they are to be properly supplied with tea, gruel, broth, and other necessaries; but are to receive no regular allowances of provisions."[14]

Wylly assigned "a well-built stone house, consisting of two apartments," to every enslaved man on taking a wife. These dwellings were not really stone buildings but rather of hybrid construction, slave structures (measuring 15' × 20' or 27'6" × 17') still extant having external walls of cut stone, limestone rubble, and tailings all set in thick lime mortar. Larger stones reinforced corners and openings, smaller stones or rocks filling areas in between. Wall faces were stuccoed inside and out, the exterior being color washed.[15] Considering the abundance of lime mortar employed and the slow curing time of this material, it is probable, though not absolutely certain, that once building corners had been defined, timber formwork was erected, into which the mixture of lime and limestone pieces was placed, local builders currently calling the walling material used at Clifton "tabby."[16] Overall Wylly's layout embodied the values of an owner whose identity with the Anglo-American slave-owning class of the southeastern coastal plain and familiarity with Sea Island planning traditions remained steadfast even though he was forced from his native country by war and revolution.[17] Ultimately Wylly was driven out of office and sent into exile again by virulently anti-abolitionist members of the Bahamian Assembly suspicious of his real motives. Rather than repatriating he sailed off to St. Vincent, taking his slaves with him. Clifton Great House burned in the mid-nineteenth century. Road works subsequently (1907) destroyed one line of slave houses. The other row remained occupied down until the 1940s. Fragile and near collapse, three were fully restored in 2013, these and ruins of the main house now presenting tangible evidence for the way Loyalist settlers briefly re-created the privileged, semifeudal, planter-dominated world they had lost during the American Revolutionary War—irretrievably as it turned out.

The uncomfortable fact that many Loyalists maintained business and extended kinship associations with their former homeland, exchanging slaves, goods, and agricultural and technical knowhow, is exemplified by Roger Kelsall, late of Beaufort and Savannah, who despised the new American republic yet stayed in touch with his former partner James Spalding of St. Simon's Island, Georgia. Kelsall sent "black seed" from Exuma to James Spalding Sr. in 1785 or 1786. Years later (in 1828) the latter's son Thomas wrote: "all the varieties of the long staple, or at least the germs of those varieties came from that [i.e., Kelsall's] seed."[18] Thomas Spalding also stated that technical information forwarded by unnamed Bahamian correspondents helped Alexander Bisset build serviceable foot gins at St. Simons in 1788. Powered by animals, water, or wind, Joseph Eve's more efficient gin was developed on Cat Island (southern Bahamas), a New Providence planter reporting in 1791 that one of "Mr. Eve's machines "which had been in use for three years" ginned 360 pounds of cotton per day. In May 1796 Spalding himself was soliciting orders on Eve's behalf through advertisements placed with the *Georgia Gazette,* supplying machines imported directly from Nassau at the extravagant cost of fifty guineas each.

Ironically, while cotton made the Sea Islands rich, it brought financial ruin to the Bahamas. Soil exhaustion, poor agricultural management, lack of experience, misunderstanding of unfamiliar environments, the "chenille worm" (which defoliated cotton plants), competition from the United States, and precipitous declines in yields caused desertion of once-productive holdings decades before emancipation. Dated New Providence, August 1780, a letter published by the *Bahama Gazette* records, "success of planters in these islands, has been as various as their systems; and, in several instances, the absence of knowledge in tropical agriculture has rendered almost entirely abortive every exertion of industry:—some lands planted in cotton have not yielded 100 lb clean cotton to the acre, whilst others have yielded 250 lb and 300 lb."[19] Traveling in 1803 Daniel McKinnen heard troubling reports from Long Island: "three years ago eight or ten plantations were entirely quitted, and thirteen others partially given up."[20] Interviewing John Kelsall he learned the latter had lately removed his slaves from Little Exuma to Rum Cay, which was otherwise deserted, Kelsall slaves numbering around 160 now raking salt and raising cattle rather than growing cotton.

Repeated reversals of similar kind created a secondary diaspora, slaves and their owners relocating, sometimes together, more often separately. Answering government questionnaires in 1800, John Kelsall wrote: "the vicinity of the American states and the success that has attended the cultivation of cotton there, will doubtless operate with many, and from a recent instance of permission given by the legislature of South Carolina to two individuals of this colony, to remove their Negroes into that state, there is no reason to doubt that the states will adopt the obvious policy of receiving all who may be inclined to go hither."[21] Kelsall did not say that his brother-in-law, John Moultrie, son of the last royal lieutenant governor of East Florida, was one of those persons granted permission to repatriate enslaved people.[22] The younger Moultrie's principal Crooked Island holding had "a considerable number of Negroes" quartered about the place, these probably coming from Nightingale—a rice plantation on the lower Pee Dee River in South Carolina—through inheritance from his maternal grandfather, George Austin, who had amassed considerable lowcountry property.

Knowing the Hermitage on Little Exuma must be sold to satisfy debts, William Kelsall's widow, the redoubtable Mary Elizabeth, petitioned to bring about sixty enslaved people into South Carolina from the Bahamas in 1805, the inference being that these individuals had been removed from the state. Perhaps she learned from her DeSaussure family connections that experienced workers would command premiums if sold by Charleston or Beaufort slave dealers.[23] Having passed into the ownership of his creditors, Lieutenant Governor Moultrie's own slaves were apparently parceled out among sugar growers of Demerara—a South American colony (now Guyana) notorious for the brutality of its plantation regimes—newly acquired by Britain from the Dutch. The ebb and flow of enslaved labor continued down until emancipation came fully into effect throughout Britain's West Indian colonies during the 1830s, ensuring that agricultural and attendant skills were transmitted from place to place.

◦ After Emancipation ◦

Final dissolution and collapse of plantation-based agriculture brought new freedoms and widespread poverty throughout the Bahamas. Across the Out Islands, individuals white and black fell back on old standbys—subsistence agriculture, fishing, sponging, salt raking, cattle ranching, wrecking, and smuggling. A solitary farmstead established on Little Stirrup Cay (Berry Islands) during the 1830s to supply passing ships with meat and greens; isolated hamlets overlooking rugged shores of South Abaco founded by emancipated slaves or liberated Africans drawn here by pineapple and sisal cultivation during the 1870s; along with numerous other examples such as Bannerman Town, described above, show formed materials still remained useful for individuals living near the margins of colonial society—indispensable perhaps among small struggling communities of farmers and fishermen, fending off starvation (not always successfully) as best they could. Several dozen roofless dwellings constituting an abandoned village called the Cove near Crooked Island's southeastern extremity are good examples. Spilling down an exposed headland above the Going Through, a narrow channel separating Crooked and Acklins Islands, which Columbus on his first voyage in October 1492 found "so shallow that I could not enter nor sail toward it," individual houses are rather squat and very sturdy.

The Cove, Crooked Island, the Bahamas,
exterior corner of typical tabby house.

One measures 25' × 16'6" overall and is divided into two rooms of unequal size, with the larger entered by doorways front and back, flanked right and left by small windows. Battered exterior walls are six feet high and eighteen inches wide at ground level reducing to twelve inches at top, corners rounded, and all surfaces coated with layers of hard stucco. Impressions of timber uprights framing openings show about half the site's dwellings were cast, their walls made from limestone rubble, small rocks, and lime mortar poured into timber forms. The rest are made of randomly unworked pieces of stone, stacked and bound with similar mortar.

When the Cove was settled is undetermined, cadastral maps giving the site's ownership history missing relevant pieces. Neighboring properties were all granted to southern Loyalists—the majority from Georgia and South Carolina—over the 1790s, suggesting the Cove was also Loyalist owned near the beginning of the nineteenth century. Clustered closely together dwellings themselves suggest slave occupation, but the settlement may well include buildings built both before and after emancipation, habitation probably extending into the early twentieth century, when hardship and Nassau's promise of regular work caused many families to leave the southern islands. Those who stayed gradually abandoned settlements grown up around former great houses built on high ground for the sake of security, populations (if they stayed) moving to new villages in lower, more accessible places as public works—paved roads, schools, and clinics—improved. Recently most older houses up in Crooked Island's Blue Hills have burned or been cleared away. On Acklins similar tabby buildings are now empty shells seen silhouetted against the sky from its principal highway or standing derelict near the beach at Pompey Bay.

⚬ Tabby and Bahamian Public Works ⚬

Few of these essentially vernacular buildings are either recorded or documented. Conversely specifications drawn up between 1897 and 1913 preserved by the Bahamian Department of Works show that analogous construction was not confined to folk housing near the beginning of the twentieth century. Framed in somewhat stilted Victoria language, these include precise instructions for building a variety of public facilities including a new residency at Fresh Creek, Andros; a police office on Ragged Island; the new residency at Watling's Island; a lighthouse and keeper's dwelling on Duck Cay, Cherokee Sound (Abaco); and lighthouse, kitchen, and tank, South East Point, Eleuthera. In each case it was specified that exterior walls were to be of the "best tabby work or rubble rock," minor variants of the same formula being repeated for scheme after scheme.

Located on a high promontory, Eleuthera's South East Point Lighthouse is among the most attractive and unaltered of these projects. Extending over thirteen pages, contract documents include detailed instructions regarding construction of masonry walls and carpentry required for installation of floor, roof, and steps, scale drawings illustrating the main building's plan, section, and roof frame.[24] The keeper's kitchen is also described, a small sketch giving its plan and location west of the lighthouse. What is more important, a penciled note on the same sheet, dated May 27, 1901, establishes that both lighthouse and kitchen were then "in course of erection." An attached notes states the contract sum for the work was £170 and names Joseph H. Cox as contractor. Besides specifying that "the building shall be erected on a spot pointed out to the Contractor," relevant documents say nothing about the site or its surroundings, a deficiency remedied by the *West Indies Pilot* for 1913, which informed mariners: "from Palmetto Point the eastern shore trends southwards 31 miles, to Eleuthera Point and is slightly indented. It is foul about the whole way, and closely skirted by a reef with deep soundings about one half to one mile outside it."[25] But the *Pilot* makes no mention of the four small islands extending offshore southeast of the Eleuthera Point proper nor the dramatic headland on which the present lighthouse stands.

Drawings depict a square, three-stage structure very much as seen today, standing about twenty feet high at maximum. The lowest level (measuring 16' × 16' overall in plan) was used as a cellar. Exterior stone steps gave entry via a single door to the main floor, above which accommodated the keeper, wooden stairs giving him access to the uppermost level—an attic-like space closed off by means of a flap door in its floor. Of pyramidal form the roof enclosing this upper room was supported on knee walls about two feet high, a single dormer window (now lost) providing light and air. Specifications describe a curiously haphazard arrangement, calling for a stool on which the keeper might stand "when putting up the light" mounted overhead at the apex of the roof in a sixteen-inch hardwood frame. The apparatus—now removed—is not described. We are told: "the lantern and fittings will be supplied to [the] Contractor in Nassau—he is to carry it to the spot and securely fix in position, bedded in white lead and the Joist or seat on roof made watertight & painted, all to be neatly painted and level, so that light may show properly & that lamp may be put up from inside roof."

Wall construction followed techniques utilized for public buildings designed and executed by government agencies throughout the colony reliant more on local building materials and indigenous building skills than imported ones. However, specifications for the various works cited reveal the Department of Public Works had discovered small quantities of Portland cement (probably shipped from England) added to traditional lime-based mortars improved "workability" and significantly reduced time required to set, thus speeding the construction process. At South East Point it was specified:

> The walls shall be of the best rubble or tabby work with sufficient long stones placed as headers to tie stone a together and on completion it shall be plastered both out-side and inside and receive three coats of whitewash.
>
> The mortar shall be mixed in the proportion of 3 parts lime to 2 parts sand and when just about to be used, the mortar shall be mixed with Portland Cement in the proportion of 5 measures of mortar to one measure of cement.
>
> The cement for this work will be supplied to the Contractor in Nassau, free and he is to convey it to the spot. The plastering will be of cement mortar.[26]

As common in multistory masonry construction, walls were designed to diminish in width with height (thereby reducing materials and foundation loads) from three feet at the lowest level to two feet at the main level and eighteen inches at the uppermost room "in the roof." Wall height also varied, beams and joists reducing clear headroom to a mere four feet in the cellar and seven feet in the main room above. Walls of each story were centered on walls of the story below, this producing an agreeable stepped profile on the exterior, "spalls" creating chamfered plinths, which, remarkably, still remain sharp and in good condition. Exposed surfaces in the cellar (where original plaster has decayed) show the contractor followed stipulations, the walls exhibiting a high proportion of relatively large, roughly squared limestone blocks set with smaller pieces of rock in quantities of lime mortar. The work is well fashioned—larger stones are laid horizontally and corners precisely made. This strongly suggests the stone-mortar mix was placed in removable timber forms raised incrementally as the building rose higher.

Names of resident keepers have not as yet been discovered. If accommodated in the lighthouse itself (as provision of a kitchen and 8' × 6' × 6' deep water tank implies) their quarters—notwithstanding the site's great natural beauty—were minimal at best, the lighthouse's main space encompassing just one hundred square feet encumbered on one side by open timber stairs accessing the upper story. With its "proper hearth and chim-ney," the adjacent kitchen was no more generous, measuring 10' × 7' overall with fifteen-inch-thick rubble walls pierced by a single small, unglazed window and battened door at the eastern end, the floor being of "beaten earth neatly levelled off 1ft, above ground level."

Today (though not for much longer if current development plans are realized), the lighthouse is unused, and visited by few locals or tourists, near impassable unpaved roads

Eleuthera, the Bahamas, South East Point
Lighthouse (1911), exterior view.

from Bannerman Town discouraging all but the most determined. When it was built, a lively coastal trade, government dock, fishermen, and occasional appearance of "wrackers" made it less lonely. All this activity is long gone, the lighthouse and other remote buildings like it demonstrating that tabby and materials called tabby were utilized by central government as the means of stretching scarce resources in places where communications were difficult, populations tiny, money very tight, and finished building products brought by sea from Britain or the United States.[27] More than four hundred years after

Columbus carried his knowledge of tapia construction to Hispaniola, Bahamian tabby remained a material of convenience. At this edge of the Atlantic World, old building methods were- like Sea Island tabby- influenced by multiple cultural traditions.

The roles southeastern planters, enslaved workers, and their descendants played are largely unexplored. It is certain that the process of cultural transfer came almost full circle more than once—around 1893, when George Adderl[e]y, a Bahamian of African descent who, it is said, burned shell for lime, built a tabby house at Crane Point, Key Vaca, Florida; or earlier, when disappointed white Loyalists—the Armstrongs and Wyllys of Abaco, Long Island (Bahamas), and New Providence—came under the pervasive influence of Thomas Spalding's architectural ideas after repatriating to St. Simon's Island about 1800. And we can be sure that across the Bahamas and the Antilles and on the Sea Islands themselves, more is to be found of the building heritage described in this book.

Today the greatest challenge is to record what remains before it entirely disappears, to stay ahead of destruction caused by wind, tide, and intrusive vegetation on those fortunate islands retaining their near-pristine state or less fortunate ones where natural environments and historic landscapes are threatened by population growth, questionable zoning practices, overdevelopment, and climate change.

Extensive and repeated flooding across the Sea islands is a new reality as are storms of unprecedented magnitude slamming into the Bahamas year after year. The full significance of these events has yet to be assessed. Nevertheless, it is already clear that loss of historic buildings will likely proceed at an accelerated pace in Atlantic coastal zones until such time as meaningful environmental protection is enacted at both national and global levels. Three or more decades ago when I first learned of tabby's long antecedents from Marcus Binney, then architectural editor of the British periodical Country Life, such destruction was unimaginable. Now warning signs are all too obvious, manifested, to give just two examples, by irreversible salt intrusion into historic tabby structures located along Beaufort's waterfront and total destruction by hurricane Matthew of late eighteenth century slave house at Clifton Plantation, New Providence in 2016. More disasters will doubtless occur, some preventable others not as rapidly changing weather patterns, vibration from ever increasing traffic or depleted aquifers cause fragile historic building fabrics to fall apart. My hope is that dedicated field archaeologists now investigating historic resources of the Sea Islands will inspire new generations to ensure that knowledge of past building technologies of the kind examined by this book is not lost, but expanded and preserved.

APPENDIX

Notes on Selected Sources

For anyone interested in southeastern tabby building, a small volume edited by E. Merton Coulter entitled *Georgia's Disputed Ruins* remains an indispensable work. Published in 1937 at time when tabby structures along the Georgia coast were reduced to "dismantled crumbling walls compassionately cloaked with a thousand profuse and graceful creepers," the book brought together three pioneering if disparate accounts.[1] The first, "Certain Tabby Ruins on the Georgia Coast" by Marmeduke Floyd, argued that ruined tabby buildings erected near Darien, Georgia, represented nineteenth-century sugar mills rather than sixteenth- or seventeenth-century Spanish missions as then commonly thought.[2] James A. Ford reached the same conclusion in the volume's second part, describing excavation in 1934 of a ruined tabby sugar mill at Elizafield in Glyn County, Georgia. Finally, Thomas Spalding's "Observations on the Method of Planting and Cultivating the Sugar-Cane in Georgia and South Carolina," first published by the Agricultural Society of South Carolina in 1816, was reprinted along with redrawn versions of his original plates.

Time has shown these are key documents for investigators, including myself. Floyd's syntheses of Spalding's writings—published and manuscript—are of particular value since the originals are difficult to find. More historic references to tabby have surfaced in recent years and are quoted at some (perhaps too great) length here, but thanks to the thoroughness of his research these expand rather than replace Floyd's commentary regarding the fabrication, constituents, and properties of tabby.[3] However, my geographical focus differs from his and that of more recent scholars—notably William M. Kelso and Lauren B. Sickels-Taves, scholars whose respective publications concentrate on structures from coastal Georgia. Here Beaufort County, South Carolina, is the starting place, its exceptional corpus of tabby structures demonstrating how lowcountry builders capitalized on indigenous resources and shaped distinctive construction systems. It is unfortunate that the forts—among the earliest and largest of these structures—are now reduced to ruin or have disappeared. For information about them, I have drawn on South Carolina's Colonial Records and materials deposited by the U.S. Engineer Department at the United States National Archives.

Dr. John Archibald Johnson (1819–92) was among the few former planters who returned to Beaufort after the Civil War. Along with fellow members of what had become a discredited class, he found himself excluded from civic affairs, despoiled of his former property, and facing an uncertain future. Only the antebellum past remained an undisputed possession, a past that by the 1870s was gaining legendary proportions in the face of Beaufort's shabby appearance, commercial stagnation, and rampant political corruption. Judging the time ripe for reassessment, Johnson prepared sixteen articles for the *Beaufort Republican*. Starting on January 16, 1873, the weekly series presented views of places, buildings (including forts), and events drawn from his own memory. While these cannot compensate for archival losses, they do preserve aspects of local history unavailable elsewhere.

For background information concerning local settlement, political, and social developments before the Civil War, I quarried the *History of Beaufort County* (vol. 1, 1996) by Lawrence S. Rowland, Alexander Moore, and George C. Rogers Jr. Periodicals including the *Southern Agriculturalist* and *American Farmer* were mined for information about agricultural matters. If something of a misnomer, Suzanne C. Linder's *Historical Atlas of the Rice Plantations of the ACE Basin* is packed with biographical information concerning planters of Beaufort County and neighboring Colleton County. *The Story of Sea Island Cotton* (2005) by Richard Dwight Porcher and Sarah Fick expands the literature about cotton itself and includes an important survey of Sea Island plantations. For lowcountry building generally, principal works used include volumes from the *Architecture of the Old South* series by Mills Lane; and *Beaufort County Above-Ground Historic Resource Survey* (1998) prepared by Bruce G. Harvey, Colin Brooker, David B. Schneider, and Sarah Fick. Stoney's classic *Plantations of the Carolina Low Country* (1938, 1964) makes brief reference to Beaufort County. Nevertheless its text is pertinent and perceptive, treating plantation architecture as architecture rather than the stepchild of genealogy—as often the case with later imitators. For Georgia's early tabby building, I relied on William Kelso's excellent summary in *Captain Jones's Wormslow* (1979). Published collections of HABS materials include Linley's *Georgia Catalogue* and Vlach's *Back of the Big House* (1993), covering plantation building types from across the Southeast besides offering detailed, sometimes provocative, commentary.

Regarding tabby's Old World antecedents, research fueled by the need to conserve long-neglected or poorly restored tâbiyah/tapia structures in Spain is opening new perspectives, with papers of the International Congress on Construction History (Madrid, 2003; Cambridge, U.K., 2006, etc.) and *Arqueologia de la Architectura*—an online periodical published via the University of Granada—providing an accessible introduction to this groundbreaking work. Popular descriptions in English of Granada itself first appeared soon after 1825, when the American writer Washington Irving (1783–1859) began occupying rooms of the Alhambra and rambling through this great Islamic palatine complex at will. Exemplifying a now-discredited Orientalist viewpoint, his subsequent writings (notably *The Alhambra,* 1832) enveloped the site and its urban surroundings in romance (largely fictional), fantasy, and sensuousness, coloring publications about the so-called Moorish monuments of Spain for generations.[4] Barrucand and Bednorz's *Moorish Architecture in Andalusia* (1992) notes these attitudes were offset (if sometimes fed) by more serious works including the monumental *Plans, Elevations, Sections and Details of the Alhambra* by Goury and Jones (1842), with its exquisitely detailed drawings that defy reproduction even in the digital age. I used my battered original copy when describing this palace city and its ubiquitous tâbiyah construction. A run of *Cuadernos de la Alhambra* bought in Granada is also referenced along with Oleg Grabar's *Alhambra* (1978). Although ignoring recent archaeological investigation, the latter author goes far beyond studies focused more on the site's decorative than architectural qualities typified by works of Antonio Fernandez-Puertas. Now dated, collected papers of Leopoldo Torres Balbas reprinted by the Instituto de Espana (1981) and his *Ciudades Hispano-Musulmanas* (1985) remain useful, especially when exploring urban defenses.

Accounts of "Barbary" and lands bordering the Sahara by James Grey Jackson (1809, 1820) are remarkably erudite for their date and of particular interest since they were read by Thomas Spalding, the sparse information they yield about building being important.[5] An earlier account given by John Windus of the embassy sent by King George II to the "Emperor of Fez and Morocco" at Meknes in 1721 is also remarkable, describing tâbiyah monuments erected on the grandest scale by the murderous, most likely insane, Alawite ruler Mawlay Isma'il, and it includes a fold-out plan of the once-sumptuous, but now-ruined, al-Badi' palace at Marrakesh.

Among more recent contributions, Paul Berthier's very scarce account (1966) of Morocco's medieval sugar industry and excavation of its sugar mills is indispensable.[6] I'm grateful to Bouquiniste Chella, Rabat, who, after combing local sources without success, eventually located a copy for me in Essasouria and made sure that it was in my hands before I left the country. Publications by French scholars working in Morocco during the early twentieth century are useful—especially in cases where archaeological excavation occurred, although stylistic analysis derived from art-historical perceptions bred in Western academies can be jarring when applied to Islamic sites.

At a different level, I found picture postcards of architectural subjects valuable if humble resources when trying to determine the degree to which tâbiyah monuments in Morocco's imperial centers are restored, rebuilt, or otherwise altered, city walls in particular having become a focus for modern activities that—in the name of preservation—hide original building sequences and techniques under new surface finishes. Images featuring *"Popular Scenes and Types"* published before 1910 in both France and its "Protectorates" may indeed have emphasized the "timelessness and backwardness of Arab North Africa" besides legitimizing French dominance of the region as Slavin has stated.[7] But mass-produced photographs of this kind include views of traditional building techniques that few savants of the period bothered to notice.

Fernand Braudel did not mention tâbiyah, tapia, or tabby. No matter—his *The Mediterranean and the Mediterranean World in the Time of Philip II* (French edition, 1966; English translation, 1972) conjures the currents of history swirling around Spain and Spanish expansion into the New World with an immediacy never surpassed. I have used George Kubler's *Arquitectura Mexicana del siglo XVI* (Mexican architecture in the sixteenth century) as a field guide, notwithstanding the weight of the well-illustrated Spanish-language edition (Mexico City, 1984). Sidney David Markman's *Colonial Architecture of Antigua Guatemala* (1996) cannot be done without either, it providing detailed information about the abandoned city's buildings, including descriptions and a history of the now-rebuilt Capitanía summarized in chapter 1. Graziano Gasparini's, *Las*

Fortificaciones del Periodo Hispanico en Venezuela (1985) opens new vistas for architectural research among inadequately surveyed regions of northeastern South America.

John Archer's *The Literature of British Domestic Architecture 1715–1842* (1985) was found valuable when unraveling shifting trends in British building generally and rural building in particular, John Newman and Nikolaus Pevsner's *Buildings of England* series providing county-by-county guides. François Cointeraux ranks as the preeminent source for understanding a short-lived enthusiasm that pisé de terre enjoyed in Europe and ultimately the United States during the early nineteenth century.[8] I have used my own original copy of a work rarely encountered outside national libraries. Facsimiles of reasonable quality have recently appeared while exhaustive commentary on the subject can be found under "pisé" in the *Cyclopaedia or Universal Dictionary of Arts, Sciences and Literature,* volume 27, published by Abraham Rees (London, 1819).

For Florida's tabby building, publications by Albert Manucy, notably his *Houses of St. Augustine 1565–1821* (1962, 1978), need no introduction. Concerning the Bahamas *Islanders in the Stream* (1992) by Michael Craton and Gail Saunders is an unassailable piece of scholarship. Lydia Austin Parrish did not live long enough to publish a projected volume provisionally entitled *Records of Some Southern Loyalists,* which concerns families, the majority from Georgia and South Carolina, who settled across the Bahamas after the American Revolution. Luckily her notes collected during the 1940s and 1950s survive, providing information about the Kelsalls of Beaufort County, Georgia, and Little Exuma, whose lifeways epitomize those of contemporary Loyalist planters. Variant copies held by the Department of Archives, Nassau, and Georgia Historical Society, Savannah, were consulted. I have followed with profit journeys described by Daniel McKinnen in his "General Description of the Bahama Islands" (1806). McKinnen traveled when plantation life across the southern islands was near its end and has presented pictures of planters facing failure, disillusion, and exile who still found time to show him what remained of their fields and gardens besides native plants and other natural productions.

Despite their prejudices, misrepresentations, unfair comparisons, and occasional fantasies, I admit a partiality for travelers' accounts bringing fresh perspectives to both familiar and forgotten places. La Rochefoucauld-Liancourt, whose thinking was strongly influenced by the European Enlightenment, is cited, the first French edition of his *Voyages* (Paris, 1799 or 1780) containing valuable information about Beaufort, South Carolina, omitted by English translations. The Reverend Jeremiah Evarts seems a more dyspeptic, far less cosmopolitan character, but his "Diary" (1822), with its undisguised abolitionist sentiments, is refreshing for the period. Another clergyman, Reverend Abeil Abbott, thought everything he saw between Savannah and Charleston too pleasing, charming, genteel, or well ordered for modern readers, his views of happy, contented, well-nourished, well-housed "Negroes" giving a Potemkin-like view of the country he passed through. Nevertheless verbal pictures of plantation layouts, gardens, and building methods found curious or worth emulating—cob, tabby, wattle and lime daub—recorded by his diary (ca. 1833) are useful.

Not really a travel book, William Elliott's *Carolina Sports* (1846) evokes the natural world he knew along the southeastern Atlantic coast, his description of hunting the elusive "devil fish"—in actuality a harmless species of manta ray—off St. Helena, Bay Point, and Hilton Head allowing us, like Elliott himself perhaps, temporary escape from the hard, demeaning, sometimes savage realities of day-to-day plantation life during the antebellum era chronicled by Theodore Rosengarten's *Tombee* (1986), a seminal work that includes the journal of Thomas B. Chaplin Jr., largely compiled near the southern tip of St. Helena Island, only two or three miles from where this contribution was completed.

NOTES

Introduction

1. This is a translation of La Rochefoucauld-Liancourt's original French text (*Voyages*, 4:119–22) given in Marcus Binney, "Planter's Summer Retreat," *Country Life*, March 27, 1980, 916.

2. Rowland et al., *History of Beaufort County*, 44.

3. A tabby wall standing on the Paul Grimball House site on Edisto Island (Charleston County) has been associated with Grimball's main residence, which burned in 1686. If so, this would be the earliest tabby known from South Carolina. According to a National Register nomination on file at the South Carolina Department of Archives and History, Columbia, S.C., no investigation has yet been undertaken to verify the identity or age of the wall in question.

4. There are stray references to earlier brick-making activities by Colonel Thomas Talbird (presumably at his plantation near Clarendon, Port Royal Island) and on St. Helena Island.

5. McCash and McCash, *Jekyll Island Club*, 58.

6. Johnson, "Beaufort & the Sea Islands," 102.

7. Rose, *Rehearsal for Reconstruction*, 60.

8. Noah Brooks, Beaufort, June 1, 1863, South Carolina Historical and Genealogical Magazine, 64:136.

9. *Free South*, January 27, 1863.

10. Welden, *Conservation and Preservation of Tabby*, 83.

11. Bede, Elizabeth A. "Conclusions" in Weldon, Jane Powers (Ed). "Conservation & Preservation of Tabby," 1998: 83, 84.

12. Craton and Saunders, *Islanders in the Stream*, 2:181.

13. Marder. "Tabby Resources in Florida," 30.

14. Granger, *Savannah River Plantations*, 439.

15. Unsourced newspaper clipping dated September 23, 1824, pasted into the *Daniel Webb Plantation Book, 1815–1860* (Colleton County, S.C.), South Carolina Historical Society, Charleston.

16. Rosengarten, *Tombee*, 632.

17. This would make it a Category 3 storm on the Saffir-Simpson scale, capable of producing "devastating damage."

18. Beaufort Coroner's Report, August 1893, 251; August 1893, 275 (microfilm, Beaufort Public Library, Beaufort, S.C.). Mortality figures are based on actual body counts. The true number of deaths was probably higher.

19. Savannah Morning News, August 19, 1893.

Chapter 1. Old World Antecedents and Their Diffusion

1. Spalding, "On the Mode of Constructing Tabby Buildings," 619; Floyd, "Certain Tabby Ruins," 73.
2. Thomas Spalding, "Sapelo," Darien, Georgia, July 20, 1816, letter to unknown correspondent, Georgia Historical Society, Savannah, Ga.
3. Shaw, *Travels in Barbary,* 48. See Marçais and Marçais, *Monuments arabes,* 113–35, 201–6, for description of Tlemcen's pisé (tâbiyah) walls.
4. John Windus saw this building technique in 1721; his account when published in 1725 was among the first such descriptions from Morocco available to English readers.
5. Armstrong, *History of Minorca,* London, L. Davis on C. Reymers, Second Edition, 1756: 58–59; James, *Herculean Straits,* Vol. 2: 203, 403.
6. Lane, *General Oglethorpe's Georgia,* 1:278.
7. Carroll, *Historical Collections,* 1:33.
8. Kelso, *Captain Jones's Wormslow,* 61; Stoney, *Plantations of the Carolina Low Country,* 12.
9. Sickels-Taves and Sheehan, *Lost Art of Tabby Redefined,* 4.
10. Elbl, *Portuguese Tangier,* 167n67.
11. Ibn Khaldûn, *Muqqadimah,* 2:359–60.
12. Waldron, Arthur *The Great Wall of China from History to Myth,* Cambridge, Cambridge University Press, 1990: 13.
13. Glick, *Islamic and Christian Spain,* 224–25.
14. I have used Clifford Edmund Bosworth, *The New Islamic Dynasties: A Chronological and Genealogical Manual* (New York: Columbia University Press, 1996), as guide to the names, regnal dates, and dynastic affiliations of Morocco's medieval rulers many of whom also ruled—however briefly—in southern Spain.
15. James Grey Jackson, *Account of the Empire of Morocco and District of Suse,* London, W. Blumer and Co. (1809), 58.
16. The site's northern extremity—a rocky outcrop overlooking the Atlantic—was developed as a military camp by the Almohads. Excavation has revealed a palace enclosed by substantial (1.4 m thick) tâbiyah walls, fragments of which still stand.
17. UNESCO, *Rabat capitale modern et ville historique: Un patrimoine partage proposition d'inscription su a liste Mondial soumise por le Royaume du Maroc,* June 20, 2011, 115.
18. The foreman in charge of restoration work here was adamant that no Portland cement was ever used in such modern mixes.
19. Charles Allain and Gaston Deverdun, "Portes anciennes de Marrakech," *Hesperis* 44 (1957): 117.
20. UNESCO, *Rabat capitale modern et ville historique,* 107–10.
21. *Encyclopedia of Islam,* new ed., 2:812, 822.
22. Fa and Finlayson, *Fortifications of Gibraltar,* 14–16.
23. *Répertoire Chronologique d'Épigraphie Arabe* 15:99–100, no. 5759.
24. Henri Basset and E. Lévi-Provençal, "Chella, une nécropole mérinide," *Hésperis: Archives bèrberes et bulletin de l'Istitut des hautes études marocaines* 2 (1922): 1–92.
25. Fa and Finlayson, *Fortifications of Gibraltar,* 15.
26. Samuel Marquez Bueno and Pedro Gurriaran Daza, "Recursos formales y constructivos en la arquitectura militar almohade de al-Andulus," *Arqueologia de la Arquitectura* 5 (enero–diciembre 2008): 115–34.

27. Basset and Lévi-Provencals, "Chella," 30.

28. Formwork dimensions given here were taken by myself from unrestored wall segments visible in 2016, these features disappearing from view as the result of renewed restoration work one year later.

29. Henri Gaillard, *Une Ville de l'Islam Fès* (Paris: J. André 1905), 68–70, stated the first burial here was that of Sultan Abd el Aziz (r. 1367–72), who died fighting in Algeria.

30. A photograph taken in 1880 shows this square tâbiyah-and-brick building then had a tiled, pyramidal roof. Subsequently, rear and side facades collapsed, doubtless bringing down the roof as they fell.

31. Gaillard, *Une Ville de l'Islam Fès,* 68–70.

32. Berthier, *Ancien Sucreries,* Vol. 1, 55–56, cited the historian El Oufrani as the source of this information, which is repeated by Gray, *Angelican Gardens of Marrakech,* London Frances Lincoln Ltd. 2013: 32.

33. In 2017 the superintendent of building at al-Badi' told me that forms (*luh* as he called them) used in construction of the palace walls were typically 2 m long × 70 cm high × 80 cm wide.

34. Windus published the drawing as "Palace at Fez," but in actuality it depicts al-Badi' in Marrakech. See Jean Meunier, "Le grand riad et les batiments Saadiens du Badi," *Hesperis* 44 (1927): 129–34, plates 1–3.

35. At ground level on the west pavilion's north side what appears to be an unrestored wall segment was cast in vertical increments measuring about three feet in height. A timber needle—circular in section—remains in situ.

36. Antonio Almagro, "Analisis arquologico del pabellon occidental del palacio al-Badi' in Marrakech," *Arquelogia de la Arquitectura* 11 (2014): 189–211.

37. Navarro Polazon et al., "Aqua, arquitectura y poder," *Arquelogia de la Arquitectura,* no. 10 (2013): n.p.

38. See Navarro Palazon, *Agdal,* for details.

39. Gray, *Gardens of Marrakech,* 18, stated the Almohad tank measures 205 m × 180 m.

40. Chouki el Hamalm, *Black Morocco,* 2013.

41. The mill is situated near a settlement called Ida Ougourd about thirteen kilometers east from Essaouira via Diabet (GPS N31_ 26.421' W9_ 37.099').

42. Berthier, *Ancienne Sucreries,* 1:163.

43. Edith Wharton, *In Morocco,* (1920):124.

44. Berthier, *Ancienne Sucreries,* 1:77.

45. Wharton, *In Morocco,* 157.

46. Berthier, *Ancienne Sucreries,* 1:132.

47. Windus, John. *A Journey to Mequinez the residence of the present Emperor of Fez and Morocco on the occasion of Commodore Stewarts's Embassy thither for the Redemption of the British Captives in the year 1721,* (London: Jacob Jonson, 1725): 24.

48. In eighteenth-century parlance *terras* was a term applied to hard, lime-based cements, often used for waterproofing cisterns.

49. Windus. *A Journey to Mequinez:* 112–14.

50. Windus, *Journey to Mequinez,* 174–75.

51. "Were these the vaulted granaries, or the subterranean reservoirs under the three miles of stabling which housed the twelve thousand horses," Edith Wharton (*In Morocco,* 66) wrote in 1920, a question that still evokes conflicting answers from local guardians and guides.

52. Brown, Kenneth L. *People of Salé- Tradition and Change in a Moroccan City 1830-1930*. Harvard University Press, Cambridge, 1976:44.

53. Sultan Muhammed IV apparently sent skilled workers to Gibraltar to assist with building a tâbiyah traverse for the Royal Engineers in 1859 or 1860.

54. Curtis, "Type and Variation: Berber Collective Dwellings of the Northern Sahara," 198–99.

55. In 2017 the superintendent of building conservation at al-Badi used the same term "luh" when describing for my benefit the formwork used in the building's original construction.

56. Braudel, *Mediterranean and the Mediterranean World*, Vol. 1, 468–67.

57. Torres Balbas, *Ciudades Hispano-Musulmanas*, 491; Graciani García and Rodriguez, "Typological Observations on Tapia Walls," 1104.

58. Bosworth, *New Islamic Dynasties*, 22–24.

59. Grabar, *Alhambra*, 41.

60. Rosenthal, *Palace of Charles V in Granada*, 3–5.

61. Pavón Maldonado, "Contribución al estudio del arabismo de los castillos de la Penínsular Ibérica," 220.

62. Shaw, *Travels in Barbary*, 48; Marçais and Marçais, *Monuments arabe*, 119, described the material used for Tlemcen's walls as "pisé soigneusement batu."

63. Located near the modern ronda del Marrubial. Measurements were taken by the author in August 1990, soon after this wall had been cleared and excavated. For a detailed discussion of formwork design, see Gil-Crespo, Ignacio-Javier, "Islamic fortifications in Spain built with rammed earth," *Construction History*, Vol. 31 No. 2 (2016): 1–22.

64. Fernandez-Puertas, *Alhambra*, 1:192.

65. See Bazzana, "Eléments d'archéologie musulmane dans al-Andalus."

66. Graciani García and Rodriguez, "Typological Observations on Tapia Walls."

67. Jaquin, Augarde, and Gerrard, "Historic Rammed Earth Structures."

68. Villegas Cerredo, *Homenaje*, 27, and Quesada-Garcia and Garcia-Pulido, "Segura de la Sierra," 115, also referred to materials exhibiting this technique called *tapias calicostradas*.

69. Henri Terrasse, "Gharnata," in *Encyclopedia of Islam*, new ed., 2:1012–20.

70. Quotation from Fernandez-Puertas, *Alhambra*, 1:285; Jules Goury and Owen Jones, *Plans Sections and Details of the Alhambra*.

71. Torres Balbas, *Las torres*, 200–201. In 1990 the tower appeared abandoned. Since then it has been heavily restored, this work concealing structural features here described.

72. Still in situ at the end of the sixteenth century when mentioned by de Molina, *Nobleza*, 138, the slab is stored at the Palace of Duenas, Seville.

73. D. Villegas, M. Cámara, and V. Compán, "Análisis estructural de la torre de homenaje de la Alhambra de Granada," *Informes de la Construcction* 66, extra 1 (2014),

74. Fernandez-Puertas, *Alhambra*, 1:201–4.

75. For photograph of towers before modern restoration, see Torres Balbas, *Ciudades Hispano-Musulmanas*, 559.

76. In 2010 several broken towers were partially reconstructed with corten steel sheets. Whatever the aesthetic merits of this clever modernist approach, the scheme was condemned by UNESCO for the incompatibility and intrusiveness of its material usage. Photographs taken in 1990 illustrate the present account.

77. Torres Balbas, "Las torres en Sevilla."

78. Graciani García and Rodriguez, "Typological Observations on Tapia Walls."

79. Torres Balbas, *Ciudades Hispano-Musulmanas*, 588 (photograph), 589 (plan).

80. Quesada-Garcia and Garcia-Pulido, "Segura de la Sierra," 109–22.

81. Grabar, *Alhambra,* 169.
82. A structural survey dated 1686 (Archivo General de Simancas) shows the entire tower riven by vertical cracks. Brooker "Conservation and Repair of Tabby," 65.
83. Grabar, *Alhambra,* 143; Goury and Jones, *Plans Sections and Details of the Alhambra,* plate 7.
84. Grabar, *Alhambra,* 145.
85. Fernandez-Puertas, *Alhambra,* 1:220, fig. 122; 223. On the site plan given by Goury and Jones, the bastion is designated no. 15; the *Torre del Homenaje,* no. 16.
86. James, *Herculean Straits,* 2:303.
87. The memo is Stehelin, "Gibraltar," 1861: 25. The Grand Battery was itself built on reused (probably tâbiyah) walls dating back to the eleventh century A.D. Fa and Finlayson, *Fortifications of Gibraltar,* 56.
88. Stehelin, "Gibraltar," 1861: 25.
89. Stehelin, "Gibraltar," 1861: 26–27.
90. The Gibraltar Museum kindly provided this information.
91. Braudel, *Mediterranean and the Mediterranean World,* 2:788.
92. John Dunton, *A True Journal of the Sally Fleet with the Proceedings of the Voyage* (London: John Dawson, 1637).
93. Olfert Dapper, *Description d'Afrique* (Amsterdam: Wolfgang, Waesberge, Boom and van Someren, 1686), 141.
94. McAndrew, *Open-Air Churches,* 389.
95. Toussaint, *Colonial Art in Mexico,* 29, 84–85, 125–28.
96. Braudel, *Mediterranean Vol. 1,* 87.
97. Deagan and Cruxent, *Archaeology at La Isabela,* 126.
98. Hoffman, *Spanish Crown and the Defense of the Caribbean,* 55–56.
99. I am told that seventeenth-century tapia can still be found in the domestic architecture of Cuba (Dr. Alicia Garcia, Havana, Cuba, personal communication, 2011).
100. Markman, Sidney David. *Colonial Architecture of Antigua Guatemala.* American Philosophical Society, Philadelphia, 1966:26.
101. Markman, *Colonial Architecture of Antigua Guatemala,* 24, 26, 99, fig. 24.
102. Markman, *Colonial Architecture of Antigua Guatemala,* 24.
103. Markman, *Colonial Architecture of Antigua Guatemala,* 100.
104. Stephens, *Incidents of Travel, Chiapas and Yucatan Vol. 1,* 269; Markman, *Colonial Architecture of Antigua Guatemala,* 203–6.
105. Kubler and Soria, *Art and Architecture,* 66.
106. Manucy and Torres-Reyes (1973:40) noted, "the walls connecting the bastions, though also of *mamposteria* looked like stone. The masons had lined the forms with stones dressed on the outer face to about six by twelve inches."
107. Gasparini, *Las Fortificaciones,* 60.
108. Portuguese City of Mazagan (El Jadida), UNESCO World Heritage nomination, inscribed July 7, 2004.
109. Arana and Manucy, *Building of Castillo de San Marcos,* 12.
110. Albert Manucy, *The Houses of St. Augustine, 1565–1821* (Tallahassee, FL: St. Augustine Historical Society, 1962), 71.
111. Bartram, John. "Diary of a Journey through the Carolinas, Georgia and Florida, July 1, 1765–April 10, 1766." Philadelphia, 1942: 52, 55, cited in Manucy, *Houses of St. Augustine,* 32.
112. Manucy, *Houses of St. Augustine,* 31–32, 146 n46, n47.
113. John Bartram *Diary of a Journey,* edited by Francis Harper, 54.

114. Manucy, *Houses of St. Augustine,* 48–75.

115. McAndrew, John. *The Open Air Churches of Sixteenth-Century Mexico.* (Harvard University Press) Cambridge, MA. 1969: 324.

116. Meraz and Guerrero Baca, "Calpan," 40–46.

117. Spalding, Thomas. "On the mode of Contructing Tabby buildings." *Southern Agriculturalist,* Vol. 3 No. 12 (December 1830): 618.

Chapter 2. Tabby in Military Building of the Southeast Atlantic Coast

1. A. S. Salley, ed. *Journal of the Commons House of Assembly, 1692–1735,* 21 vols. (Columbia, S.C., 1907–1946), November 15, 1726–March 11, 1727: 77.

2. Crane, *Southern Frontier,* 246.

3. H. Smith, "Beaufort, the Original Plan and the Earliest Settlers," 144.

4. Reports by local informants of this fort's foundations being found on the south side of central Bay Street, Beaufort, are mistaken, description of a "timber crib" referring to late eighteenth-century docks made for Captain Francis Saltus and other merchants rather than any military building.

5. *Journal of the Commons House of Assembly, 1692–1735,* January 20, 1726: 77.

6. Oglethorpe to the Trustees of Georgia, February 10, 1733, in Lane, *General Oglethorpe's Georgia,* 1:4.

7. Thomas Causton to his wife, 1733, in Lane, *General Oglethorpe's Georgia,* 1:8.

8. *Journal of the Commons House of Assembly, 1692–1735,* January 24, 1734: 36.

9. British Public Record Office, South Carolina, vol. 9, F. 90, transcript, South Carolina Department of Archives and History microfilm, roll 5, 200–201.

10. No evidence was found for timber platforms when the southwest bastion was excavated in 2015.

11. Archival depositories in South Carolina, collections of the Public Record Office (London) and King's Topographical Collection (British Library, London) were searched by the writer for early drawings of Fort Frederick without result.

12. Pepper and Adams, *Firearms and Fortifications* (1986), 3.

13. Arana and Manucy, *Building of Castillo de San Marcos,* 36.

14. DelaBere, who represented St. Helena Parish, continued in office down until his death in 1739. In 1754 he was given permission of the House to absent himself from the Ninth Royal Assembly (1731–33) while working on Fort Frederick. Bond stayed in office down until 1754.

15. A.S. Salley, ed., *Journal of the Commons House of Assembly, 1736-1750,* 9 vols. (Columbia, S.C.: 1951-1962): March 30, 1743: 347.

16. *Journal of the Commons House of Assembly, 1736–1750,* March 31, 1743: 350.

17. *Journal of the Commons House of Assembly, 1736–1750,* January 21, 1739/40. Easterby Colonial Records of South Carolina Commons House of Assembly, September 12, 1739–March 26, 1741: 186.

18. Papers in the State Paper Office, London Collection of the South Carolina Historical Society. Vol. 2: 277 Charleston, S.C. Historical Society, 1858.

19. Benjamin Walker to Col. Vanderdussen in London, October 15, 1742. Collection of South Carolina Historical Society, Vol. 2, 277 (Charleston, S.C. 1858).

20. *Journal of the Commons House of Assembly, 1736–1750,* March 30, 1743: 347.

21. *Journal of the Commons House of Assembly, 1736–1750,* March 28, 1749–March 19, 1750: 147.

22. *Journal of the Commons House of Assembly, 1736–1750,* March 28, 1749–March 19, 1750: 92.

23. *Journal of the Commons House of Assembly, 1736–1750,* May 5, 1752: 280.

24. The Independent Company of Foot, a unit of British regulars, garrisoned the fort until they were transferred to Georgia in 1736. During the next two years provincial solders were stationed there. British soldiers from the Forty-Second Regiment in Georgia provided a garrison from 1738 until about 1744. Regulars of a newly raised Independent Company began garrisoning the fort in 1746, and British soldiers continued to be stationed there intermittently, until Fort Lyttelton was begun in 1758. Commons House of Assembly Journal March 4, 1757.

25. Commissioners of Fortification for the Province of South Carolina. *Journal,* 1755–70. S.C. Department of Archives and History, Columbia, S.C., November 17, 1757: 108.

26. Christopher Looby, ed., *The Complete Civil War Journal and Selected Letters of Thomas Wentworth Higginson* (Chicago: University of Chicago Press, 2000), 53–54.

27. *Colonial Records of North Carolina,* 7:24.

28. I am indebted to the Friends of Historic Jekyll Island for bringing this map to my attention and to Warren Murphey (Jekyll Island Museum) for obtaining a copy from collections of the National Park Service, Frederica, Georgia.

29. Anon. Description of Frederica *London Magazine* Vol. 14. London 1745: 395–96.

30. Linley, John. The *Georgia Catalogue Historic American Building Survey: A Guide to the Architecture of the State,* University of Georgia Press, Athens, GA 1982:321, confirms the actual dimension as 96" × 96".

31. Moore, *Voyage to Georgia,* 83.

32. Spalding, July 30, 1844, cited in Coulter, *Georgia's Disputed Ruins,* 73.

33. Kelso, *Captain Jones's Wormslow,* 57.

34. Kelso, *Captain Jones's Wormslow,* 8.

35. Drawing by Lieutenant Hesse, Huntington Library, San Marino, Calif., HM 15399.

36. Commissioners of Fortification, *Journal,* August 25, 1757.

37. Rowland et al., *History of Beaufort County,* 156.

38. Commissioners of Fortification, *Journal,* December 5, 1757.

39. Commissioners of Fortification, *Journal,* February 9, 1758.

40. N. Barnwell and John Mullryne authorized payment to John Gordon of remainder of money granted for building of Fort Lyttelton, July 24, 1758.

41. Commissioners of Fortification, *Journal,* July 31, 1758.

42. Commissioners of Fortification, *Journal,* September 14, 1758. Information from letter dated September 6, 1758 from Supervisors.

43. Commissioners of Fortification, *Journal,* September 14, 1758.

44. This refers to flooring joists for the attic story.

45. Smith, Henry A.M. "Beaufort—the Original Plan and the Earliest Settlers," *South Carolina Historical and Geneological Magazine,* Vol. 9 No. 3, July 1908:149. Smith has a quotation from Dr. Miligan in his short *Description of the Provinece of South Carolina* written in 1763.

46. Commissioners of Fortification, *Journal,* December 28, 1756.

47. Dr. Nicholas Butler, Charleston, S.C., personal communication, April 2010.

48. First Council of Safety of the Revolutionary Party, *South Carolina Historical and Geneological Magazine,* Vol. 2 No. 1: 16–17 (January 1901).

49. The main plantation house (built of brick) at Laurel Bay was heavily damaged by fire although subsequently rebuilt. On the opposite side of the river, Heyward's Old House was burned to the ground.

50. Moultrie, *Memoirs of the American Revolution,* 1:290–291.

51. Rowland et al., *History of Beaufort County,* 288.

52. General Thomas Pinkney, Charleston, S.C., July 3, 1812, *Corps of Engineers Reports July 3, 1812–October 4, 1823,* U.S. National Archives, College Park, Md., RG 77.

53. U.S. National Archives, RG 77, Drawer 146, Sheet 3. Extract from letter of Maj. Macomb, dated September 28, 1808.

54. "Report of Secretary of War on Seacoast Fortifications and Batteries," December 1811, National Archives, Designation/Call Reference I, 307–11.

55. Manuel, "Forts Hampton, Winyaw, and Marion," 43–48.

56. Mandar, 600–622, plate 8, 97.

57. Wade, *Seacoast Fortifications,* 6.

58. Captain Swift to Captain McRae, September 2, 1812. Buell Collection of Historical Documents relating to the Corps of Engineers 1801-1819 U.S. National Archives RG 77, Records of the Chief of Engineers.

59. Prentice Willard to Major McRae, Beaufort, January 16, 1813, U.S. National Archives, Buell Collection.

60. U.S. National Archives, College Park, Md., Cartographic Department, RG 77, 146-5.

61. U.S. National Archives, College Park, Md., Cartographic Department, RG 77, 146-4.

62. Johnson, "Beaufort and the Sea Islands," typescript Beaufort County Library: 88.

63. Lepionka, "Fort Lyttelton Excavations," 38-41.

64. U.S. National Archives, College Park, Md., Corps of Engineers Reports, RG 77, entry 221, 464.

65. U.S. National Archives, College Park, Md., RG 77 66-1.

66. Manuel, "Forts Hampton, Winyaw, and Marion," 45.

67. "Tower on Tybee Island at the Mouth of the Savannah River Georgia." Surveyed and Drawn by Lieut. Mansfield of the Corps of Engineers, 24 January 1835, U.S. National Archives, RG 77, Cartographic Division, College Park, Md..

68. *Tower on Tybee Island at the Mouth of the Savannah River Georgia Surveyed and Drawn by Lieut. Mansfield of the Corps of Engineers, 24 January 1835,* U.S. National Archives, Cartographic Division, College Park, Md., RG 77 68-3. Another Martello tower was erected near the ruins of Fort Johnson on James Island, S.C., in the 1820s, where construction was of brick. Everything is now reduced to rubble (Trinkley et al., *Property Nobody Wanted,* 88–89).

69. Sheila Sutcliffe, *Martello Towers* (Cranbury, N.J.: Associated University Presses, 1973).

70. Henry Middleton to Brigadier General I. G. Swift, U.S. National Archives, College Park, Md., "Corps of Engineers Reports, July 3, 1812–October 4, 1823," 73.

71. Henry Middleton to Brigadier General I. G. Swift, U.S. National Archives, College Park, Md., *Corps of Engineers Reports, July 3, 1812–October 4, 1823,* 69.

72. Mansfield, U.S. National Archives, College Park, Md., RG 77 68-3.

73. Lane, *Architecture of the Old South, Georgia,* 74, 82–85.

74. South Carolina Department of Archives and History Columbia, S.C., 1825-5-09, *Copy of the Report from the Keeper of the Beaufort Arsenal, July 1, 1825.*

75. *Copy of the Report from the Keeper of the Beaufort Arsenal,* July 1, 1825.

76. *Report and Resolutions of the South Carolina Legislature, 1857* cited in G. B. Smith, "Beaufort Volunteer Artillery."

Chapter 3. Tabby Making: Materials and Fabrication

1. Macomb, "Observations on the Art of Making the Composition Called Tapia," 128.
2. Macomb, "Observations on the Art of Making the Composition Called Tapia," 128.
3. James Julius Sams, undated typescript titled *Dathaw*, Sams Family Collection, Beaufort, S.C.
4. Thomas Spalding, "On the Mode of Constructing Tabby Buildings and the Propriety of Improving Our Plantations in a Permanent Manner," *Southern Agriculturalist,* December 1830 (Charleston, S.C.) Vol. 3 No. 12: 617–623.
5. Sickels-Taves and Sheehan, *Lost Art of Tabby Redefined,* 55, called tabby of this variety "Spalding Tabby."
6. K. Jones, *Port Royal under Six flags,* 65, citing Jeannette Thurber Connor, ed., *Colonial Records of Spanish Florida II* (Deland: Florida State Historical Society, 1930), 1:171.
7. Floyd "Certain Tabby Ruins," 53.
8. Lawson, *History of North Carolina,* 82.
9. Bruce, *Memoirs of Peter Henry Bruce,* 438.
10. South Carolina Commissioners of Fortification, *Journal,* June 16, 1757.
11. U.S. National Archives RG 77, Miscellaneous Papers, Engineer Department, Washington, D.C., Draft of letter to Secretary of War, March 1821.
12. See Elijah Paine Jr., *Report of Cases Heard and Determined by the Circuit Court of the United States for the Second Circuit* (New York: R. Donaldson), 1827, 1:314. This document describes the "tappy" as comprising "equal proportions of sharp sand, fresh lime and oyster shells, with water sufficient to produce adhesion" (1:324).
13. Spalding, "On the Mode of Constructing Tabby Buildings," *Southern Agriculturalist*, Vol. 3 No. 12, December 1830: 619.
14. Gilman, *Economical Builder,* 6. Writing in 1796 La Rochefoucauld-Liancourt (*Voyage,* 4:121) said the cost of tabby in Beaufort was then ten pence per cubic foot; however is not clear exactly what unit of measurement the French visitor used.
15. Commissioners of Fortification, *Journal,* August 25, 1757.
16. Commissioners of Fortification, *Journal,* October 6, 1757, October 27, 1757.
17. Commissioners of Fortification, *Journal,* July 24, 1758.
18. Commissioners of Fortification, *Journal,* August 3, 1758.
19. Commissioners of Fortification, *Journal,* September 1, 1758.
20. Commissioners of Fortification, *Journal,* May 16, 1757.
21. Fleetwood, *Tidecraft,* 102.
22. *Diary of Charles Drayton I,* 1799-1805 Drayton papers 1701–2004 College of Charleston, S.C. MSS 0152.
23. Powell, *Ninth Annual Report of the United States Geological Survey,* 229.
24. Coulter, *Georgia's Disputed Ruins,* 73 (Spalding to N. C. Whiting, New Haven, July 30, 1844).
25. Spalding to H.C. Whiting, New Haven July 30, 1844. Cited in Coulter, E. Merton. *Georgia's Disputed Ruins* (Chapel Hill. University of North Carolina Press, 1937), 75.
26. Sickels-Taves and Sheehan, *Lost Art of Tabby Redefined,* 24.

27. Coulter, *Georgia's Disputed Ruins,* 73 (Spalding to N. C. Whiting, New Haven, July 30, 1844).

28. Thomas B. Chapman Jr. mentioned building a lime kiln "12 feet across" on St. Helena Island but gives no details (Rosengarten, *Tombee,* 540).

29. Drayton, *Diary,* November 1798, November 1801.

30. An early reference to indigenous African practice was given by d'Almeida (1505), who observed at Kilwa (Kenya), "Lime is prepared here in this manner: large logs of wood are piled in a circle and inside them coral limestone is placed; then the wood is burnt." Garlake, *Islamic Architecture East African Coast,* 16.

31. Commissioners of Fortification, *Journal,* June 23, 1757.

32. Spalding, "On the Mode of Constructing Tabby Buildings," 619.

33. Sickels-Taves and Sheehan, *Lost Art of Tabby Redefined,* 22–23.

34. The majority of vertical wall cracks now visible resulted from the Charleston earthquake in 1886. See Clarence Edward Dutton, *The Charleston Earthquake of August 31, 1886,* in United States Geological Survey, Ninth Annual Report of Director, Washington, D.C., GPO, 1889, 298.

35. Commissioners of Fortification, *Journal,* March 2, 1758. In Charleston, forms utilized for an early horn work erected during the 1760s (now represented by shapeless fragments standing in Marion Square) were only sixteen inches high, tied with one-inch-diameter "pins" positioned about 3'7" on center.

36. Macomb, "Observations on the Art of Making the Composition Called Tapia," 130–31.

37. Jeffrey W. Cody, "Earthen Walls from France and England for North American Farmers, 1806–1870." 6th International Conference of Earthen Architecture, La Cruces, New Mexico, Oct. 14–19, 1990 (Getty Conservation Institute, Los Angeles, 1990), 35–49.

38. Thomas Walker was likely the same Thomas Walker, "Carpenter," who traveled to Georgia at the trustees' expense in 1735–36 and occupied Lot 10 N in Frederica (see *Early Settlers of Georgia,* 55).

39. Gwilt, *Encyclopedia of Architecture* 1867: 1332.

40. Macomb previously cited (Observations etc. 130–31).

41. Return of Works Done at Fort Lyttelton. November 16, 1758. Commissioners of Fortification, *Journal.*

42. Rees, *Cyclopaedia,* vol. 27, "Pise."

43. Coulter, *Georgia's Disputed Ruins,* 74 (Thomas Spalding to N. C. Whiting, New Haven, July 30, 1844).

44. Thomas Spalding to unknown correspondent, 1816, Georgia Historical Society, Savannah.

45. Thomas Spalding to unknown correspondent, 1816, Georgia Historical Society, Savannah.

46. Around Darien, Georgia, and in other parts of coastal Georgia, Spalding's influence was far greater, numerous tabby structures having been built using twelve-inch-high formwork as he recommended. Sickels-Taves and Sheehan, *Lost Art of Tabby Redefined,* 55, called tabby of this variety "Spalding Tabby."

47. Thomas Spalding, "On the Mode of Constructing Tabby Buildings (etc.)." *Southern Agriculturalist,* December 1830, Vol. 3 No. 12: 618. For La Rochefaucault on tabby, see his "Voyages" *Manière de Bâtir en Taby.* Vol. 4: 119–122.

48. Author wishes to thank Craig M. Bennett Jr. (Bennet Preservation Engineering) for bringing this site to his attention.

49. M. Crook and O'Grady, "Spalding's Sugar Works," 15–22. Although substantial portions of

the sugar house were still standing in 1980 when recorded by the authors cited, the entire building has since collapsed.

50. Schafer, *Zephanieh Kingsley II,* 164–65, 170. Following introduction of milling machinery in 1829, fifty hogsheads of sugar were shipped from White Oak. Previously Kingsley had transported his cane to Mr. Carnochan's mill (presumably at the Thickets) for processing.

51. Dr. Mary Socci, personal communication, December 2010. Among excavated materials, chunks of lime mortar suggest the lower floor was cast in tabby.

52. Thomas Spalding to N.C. Whiting July 30, 1844. Cited Coutler, "Disputed Ruins," 75.

53. Johnson's remarks were published in the *Beaufort Republican.* I have used an undated typescript of his manuscript held by Beaufort County Library, "Beaufort and the Sea Islands," which may be the original.

54. Alexander Macomb, "Tapia," *American Farmer* (Baltimore) 8, no. 45 (January 28, 1827): 953–54.

55. Gilman, The Economical Builder, 4.

56. Commissioners of Fortification, *Journal,* April 6, 1758.

57. Letter dated "Beaufort, December 2, 1757, from Thomas Wigg, Nath Barnwell, John Mullryne and John Gordon to Commissioners of Fortification in Charles Town." in *Journal of the Commissioners of Fortification for the Province of South Carolina 1755–1770.*

58. Commissioners of Fortification, *Journal,* June 15, 1758.

59. Commissioners of Fortification, *Journal,* June 26, 1758.

60. Gilman, *The Economical Builder,* 4.

61. Macomb, "Observations on the Art of Making the Composition Called Tapia."

62. Gilman, *The Economical Builder*, 4.

63. Floyd, "Certain Tabby Ruins," 70.

64. U.S. National Archives, Cartographic Division, College Park, Md., RG 77, drawer 67, sheet 2.

65. U.S. National Archives, College Park, Md., Cartographic Division, Records of the Office of the Chief of Engineers, *Historical Sketches of Fort Johnston in North Carolina Written April 1819 by Memory Only*, U.S. National Archives, RG 77, Map File DR 62-A.

66. U.S. National Archives, College Park, Md., Cartographic Division, Records of the Office of the Chief of Engineers, *Historical Sketches of Fort Johnston in North Carolina Written April 1819 by Memory Only*, U.S. National Archives, RG 77, Map File DR 62-A.

67. Brigadier General Joseph G. Smith, *Letterbook 1804–1809,* 1–6, U.S. National Archives, RG 77.

68. Busch, *Beaufort Baptist Church.*

69. *Beaufort Gazette,* March 23, 1917.

70. Sams, "Dathaw."

71. Macomb, "Tapia," 953–54.

72. George Jenkins, letter to editor, *American Farmer* 9, no. 49 (February 22, 1828): 391.

73. George Jenkins, letter to editor, *American Farmer* 9, no. 49 (February 22, 1828): 391.

Chapter 4. Tabby Construction Details and Operational Procedures

1. Sickels-Taves and Sheehan, *Lost Art of Tabby Redefined,* 111.

2. Fischetti, "Tabby," 58.

3. Tabby foundation strips supporting exterior walls of east and west wings at the Sams House are approximately 2' deep × 1'6" wide.

4. Alexander Macomb, *American Farmer* Vol. 8 No. 45, (January 26, 1827), 353.

5. James Simons, *South Carolina State Gazette,* June 18, 1800, 3.

6. In eighteenth-century London, common or stock bricks usually measured 9" × 4½" × 2½". These figures are used in calculating the wall thickness required by the London Building Act, 1667, given in Cuickshank and Wyld, "London, the Art of Georgian Building," Architectural Press, London 1975:31.

7. Spalding to N.C. Whiting, New Haven, July 30, 1844. Cited Coulter, *Georgia's Disputed Ruins,* 1937: 75-76.

8. Gilman, *The Economical Builder,* 4–5.

9. Spalding to N.C. Whiting, New Haven, July 30, 1844. Cited in Coutler, *Georgia's Disputed Ruins,* 1937: 74.

10. La Rochefoucauld-Liancourt, *Voyage,* 4:121.

11. The southern chimney base at Montpelier measured about 8' × 3'10" in plan (excluding a small pier on one side).

12. Alexander Macomb, "Answer to Inquiries as to the Manner of Building and Attaching Chimneys," *American Farmer* Vol. 8 No. 49 (February 15, 1827): 391.

13. Manucy, *Houses of St. Augustine,* 44n8, 171.

14. Thomas Spalding, Darien, 20 July 1816 to unnamed correspondent. Typed copy of unlocated original, Georgia Historical Society, Savannah, GA.

15. Coulter, *Georgia's Disputed Ruins,* 130. A photograph used as the frontispiece for the work gives a general view of slave dwellings at the Thickets. Floyd's manuscript notes (Georgia Historical Society, Savannah, M. H and D. B. Floyd Coll., no. 1308, folder 124, fol. 58) amplify the published description, stating the houses measured 14'2" × 33'4" with a central brick chimney, back-to-back fireplaces heating the two cells each of which was subdivided into one large and two very small rooms.

16. Albert Manucy, "The Fort at Frederica," *Notes in Anthropology* 5 (Dept. of Anthropology, Florida State University, 1962): 71, 75. Two brick vaults are now visible, one original the other reconstructed ca. 1906.

17. *Tower on Tybee Island at the Mouth of the Savannah River Georgia Surveyed and Drawn by Lieut. Mansfield of the Corps of Engineers, 24 January 1835,* U.S. National Archives, College Park, Md., Cartographic Department, RG 77, 68-3.

18. Cruickshank and Wylde, *London: The Art of Georgian Building,* 192–93.

19. Henry Holland, "Pisé," *American Farmer* 3 (1821): 35. For a color wash resembling Portland stone, the English landscape architect Humphrey Repton recommended one part lamp black, three parts yellow ochre, five parts lime in a medium of milk (George Carter, "Coloured Houses," *Country Life,* February 6, 1992, 116).

Chapter 5. Tabby Brick, Wattle and Daub, and Cements

1. Powell, *Ninth Annual Report of the United States Geological Survey,* 229.

2. Archaeological evidence suggests it was built ca. 1810 to replace an earlier tabby kitchen built before the Revolution at a time when the Phase I of the William Sams House was under construction.

3. Salley, *Minutes of the Vestry of St. Helena's Parish,* 62.

4. Marder, "Tabby Resources in Florida," 28.

5. Thomas, *St. Catherines*, 24.

6. Kelso, *Captain Jones's Wormslow*, 72.

7. Abbot, Abiel. *Journey to Savannah*. Cited Trinkley, "Further Investigations of Prehistoric and Historic Lifeways Callawassie and Spring Islands," 149.

8. Brooker, "Callawassie Island Sugar Works," 147–49, fig. 40.

9. R. Crook, "Gullah-Geechee Archaeology," 18.

10. Macomb, Alexander. *American Farmer,* Vol. 8 (January 26, 1827): 353-354.

11. The formulations given are taken from Parker's original patent. Derived from London Clay, an Eocene deposit extensively used in brick manufacture, the nodules were almost worked out by 1830 but are now, I find, common on the foreshore near Warden Point, Isle of Sheppey. Parker, James, 1796 Patent #2120.

12. Summerson, *Georgian London*, 184.

13. Abbot, *Journey,* 305.

14. Mathews and Grasty, *Limestones,* 272.

15. I must thank Mr. Gerry Crosby (Wildwood Contractors Inc., Walterboro, S.C.), and the City of Charleston for inviting me to consult on this project.

16. A manuscript sketch of Beaufort's defenses (U.S. National Archives) dated 1808 locates an otherwise unknown "Navy Yard" either beside or close to the wide creek bounding Block No. 13 on its west side. In 1808, Captain Francis Saltus built five gunboats for the U.S. Navy in Beaufort, Nathaniel Ingram acting as supervising navy agent.

17. In 1911 Beaufort Council awarded construction of another seawall located at the foot of Charles Street to R. R. Legare. The mix specified was "1 Part Cement, 2½ parts Sand and 4 parts Shell," the price having increased to "$3.25 per linear foot." See Beaufort Council, *Minutes,* October 16, 1911, microfilm, Beaufort County Library.

18. Ashurst and Dimes*, Conservation of Stone,* 2:82.

19. Mathews and Grasty, *Limestones* 328, 324.

20. All first-floor timbers were missing when the interior was photographed by Jack Boucher in April 2003, a small number of joists but no floorboards remaining at second-floor level. An unusual (for Beaufort County) brick oven built into the building's northeast corner was found completely robbed.

21. *Preservation Progress* 56, no. 3 (Preservation Society of Charleston, 2012): 14. Dr. Nicholas Butler (personal communication, February 2013) kindly provided information concerning the identity of Fort Darrell.

22. Spalding, "On the Mode of Constructing Tabby Buildings etc.," *Southern Agriculturalist* Vol. 3 No. 12, December 1830:618.

23. Coulter, *Georgia's Disputed Ruins,* 127–29.

24. Sullivan, Buddy. "Tabby: A Historical Perspective of an Antebellum Building Material in McIntosh County, Georgia" in *The Conservation and Preservation of Tabby: A Symposium on Historic Building Material in the Coastal Southeast.* February 25–27, 1993 Jekyll Island, GA, 1998, Jane Powers Weldon, ed., Georgia Department of National Resources, Atlanta, 1998: 45.

25. The late Gordon Mobley, manager of Spring Island, gave me this information but, for obvious reasons, was reluctant to provide full details about the material's source.

26. Barnwell, Joseph W. "Life and Recollections," typescript 1929: 105.

27. Noah Brooks, letter dated Beaufort, June 17, 1863, published in *South Carolina Historical and Geneological Magazine,* 64:136.

Chapter 6. Tabby Building in Beaufort Town, South Carolina

1. Smith, A.M. "Beaufort—the Original Plan and the Earliest Settlers." *S.C. Historical and Geneological Magazine*, Vol. 9 No. 3, July 1908: 142.

2. I examined the original of this document at the old Public Record Office, Chancery Lane, London, before its removal to the U.K. National Archives, Kew. Undated, this is a professional survey probably made soon before or just after the town's foundation. Local historians mistakenly claim it bears Queen Anne's cipher, however it bears a standard crowned stamp used by the Public Record Office during the Victorian period.

3. Edward T. Price, "The Central Courthouse Square," in Upton and Vlach, *Common Places* (1986), 130.

4. Crane, *Southern Frontier,* 165.

5. *Commons Journal,* March 17, 1724: 46.

6. *Journal of Commons House Assembly of South Carolina.* June 13, 1724. A.S. Salley Editor. General Assembly of South Carolina. 1944: 37.

7. Bull's message to the S.C. Commons House of Assembly is dated February 23, 1737/38.

8. Rowland et al., *History of Beaufort County,* 153; Charleston, S.C., Register of Mesne Conveyances (RMC) Charleston, S.C. Book F-4:297.

9. Charleston, S.C., RMC, Book F-4:297.

10. *Loyalist Transcripts,* 52:490.

11. *Loyalist Transcripts,* 52:476.

12. Confiscated by federal authorities, the property was purchased by Sylvanus Mayo for $3,100 at auction on January 21, 1863. Mayo (called a "wholesale grocer") also purchased the water lot fronting this property. *Sanborn Insurance Maps of Beaufort,* microfilm, Beaufort County Library, Beaufort, S.C.

13. While the footprint of the main building may still exist, its cistern and traces of outbuildings were destroyed by new building in 2000.

14. The Beaufort River is brackish and unsuitable for drinking purposes. Wyers Pond (now filled), shown by early navigational charts and maps in the western part of town, was probably used as a freshwater source by mariners and residents, but the vicinity became notorious for fevers during the early 1800s.

15. *Merchant's Account Book, 1785–1791,* Historic Beaufort Foundation, Beaufort, S.C.

16. I owe this reference to Sarah Fick, Historic Preservation Consultants, Charleston, S.C.

17. Nancy Rickter Webb, whose great-grandfather established a cotton gin here after the Civil War, kindly confirmed the photograph's identity and location.

18. Lengnick, *Beaufort Memoirs,* 10.

19. Lengnick, *Beaufort Memoirs,* 11.

20. Description of roof form from Simon and Lapham, *Charleston,* 20.

21. Salley, *Minutes of the Vestry of St. Helena's Parish,* 60.

22. Salley, *Minutes of the Vestry of St. Helena's Parish,* 62.

23. Siebert, *Loyalists in East Florida,* 1:23–24.

24. Siebert, *Loyalists in East Florida,* 1:130.

25. Joseph Brevard, *The Public Statute Law of South Carolina,* Vol. 1-3 Charleston, S.C.:702.

26. Information is based on a map (U.S. National Archives, College Park, Md., Cartographic Division, RG 58, item 25) apparently prepared by Thomas Fuller, ca. 1799, which, as far as I know, now exists only as a copy redrawn from a lost original.

27. Rowland et al., *History of Beaufort County,* 263.

28. Rowland et al., *History of Beaufort County,* 259.

29. Joseph Brevard, *The Public Statute Law of South Carolina,* Vol. 1-3, Charleston, S.C., Joseph Hoff, 1814.

30. Tucker, *Jeffersonian Gunboat Navy,* 62. The "Federal Shipyard" located on Black's Point is identified on a manuscript sketch of Beaufort's defenses dated 1808, a property that (as executor for James Black's estate) Saltus came to control on January 23, 1819.

31. N. Barnwell, Will. Elizabeth herself died in 1817, her house subsequently being bought by a nephew, John Gibbes Barnwell (1778–1828). Richard Gough died in 1796.

32. Lane, *Architecture of the Old South, South Carolina,* 134.

33. Beaufort Block No. 43 remained open until 1882, when subdivided for housing; see J. C. Mayo, *Palmetto Post,* Port Royal, S.C., January 19, 1882.

34. Sold out of the house during the Depression and removed to Great Neck, Long Island, then to La Jolla, California, paneling of the southeast room was acquired and reinstalled in the 1980s.

35. The back porch was replaced in concrete block during the 1950s and again in the 1980s in order to brace and strengthen the dwelling's unstable rear wall.

36. *Country Life,* April 3, 1980, 1023.

37. Palladio, *Four Books of Architecture,* 2:45.

38. During alteration to what was the Beaufort Federal Courthouse in the late 1980s, I found that pieces of tabby wall belonging to Barnwell Castle still survive mutilated but in situ below the present building.

39. Barnwell, Joseph, "Life Recollections," typescript, Beaufort County Library: 16-17.

40. Newspaper accounts attribute design of the replacement porch and other alterations to William G. Preston of Boston.

41. In Charleston early ventures were brokered through the firm Saltus and Yates, Ship's Chandlers, located on East Bay Street at the wharf called "No. 2 Crafts 4th. Range" when first recorded by the Charleston directory of 1796.

42. SCDAH, *General Assembly Petitions,* 1796-14-01.

43. SCDAH, *General Assembly Petitions,* 1796-14-01 and duplicate.

44. SCDAH, Senate Committee Report 17961701; Committee Report 1796, no. 97; Petitions, 1800.

45. Frank A. Ramsey, Special Correspondent (Charleston) *News and Courier,* 20 June 1954.

46. Beams, salvaged from another project, were inserted by Roger Pinckney, who recalled when interviewed by the author (1996) this dangerous procedure came perilously close to disaster. Regrettably, the ell's original cobbled basement floor (the only example known from Beaufort), partially exposed by excavation in 1995, was first robbed and then covered with concrete in 2001.

47. Two original demi-lune window frames were found loose in the roof space in 1969.

48. Nichols, *The Early Architecture of Georgia,* 1957: 49.

49. After the Civil War, the original structure was partially rebuilt at its northern end, its gable removed and replaced by an upper-story connector spanning over a passageway.

50. Marquis de Lafayette probably landed here on his visit to Beaufort on March 18, 1825, having arrived aboard the *Henry Schulz,* a steamship owned by Captain Lubbock.

51. Raines, *Six Decades in Texas,* 1.

52. Daner Collection, Historic Beaufort Foundation, Beaufort, S.C.

53. *Journal of the Proceeding of the Trustees of Beaufort College,* Beaufort County Library, Beaufort, S.C., August 21, 1801.

54. Lane, *Architecture of the Old South, South Carolina,* 173.

55. *Journal of the Proceeding of the Trustees of Beaufort College,* Beaufort County Library, Beaufort, S.C.

56. Salley, *Minutes of the Vestry of St. Helena's Parish,* 226.

57. Rowland et al., *History of Beaufort County,* 286, stated that during the 1817 epidemic "one seventh of the town's population and a much higher percentage of students died."

58. *Journal of Proceedings of the Trustees of Beaufort College.* May 15, 1822.

59. Salley, *Minutes of the Vestry of St. Helena's Parish,* 225.

60. Salley, *Minutes of the Vestry of St. Helena's Parish,* 227.

61. Johnson, Dr. John A. *Beaufort and the Sea Islands* (typescript n.d.) 24.

62. *Beaufort Gazette,* March 23, 1917, article by Dr. John Archibald Johnson, "Beaufort and the Sea Islands."

63. Information is based on an investigation that involved crawling under the floor of the present church along trenches previously made by heating contractors. Cut down to facilitate rebuilding, early brick walls enclosing the church still exist.

64. A. S. Salley, Jr., ed., *Minutes of the Vestry of St. Helena's Parish SC. 1726–1812* (Columbia, S.C.: State Company, 1919), 219.

Chapter 7. Tabby in the Domestic Architecture of the Sea Islands before the American Revolution

1. Rowland et al., *History of Beaufort County,* 153–54. According to Governor Glenn, 1,764 barrels of beef valued in total at £11,466 S.C. currency were exported from Charleston during the year 1747–48 (Glenn to Duke of Argyll, cited Weston, *Documents,* 1856, 71).

2. Charleston, S.C., RMC Book C-C, 300.

3. Dabbs, *Sea Island Diary,* 72–73.

4. *St. Helena Cemetery Survey,* Site U-13-ST H 28. The site is located off Capers Island Road on private land half a mile northeast of David Green Road.

5. Earl of Egmont, Journal, *Colonial Records of Georgia,* 5:248.

6. Colin Brooker. "The Major William Horton House, Jekyll Island, Georgia: Preservation, Interpretation and Display," (Report on file, Jekyll Island Museum, Jekyll Island, Ga.), 1-25.

7. Bryan, *Journal of a Visit to Georgia,* 25–26.

8. Full possession of Jekyll Island was acquired by Chistophe Anne Poulain du Bignon in 1794. In 1814 the house was plundered by British sailors and marines, who carried off twenty-seven slaves. Union soldiers occupied the island in 1862, destroying Confederate fortifications and shipping salvaged materials to Beaufort, S.C. According to John Eugene du Bignon, "the old manor house was burned in the [Civil] war," probes indicating that brick foundations were also robbed. Extensive restoration and reconstruction commenced in 1896, financed by members of the Jekyll Island Club.

9. Jekyll Island Museum.

10. *Colonial Records of Georgia,* 1:362.

11. *Colonial Records of Georgia,* 25:97. John Pye to Harman Verelst, August 27, 1746.

12. Carrol, *Historical Collections,* 1:386.

13. La Rochefaucault-Liancourt, "Voyages," Vol 4 l'an VII, 140.

14. Rowland et al., *History Beaufort County,* 165.

15. Dabbs, *Sea Island Diary,* 60.

16. SCDAH, Charleston Wills, RR 1767–71.

17. Rowland et al., *History of Beaufort County,* 166.

18. Fragment of Account Book attributed to John Mark Verdier- Historic Beaufort Foundation, Beaufort, S.C.

19. J. W. Barnwell, "Recollections," 5.

20. This White Hall is not to be confused with Whitehall Plantation on Lady's Island.

21. Exact building sequences at White Hall remain to be determined. A probable construction date "in the late 1780s" is given by Cole, in *Historic Resources of the Low Country,* 172.

22. Gowans, *Mansions of Alloway Creek,* 367–373.

23. La Rochefoucauld-Liancourt was entertained by General Barnwell and his sons, which may mean that the duke's published comments about cotton cultivation and processing in Beaufort were based on what he saw at Retreat.

24. Originally called Dathow Island, the name was altered to Dataw by Alcoe S.C. (Alcoa) following their purchase and development of the property in 1983.

25. Sams, James Julius. "Dathaw," n.d. 5.

26. Drawings by Eugenia Smith are privately held by descendants of the Sams family, Beaufort, S.C.

27. Each base measures about 7' × 3'4" in plan.

28. Trinkley and Hacker, *Investigation of the St. Queuntens Plantation,* 56.

29. Swanson, *Remaking Wormsloe,* 62.

30. Elliott, "On the Cultivation and High Prices of Sea-Island Cotton," 157.

31. Elliott, "On the Cultivation and High Prices of Sea-Island Cotton," 127.

32. Rosengarten, *Tombee,* 6.

33. Rosengarten, *Tombee,* 66.

34. Swanson, *Remaking Wormsloe,* 62.

35. Peter H. Wood, "Slave Labor Camps in Early America: Overcoming Denial and Discovering the Gulag" (1999), cited in Mack and Hoffus, *Landscape of Slavery,* 24.

36. Bishir, *North Carolina Architecture,* 149.

37. Porcher and Fick, *Story of Sea Island Cotton,* 362, 410–12.

38. Porcher and Fick, *Story of Sea Island Cotton,* 400.

39. G. Johnston, *Social History,* 52.

40. Dabbs, *Sea Island Diary,* 82. The incomplete original *Plantation Journal* attributed to Ebenezer Coffin is held by the S.C. Historical Society, Charleston.

41. Fleetwood, *Tidecraft,* 107.

42. G. Johnston, *Social History,* 108.

43. Tombee and Riverside were apparently parts of the same property at the beginning of the nineteenth century. Only segments of Riverside's main house survive (notably part of the south facade), showing that the building was a single pile structure measuring 20'2" in width. Exterior tabby walls were twelve inches thick and stuccoed on outer faces. Pour lines show tabby was cast in vertical increments measuring 1'9" in height. The construction date of the house is not known, but repairs said to have been made during the 1790s suggest it could go back to sometime just before or soon after the Revolution.

44. Rosengarten, Theodore. *Tombee, Portrait of a Cotton Planter.* William Morrow and Co. Inc. New York, 1986: 66.

45. When exactly the house was built has not been determined. A house is definitely shown standing on the site by a map dated 1838.

46. Brooker, "Haig Point House."

47. Page, *Historic Houses Restored and Preserved*, 139–43.

48. St. Simons Public Library, *Old Mill Days*, n.p.; Vanstory, *Georgia's Land of the Golden Isles*, 165–67.

49. Sams, James Julius. "Dathaw." Undated typescript. Sams Family Papers Dataw Historic Foundation, Dataw Island, S.C.: 10.

50. Sams, James Julius. "Dathaw." Undated typescript. Sams Family Papers Dataw Historic Foundation, Dataw Island, S.C.: 5.

51. Sams, James Julius. "Dathaw." Undated typescript. Sams Family Papers Dataw Historic Foundation, Dataw Island, S.C.: 10.

52. Elizabeth was the daughter of Thomas Barksdale and Mary Barksdale, a daughter of Arnoldus Vanderhost.

53. U.S. Census 1850, Agricultural Schedule, cited in Trinkley, *Second Phase of Archaeological Survey on Spring Island*, 34.

54. Since photographed by Jack Boucher, the south wing has been disfigured by clumsy and intrusive attempts at stabilization.

55. Stoney, *Plantations of the Carolina Low Country*, 44–45.

56. Hubka, "Just Folks Designing," 430.

57. Rochefoucauld-Liancourt, *Voyages*, 2:376–77.

58. Elliott, *Carolina Sports*, 35.

59. M. H. and D. B. Floyd Coll., no. 1308, folder 124, fol. 58, Georgia Historical Society, Savannah. For illustration see Lane, *Architecture of the Old South, Georgia*, 71.

60. Coulter, *Georgia's Disputed Ruins*, Chapel Hill, 1937: 76.

61. Lane, *Architecture of the Old South, Georgia*, 68–69, 70 (photographs taken before restoration). I am grateful to Dr. Lewis H. Larson Jr. for showing me Spalding's much-rebuilt house and identifying features surviving from the original building.

62. Holahan cited in Trinkley.

63. Civil War Diary of John Frederic Holahan. (Typed transcript of Ms. on file, Bluffton Historical Preservation Society, Bluffton, S.C.) January 1862.

64. Built of tabby brick and now almost completely robbed the south chimney measured 3'10" × 5'6" in plan. (Brooker, "Tabby Structures on Spring Island," Chicora Research Sereis 20, 1990 page 144, Figure 26 page 145.) Doubtless the north chimney was similar in if not exactly the same size.

65. HABS has confused this White Hall located near the Broad River with another Whitehall on Lady's Island, misidentifying photographs from the site under discussion.

66. Charleston Museum, Charleston, S.C. The same building is illustrated by a small colored sketch dated 1830 in Charles Heyward's *Diary* (Mack and Hoffus, *Landscape of Slavery*, 4–6). Tavis Folk (personal communication, 2011) reported that nothing tangible survives of the Rosehill house above ground.

67. Robinson, *Georgian Farms*, 92.

68. Eugene Genovese, *Roll, Jordan, Roll*, 116–17.

69. Sams, James Julius, "Dathaw," n.d. 5.

70. Nathaniel Kent, *Hints to Gentlemen of Landed Property*, London 1775: 158.

71. Maine Board of Agriculture, *Third Annual Report, 1858,* cited in Hubka, *Big House, Little House,* 194.

Chapter 8. Slave Dwellings, Settlements, and the Quest for Rural Improvement

1. Robinson, *Georgian Model Farms,* 30–31.

2. Kemble, *Journal of a Residence on a Georgian Plantation,* 288.

3. Kent, *Hints to Gentlemen of Landed Property,* 228, cited in Archer, *Literature of British Domestic Architecture,* 461.

4. John Wood, *A Series of Plans, for Cottages,* 3–4, cited in Archer, *Literature of British Domestic Architecture,* 843.

5. Farnsworth, *Island Lives,* 267.

6. Contemporary French promoters of pisé include Georges-Claude Goiffon, whose "L'Art du Maçon Piseur" was published in *Le Jai,* Paris, 1772, and Jean-Baptiste Rondelet, who published volume 1 of his *Traité de l'art de bâtir* in 1812.

7. Holland, "Pisé," London, 1797: 387.

8. Holland. "Pisé," 1797: 393.

9. Holland, "On Cottages." In *Communications*, Vol. 1 Part 2: 97–102.

10. Holland, "On Cottages," 1797: 99.

11. Robinson, *Georgian Model Farms,* 55.

12. Salmon, "Improved Moulds."

13. Archer, *Literature of British Domestic Architecture,* 985, 682n11.

14. Rees's work was "discovered" after World War I by Clough-Williams Ellis, who quoted numerous extracts in his now classic *Cottage Building in Cob, Pisé, Chalk and Clay* (London: Country Life, 1920).

15. Archer, *Literature of British Domestic Architecture,* 421.

16. McCann, *Clay and Cob Buildings,* 20.

17. *American Farmer* 3 (March 30–April 27, 1821).

18. Lane, *Architecture of the Old South, South Carolina,* 153.

19. Spalding, Thomas. "On the mode of Constructing Tabby buildings and the propriety of improving our plantations in a permanent manner." *Southern Agriculturalist* (Charleston, S.C.) 3, no. 12 (December 1830): 617–24.

20. *Farmers Register* 6, no. 6 (September 1, 1838): 822–23.

21. Vlach, *Back of the Big House,* 159, 177. J. H. Cocke, letter to the *American Farmer,* June 4, 1821.

22. *Southern Agriculturalist,* April 1830, 167.

23. R. Crook, "Gullah-Geechee Archaeology."

24. Todd and Hutson, *Prince William's Parish.*

25. Eugene Genovese, *Roll, Jordan, Roll,* 525.

26. Brown, *The English Country Cottage,* 200.

27. Rowland et al., *History of Beaufort County,* 378.

28. College of William and Mary, Field School, 2010. Also: *Nemours Gazette,* Vol. 11 No. 3 2011: 7.

29. Rowland et al., *History of Beaufort County,* 378.

30. Jessica Stevens Loring, *Auldbrass the Plantation Complex by Frank Lloyd Wright: A Documented History of its South Carolina Lands* (Greenville: Southern Historical Press, 1992), 113.

31. Loring, *Auldbrass,* 115–17.

32. Alexander Macomb, "Inquiries," *American Farmer* 8, no. 40 (February 15, 1827): 391.

33. Tabby foundations of a rectangular outbuilding at 1110 Eleventh Street, Port Royal, are the only components of Battery Plantation now known to exist.

34. U.S. National Archives, College Park, Md., U.S. Coastal Survey and Geodesic Survey, "Peninsular of Lands End, St. Helena Island, S.C., from Surveys Executed in 1862 by the Parties under the Direction of C. O. Boutelle," manuscript draft, U.S. National Archives, Records of the Office of Engineers, RG 77 b I 48.

35. Floyd Papers, Georgia Historical Society, Savannah.

36. St. Simons Public Library, *Old Mill Days,* n.p.

37. Trinkley, *Archaeological Excavation of 38BU96,* 21.

38. Standing to a maximum height of 4'3", these tabby bases take the form of a square "U" in plan measuring 5'1" × 2'7" overall.

39. This settlement is attested by two stereoscopic views (Stereoscope Collection no. 22 and no. 26) held by Beaufort County Library, Beaufort, S.C.

40. Anon., *Southern Agriculturalist,* September 1836, 580.

41. Kemble, *Journal of a Residence on a Georgian Plantation,* 257.

42. Du Prey, *John Soane,* 263.

43. Newman and Pevsner, *Buildings of England,* 293.

44. An arc of forty-six timber-framed slave cabins probably dating to the 1820s is recorded from Bulow Sugar Works, Flagler County, Florida (Wayne, *Sweet Cane,* 101).

45. Brooker, "Architecture of the Haig Point Slave Rows."

46. Evarts, "Diary of a Journey to Savannah." Also cited Starr, Rebecca K. "A Place Called Daufuskie, Island Bridge to Georgia 1520–1830." M.A. Thesis, Department of History, University of South Carolina, Columbia, S.C., 1984: 68.

47. See Eugene Genovese, *Roll, Jordan, Roll,* 545.

48. Vlach, *Back of Big House,* 139.

49. This chimney is too far distant from the main house to have belonged to the owner's kitchen, but big enough to serve a largish group of enslaved individuals. Compare with Olmstead, "Cotton Kingdom: A Traveller's Observations on Cotton and Slavery in the American Slave States." Edited by Arthur Schlesinger, New York, Alfred A. Knopf 1953: 81. Cited by Vlach *Back of the Big House,* 144.

50. Individual chimney bases measured approximately 6'7" north/south × 2'6" east/west.

51. Brooker, "Architecture of the Haig Point Slave Rows," 1989: 231; 232.

52. Granger, *Savannah River Plantations,* 16–18.

Chapter 9. Workplaces and Gardens: Processing and Storage Facilities

1. Steam engines at Old Combahee Plantation in Prince William's Parish were "carried off" in September 1865; boilers and a pumps were still there in February 1866, all the rest of the machinery "except the driving wheel" gone. Loring, *Auldbrass,* 115–16.

2. Stephenson, Harvey, and Poplin, *Archaeological Survey of the Burlington Plantation Tract,* 24; Carroll, *Historical Collections,* 1:386.

3. Du Tertre, *Histoire Générale des Antilles Francois,* 2:105.

4. As relocated and rebuilt, two brick indigo vats from Otranto Plantation, Berkeley County, S.C., each measure about fourteen feet square. They are positioned end to end with one vat raised slightly higher than the other.

5. Carroll, *Historical Collections,* 1:389.

6. Rowland et al., *History of Beaufort County,* 166.

7. Rowland et al., *History of Beaufort County,* 219.

8. (Beaufort) *Free South,* June 6, 1863, cited in Rose, *Rehearsal for Reconstruction,* 247. At Hobonny on the Combahee, the threshing mill and engine deemed equal in value ($6,000) to the "family mansion house" were burned in January 1865 along with most other buildings on the estate. Union soldiers made off with fifteen thousand bushels of rice worth $15,000 (*Reports of the Committees of the Senate of the United States, First Session of the Forty-Third Congress, 1873–74*).

9. Ref. Williams Middleton to his sister Eliza M. Fisher, September 25, 1865, quoted in Loring, *Auldbrass,* 115.

10. Doar, *Rice and Rice Planting,* 12, 36.

11. While attending a conference at White Oak Plantation, near Yulee, Florida, in 1988, I saw tabby foundations thought to be of a rice mill built under the direction of either Zephaniah Kingsley after he was banished from South Carolina in 1780 or his eldest son, Zephaniah Kingsley II, who owned White Oak Plantation down until the late 1830s.

12. Fleetwood, *Tidecraft,* 102–3.

13. The supposed cotton storage barn at Chocolate Plantation on Sapelo Island is a freestanding, single-story tabby structure measuring 96' × 110', ventilated by means of vertical slots in its exterior walls.

14. Rose, *Rehearsal for Reconstruction,* 16–17.

15. Walch Papers, Bolton Archive and Local Studies Service, Bolton, Lancs. J. Townsend to Henry Ashworth, January 14, 1864.

16. "Losses Due to the Enemy," South Carolina, Freedmen's Bureau Field Office Records, 1865–1872. NARA microfilm publication M1910. Washington, D.C.: National Archives and Records Administration, n.d., transcripts by F. O. Clark, https://www.sciway3.net /clark/beaufort/Beaufortloss.html.

17. I must thank Karen Smith (S.C. Institute of Archaeology, Columbia) for bringing this photograph to my attention.

18. On January 1, 1863, Lincoln's Emancipation Proclamation was read with great celebration at Old Fort, an event marked in 2016 with designation of Smith's (Old Fort) Plantation as a National Historic Monument. For ceremonies, see Rose, *Rehearsal for Reconstruction,* 196–97.

19. Michael Trinkley, ed., "Archaeological Excavations of 38 BU96 A portion of Cotton Hope Plantation, Hilton Head Island, Beaufort County, South Carolina." Research Series 21 Chocora Foundation Inc. Columbia, S.C., October 1990:132.

20. Elliott, "On the Cultivation and High Prices of Sea-Island Cotton," Southern Agriculturalist I, April , 1828: 154.

21. Painting in a private family collection, Charleston, S.C.

22. *National Freedman* 2, no. 7 (1866): 192.

23. Porcher and Fick, *Story of Sea Island Cotton,* 195.

24. La Rochefoucauld-Liancourt, *Voyages,* 4:127.

25. Turner, *Cotton Planters Manual,* 135.

26. Kemble, *Journal of a Residence on a Georgia Plantation,* 43–44.

27. Elliot, William. "On the Cultivation and High Prices of Sea-Island Cotton." *Southern Agriculturalist,* April, 1828: 152.

28. R.F.W. Alston, "Sea-Coast Crops of the South," *De Bow's Review,* 16 (June 1854): 609–612.

29. James E. Taylor, "History of Sea Island Cotton," *Frank Leslie's Illustrated Newspaper,* April 17, 1863: 77.

30. Wells, *Yearbook 1855 and 1856,* 23.

31. Robert Ralston, "On Cotton Planting," letter to the Agricultural Society of South Carolina, July 30, [1824]. Published in *Original Communications Made to the Agricultural Society of South Carolina; and Extracts from Several Authors on Cotton* (Charleston, S.C.: Archibald E. Miller, published by order of the Society, 1824): 41.

32. Almost one-third of St. Helena Island, comprising eight thousand acres, was bought in 1863 for $7,000 by a partnership of Boston business men led by Edward Philbrick (Rose, *Rehearsal for Reconstruction,* 215).

33. Fleetwood, *Tidecraft,* 110.

34. Kemble, *Journal of a Residence on a Georgian Plantation,* 42–45.

35. Vlach, *Back of the Big House,* 111.

36. Rowland et al., *History of Beaufort County,* 283.

37. "Losses Due to the Enemy," South Carolina, Freedmen's Bureau Field Office Records, 1865–1872. NARA microfilm publication M1910. Washington, D.C.: National Archives and Records Administration, n.d., transcripts by F. O. Clark, https://www.sciway3.net /clark/beaufort/Beaufortloss.html.

38. U.S. Census 1850, Agricultural Schedule, cited in Trinkley, *Second Phase of Archaeological Survey on Spring Island,* 34.

39. Trinkley, *Second Phase of Archaeological Survey on Spring Island,* 37.

40. Beckford, *Descriptive Account of the Island of Jamaica,* 2:28.

41. Spalding's paper is reprinted in Coulter, *Georgia's Disputed Ruins,* 229–52.

42. Superseding boiling trains of the so-called Spanish type, which consisted of a series of metal kettles set above separate furnaces (a system that remained current in Morelos, Mexico, through the nineteenth century), the Jamaica train became popular elsewhere because it minimized fuel consumption; cf. Wayne, *Sweet Cane,* 26.

43. Test excavation (1984) was not sufficiently extensive to confirm the location of this supposed chimney, which was probably robbed once sugar production on the island ceased.

44. Spalding, Thomas. "Observations on the Method of Planting and Cultivating the Sugar Cane in Georgia and South Carolina," (Charleston, S.C.: J. Hoff, 1816); Reprinted in Coulter, ed., *Georgia's Disputed Ruins* (Chapel Hill: University of North Carolina Press, 1937): 243.

45. Coulter, *Georgia's Disputed Ruins,* 109, gives a restoration of Thomas Spalding's sugar house on Sapelo.

46. Spalding, "Observations," in Coulter, *Georgia's Disputed Ruins,* 243.

47. Positive identification of the owner operating the Callawassie sugar mill has eluded investigators for many years. I am grateful to Evan Thompson for finding the reference cited.

48. Elizabeth was the daughter of Daniel Heyward (son of Thomas Heyward Jr.) who died on April 28, 1796, and Ann Sarah Trezevant who had married Daniel Heyward on November 11, 1793. Ann married her second husband, Nicholas Cruger Jr. of St. Croix, on July 1, 1779. *SC Historical and Genealogical Magazine* 3:44–45.

49. Antebellum surveys show that the approach road or track leading to Hamilton's residence

on Tabby Point was lined by rows of slave houses on both sides. These dwellings had disappeared above ground by 1982.

50. Nicholas Cruger Jr. was presumably the son of Nicholas Cruger, whose company Cruger and Beckman of Christiansted, St. Croix, employed the young Alexander Hamilton in its counting house for several years after 1768.

51. Leech, Wood, et al., *Archival Research, Archaeological Survey, and Site Monitoring Back River Chatham County, Georgia and Jasper County, South Carolina,* 166–216.

52. HABS Survey No. VI-2-c: Sugar Factory, Estate Reef Bay, St. John, Virgin Islands.

53. Thomas Spalding, Darien, July 20, 1816, to unnamed correspondent, typed copy of unlocated original, Georgia Historical Society, Savannah.

54. Edward Barnwell, "On the Culture of Sugar," *Southern Agriculturalist,* Vol. 1 No. 11, 1828, 485–89.

55. Transcripts, Hopeton Plantation Books, Floyd Papers, Georgia Historical Society, Savannah.

56. Coulter, *Georgia's Disputed Ruins,* 103.

57. Brooker, "Callawassie Island Sugar Works," 131; Brooker, Maygarden, et al., *Archaeological Data Recovery at Ashland-Belle Helene Plantation,* Vol. 3, Section 6: 1.

58. Edward Barnwell, "On the Culture of Sugar," *The American Farmer,* No. 41, Vol. 10 (Dec. 1828), 323.

59. Wayne, *Sweet Cane,* 43.

60. B. T. Sellers to Williams Middleton, cited in Loring, *Auldbrass,* 114.

61. Harford C. Eve, *Ocean Bluff Journal,* Sunday, December 10, 1911, copy, Historic Beaufort Foundation, Beaufort, S.C.

62. Daise, *Reminiscences of Sea Island Heritage,* [30].

63. Painting in a private collection, Charleston, S.C.

64. St. Simon's Public Library, *Old Mill Days,* n.p.; HABS, *Slave Hospital and Greenhouse, Retreat Plantation, St. Simons G-21.*

65. Reprinted in Hunt and Willis, *Genius of the Place,* 130.

66. Cited in Drayton Hall leaflet, "Self Guided Walks" (n.p., n.d.).

67. Spalding, "On the Mode of Constructing Tabby Buildings."

Chapter 10. Chapels and Cemeteries

1. Linder, *Anglican Churches,* 62.

2. James Gibbes, *Book of Architecture Containing Designs of Buildings and Ornaments* (London, 1725).

3. Rosengarten, *Tombee,* 567–68.

4. *Palmetto Post,* February 1865.

5. Sams, "Dathaw," 11–12.

6. Sams Family Collections, Beaufort, S.C.

7. Sams, "Dathaw," 11–12.

8. Harrison, W.P. [William Pope]. *The Gospel Among the Slaves: A Short Account of Missionary Activities Among the African Slaves of the Southern States.* Nashville, TN. Publishing House of the M.E. Church, 1893: 206.

9. K. Jones, *Port Royal under Six Flags,* 166.

10. Enclosing tabby walls are twelve inches wide and stand about twenty-four inches above present grade.

11. *St. Helena Island Cemetery Survey 1999*, Site U-13-ST H 33.

12. *St. Helena Island Cemetery Survey 1999*, Site U-13-ST H 11. Recent, unverified reports indicate that the enclosure is now completely destroyed.

13. *St. Helena Island Cemetery Survey 1999*, Site U-13-ST H 31.

14. Salley, *Minutes of the Vestry of St. Helena's Parish*, 221, 224. Vestry ordered "a plain Marble Tombstone with suitable inscription." There is no mention of payment for a tomb chest, which might explain its inexpensive construction.

15. According to a sale notice published by the *City Gazette,* Charleston, January 17, 1817, this property adjoined lands of "Mr. Philip Givens, Mr. Calder and Dr. Hamilton." The sale took place at Fairfield, Captain Francis Saltus's plantation in Prince William's Parish. Whether any business or social relationship existed between Saltus and the Kelsalls has not been determined.

16. William B. Kelsall, Will.

17. According to the *Bahamas Gazette,* Nassau, a Mrs. Mackay, "wife of Alexander Mackay formerly of Grenada," died at Exuma in November 1792.

18. Parrish, "Records of Some Southern Loyalists" (Gainesville copy), 6.

Epilogue: A Legacy of the Loyalists?—Tabby in the Bahamas

1. J. Miller and C. Brooker, unpublished interview with Mrs. and Mrs. J. Cooper, Freetown, Grand Bahama, October 2008, transcript, National Museum of the Bahamas.

2. Acting Governor Churchill to Mr. Chamberlain, Govt. House, Nassau, September 18, 1899, *House of Commons Sessional Papers,* vol. 55, received October 4, 1899.

3. *Commissioner's Report for Eleuthera,* 1950, Dept. of Archives, Nassau.

4. Letter from Spalding to N. C. Whiting, New Haven, dated July 30, 1844, printed in Coulter, *Georgia's Disputed Ruins,* 73.

5. Brooker, "Charting the Bahamas, a Preliminary Account of British Naval Surveys of the Bahamas," *Journal Bahamas Historical Society*, Vol. 31, October 2009: 43–56.

6. Keith Bishop (Islands by Design, Nassau), personal communication, March 14, 2012.

7. Bruce, Peter Henry. *Memoirs of Peter Henry Bruce Esq., a Military Officer in the Service of Prussia, Russia and Great Britain.* London, T. Payne and Son, 1782: 396.

8. McKinnen, "General Description of the Bahama Islands," 377.

9. Craton and Saunders, *Islanders in the Stream,* 2:238; *Minutes of Evidence before Select Committee of the House of Lords on the African Slave Trade, 7 May, 1850,* pp. 75–79. Given over to the cultivation of sugar, this last Kelsall Plantation, incorporating some four hundred acres, was located near Candelaria in Holguin Province, Cuba.

10. See J. E. Baxter, J. D. Barton, and S. Frye, "Learning from Landscapes: Understanding Cultural Change and Practice at Polly Hill Plantation," in *Proceedings of the 12th Symposium on Natural History of San Salvador* (San Salvador, Bahamas: Gerace Research Centre), 2009, 12.

11. Governor Rawson to Lord Carnarvon, Government House, Nassau, December 17, 1866, quoted in *London Gazette,* January 29, 1867.

12. *Bahama Herald,* July 31, 1858.

13. William's father, Alexander Wylly, a merchant of Savannah who served as clerk of the crown and common pleas, owned Colerain Plantation, on the Savannah River, which he

abandoned when political events turned ugly for former royal officials at the beginning of the Revolution (Granger, *Savannah River Plantations,* 180).

14. Wylly, *Regulations,* 12; Craton and Saunders, *Islanders in the Stream,* 300.

15. Stucco finishes were color washed with pink dye obtained by steeping logwood (*Haematox-ylum campechianum*) in water. This tree was perhaps introduced to the Bahamas by pirates, who harvested its wood in Capeche (Mexico) and along the coast of Honduras.

16. Without prompting from myself, contractors from New Providence used timber form boards when restoring and stabilizing "tabby" structures at Clifton in 2006, saying this was the traditional island method.

17. The interweaving of relationships is shown by the fact that William Wylly's nephew Alexander William Wylly married Thomas Spalding's daughter Elizabeth.

18. Thomas Spalding, Sapelo Island, April 1828. Letter to the Savannah *Georgian.* Reprinted in Ulrich B. Phillips, *Plantation and Frontier Documents, 1649–1863, Illustrative of Industrial History in the Colonial and Ante-Bellum South*, Vol. 1 (Cleveland: Arthur H. Clark Company, 1909), 269.

19. *Bahama Gazette,* New Providence, August 12–19, 1786; article signed ERRATOR dated New Providence, August 16, 1786.

20. McKinnen, "A General Description of the Bahama Islands," in Edwards, Bryan, *The History Civil and Commercial of the British Colonies in the West Indies in Four Volumes*, Vol. 4, Philadelphia, James Humphreys, 1806: 367.

21. Parrish, "Records of Some Southern Loyalists" (Gainesville copy), 358, citing Chalmers Correspondence, John Carter Brown Library, Providence, R.I.

22. "Petition of William Blacklock on Behalf of John Moultrie and William Moss Asking for Permission to Bring into the State Some Negroes from the Bahamas," S.C. Dept. of Archives and History, S150 150071.

23. Mary Elizabeth Kelsall switched from Anguilla to Bourbon cotton—a short-staple variety that she found immune to predation by the chenille worm.

24. Ministry of Works (Bahamas), Specification Book for 1897–1902, "Specification of Work Required to Be Done in Erecting a Stone Lighthouse, a Kitchen and Tank at South East Point Eleuthera," Department of Archives, Nassau.

25. West Indies Pilot, Vol. 1, Bermuda Islands, Bahama Islands and Greater Antilles. U.S. Government Printing Office, Washington D.C., 1927: 159.

26. "Specification for lighthouse, a kitchen and tank at South East Point," *Eleuthera Ministry of Works Specification Book. 1897–1902,* [1901]: 356–365 (Department of Archives, Nassau, The Bahamas).

27. In 1924 a local benefactor built an ambitious Anglican church at Pompey Bay, Acklins Island, for an exceedingly remote community now dispersed—using rubble bound with lime mortar and cement cast between timber form boards.

Appendix: Notes on Selected Sources

1. Frances Anne Kemble, *Journal of a Residence on a Georgian Plantation in 1838–1839.* (London, 1863): 365.

2. James A. Ford, "Archaeological Report on the Elizafield Ruins." In Coulter, *Georgia's Disputed Ruins*: 193–225.

3. Marmeduke Floyd, in Coulter, *Georgia's Disputed Ruins*: 193–225.

4. "The Alhambra: A Series of Tales and Sketches of the Moors and Spaniards" first published by Carey and Lea, Philadelphia, 1832.

5. James Grey Jackson, *An Account of the Empire of Morocco and the District of Suse* (London: W. Bulmer and Co., 1809).

6. Paul Berthier, *Les anciennes sucreries du Maroc et leurs reseaux hyrauliques. Etude archae-ologique of d'histoire exonomique*, 2 vols. (Rabat: Ministère de L'Education Nationale, 1966).

7. David Henry Slavin, *Colonial Cinema and Imperial France, 1919–1939: White Blindspots, Male Fantasies, Settler Myths* (Baltimore: Johns Hopkins University Press, 2001), 19–20.

8. François Cointeraux, *Ecole d'Architecture rurale*. Paris. Chez l'Autéur, March 1790.

SELECT BIBLIOGRAPHY

Manuscripts and Manuscript Collections

Abbot, Abiel, "Journals. A Yankee Preacher in South Carolina Society," (John Hammond Moore, Editor), South Carolina Historical Magazine, Vole. 68. No. 2, 51–73; No. 4, 232–254.

Bahamas Department of Works. "Specifications of Work to be Done in Building a Residency for the Commissioner at Fresh Creek, Andros Island." Specification Book, 1913, 88–104. Bahamas Department of Archives, Nassau.

Barnwell, Joseph W. "Life and Recollections." Charleston, South Carolina, March 11, 1929. Typescript, Beaufort County Public Library, Special Collections, Beaufort, S.C.

Barnwell, Nathaniel. Will, December 21, 1773. Charleston Wills, vol. 17, pp. 650–54. WPA transcript, S.C. Dept. of Archives and History, Columbia.

Bartram, John. "Diary of a Journey through the Carolinas, Georgia and Florida, July 1, 1765–April 10, 1776." Annotated by Francis Harper. American Philosophical Society, Philadelphis, Pennsylvania, 1942.

Beaufort College. "Journal of the Proceeding of the Trustees of Beaufort College." Beaufort County Library, Special Collections, Beaufort, S.C.

Beaufort Council Minutes. Microfilm, Beaufort County Library, Special Collections.

Bryan, Jonathan. *Journal of a Visit to the Georgia Islands of St. Catherines, Green, Ossabaw, Sapelo, Jekyll and Cumberland, with comments on the Florida Islands of Amelia, Talbot and St. George in 1753.* Virginia Steele Wood and Mary R. Bulard (Editors). Mercer University Press in Association with The Georgia Historical Society, 1996.

Brooker, Colin H. "Architectural and Archaeological Investigation at 802 Bay Street, Beaufort, South Carolina." Brooker Architectural Design Consultants, Beaufort, S.C., 1996. Typescript, Beaufort County Library, Beaufort, S.C.

———. "The Saltus/Habersham House: Architectural and Archaeological Investigation at 802 Bay Street, Beaufort, South Carolina." Brooker Architectural Design Consultants, Beaufort, S.C., 1996. Typescript, Beaufort County Library, Beaufort, S.C.

Busch, James W. "The Beaufort Baptist Church." Undated typescript, Beaufort County Library, Beaufort, S.C.

Carroll, B. R. *Historical Collections of South Carolina.* 2 vols. New York, Harper and Brothers, 1836.

Colonial Office, London. CO 23 *Bahamas Original Correspondence*, 1807–1900. U.K. National Archives, Kew, Surrey, UK.

Colonial Records of South Carolina. *Journal of the South Carolina House of Assembly.* 1730–47. Columbia, S.C., Department of Archives and History.

Columbus, Christopher. *The Journal: Account of the First Voyage and Discovery of the Indies.* Translated by Marc A. Beckworth and Luciano F. Farina. Vol. 1, part 1. Nuova Raccolta Colombiana. Rome, Instituto Poligrafico e Zecca Dello Stato, Libreria Della Stato, 1990.

Combe, Ét, Sauvaget, J. & Wiet, G. *Répertoire Chronologique d'Epigraphie Arabe*, 15. Cairo, Institut Français d'Archéologique Orientale, 1956.

Commissioners of Fortification for the Province of South Carolina. *Journal,* 1755–70. S.C. Department of Archives and History, Columbia.

Drayton, Charles I. "Diary, 1791–1798." Drayton Papers, College of Charleston, Special Collectins, Charleston, S.C.

Evarts, Jeremiah. "Diary, 1820." Georgia Historical Society, Savannah, Georgia, MS240.

Gritzner, Janet B. "Tabby in the Coastal Southeast: The Culture of an American Building Material." Diss., Louisiana State University, Department of Geography and Anthropology, 1978.

Hewitt. "An Historical Account of the Progress of the Colonies of South Carolina and Georgia." In *Historical Collections of South Carolina etc.,* edited by B. R. Carroll, 2 vols. Harper and Brothers, New York, 1836: 10–533.

Holahan, John Frederick. Diary, February 1862. Manuscript, Bluffton Historic Society, Bluffton, S.C.

Ibn Khaldûn, Abd al-Rahman. *The Muqaddimah: An Introduction to History.* Translated by Franz Rosenthal. 2nd ed. 3 vols. Bollingen Series 43. Princeton, NJ: Princeton University Press, 1967.

Johnson, Dr. John Archibald. "Beaufort and the Sea Islands: Their History and Traditions." Typed transcript of newspaper articles, Beaufort County Public Library, Special Collections, Beaufort, S.C.

Kelsall, William. Will, August 22, 1791. Registrar General, Nassau, Bahamas.

Leech, Richard, Jr., Judy L. Wood, et al. *Archival Research, Archaeological Survey, and Site Monitoring Back River Chatham County, Georgia and Jasper County, South Carolina.* U.S. Corps of Engineers, Savannah District. August 1994.

Lepionka, L. "Fort Lyttelton Excavations." Typescript, 1988 copy in author's collection.

Loyalist Transcripts. S.C. Department of Archives and History, Columbia, S.C. Transcript of Papers of the American Loyalist Claim Commission 1780–1835. Formerly Audit Office, Public Record Office, London, now U.K. National Archives, Kew, Surrey.

Parrish, Lydia Austin. "Records of Some Southern Loyalists." Typescript, Georgia Historical Society, Savannah; Department of Archives, Nassau, Bahamas; University of Florida, Gainesville.

Salley, A.S. (Editor). *Minutes of the Vestry of St. Helena's Parish, South Carolina 1726–1812.* Columbia, The State Company, 1919 reprinted by South Carolina Archives Department, 1958.

Sams, James Julius. "Dathaw." Undated typescript. Sams Family Collection, Beaufort, S.C. Copy, Caroliniana Library, University of South Carolina, Columbia.

Starr, Rebecca K., and Charles H. Lowe. "Daufuskie Island, South Carolina, Cultural Resources Survey, 1981." Ms. on file, S.C. Department of Archives and History, Columbia.

Truck Farm Journal of Harford C. Eve. "Ocean Bluff" begun August 18, 1911. Copy, Historic Beaufort Foundation, Beaufort, S.C.

U.S. National Archives, College Park, Md., Cartographic Division, Records of the Office of the Chief of Engineers, *Historical Sketches of Fort Johnston in North Carolina Written April 1819 by Memory Only,* U.S. National Archives, RG 77, Map File DR 62-A.

———. U.S. Coastal Survey and Geodesic Survey. "Peninsular of Lands End, St. Helena Island, S.C., from Surveys Executed in 1862 by the Parties under the Direction of C. O. Boutelle." Manuscript draft, U.S. National Archives, Records of the Office of Engineers, RG 77 b I 48.

————. U.S. Coastal and Geodesic Survey. "Whale Branch Passage between Coosaw and Broad Rivers, SC." Draft and proofs for 1916 and earlier editions.

Walch Papers. Bolton Archive and Local Studies Service, Bolton, Lancashire. J. Townsend to Henry Ashworth, January 14, 1864.

Weston, Plowden Charles Jennet: *Documents Connected with the History of South Carolina.* London, Privately Printed, 1856.

Newspapers

Bahamas Gazette. Department of Archives, Nassau, The Bahamas.

Charleston City Gazette. S.C. Historical Society.

Charleston Courier. Charleston County Library.

Georgia Gazette. Georgia Historical Society.

The New South. Beaufort, S.C. County Library.

Palmetto Post. Beaufort, S.C. County Library.

South Carolina State Gazette. Charleston County Library.

Articles and Reports

Almagro, A. "Analisis arquelogico del pabellon occidental del palacio al-Badi' in Marrakech." *Arquelogia de la Arquitectura*, 11 (2014).

Anon. "Preserving out Past." Nemours Gazette, Vol. 11. Nemours Wildlife Foundation. Yemassee, S.C. Summer 2011: 11.

Ashurst, John. "Mortars for Stone Building." In *Conservation of Building and Decorative Stone*, 2 vols., by John Ashurst and Francis G. Dimes. London: Heinemann, 1990: 77–96.

Baxter, Jane E., and John D. Burton. "Building Meaning into the Landscape: Building Design and Use at Polly Hill Plantation, San Salvador." *Journal of the Bahamas Historical Society* 28 (October 2006): 56–61.

Bazzana, A. "Eléments d'archéologie musulmane dans al-Andalus: Caractères spécifiques de l'architecture militaire arabe de la région valencienne." *Al-Qantara* 1 (1980): 233–52.

Brooker, Colin H. "Architectural Remains at 38 BU806." In *An Archaeological Survey of the Baker Field Expansion Project, Hilton Head Island, Beaufort County, South Carolina*, ed. Michael Trinkley, 39–50. Chicora Research Series 17. Columbia, S.C.: Chicora Foundation, 1989.

————. "Architecture of the Haig Point Slave Rows." In *Haig Point, a 19th Century Plantation, Daufuskie Island, SC*, ed. Michael Trinkley, 208–42. Chicora Research Series 15. Columbia, S.C.: Chicora Foundation, 1989.

————. "Callawassie Island Sugar Works: A Tabby Building Complex." In *Further Investigations of Prehistoric and Historic Lifeways on Callawassie and Spring Islands, Beaufort County, SC.,* ed. Michael Trinkley, 110–54. Chicora Research Series 23. Columbia, S.C.: Chicora Foundation, 1991.

————. "Charting the Bahamas: A Preliminary Account of British Naval Surveys, 1815–40." *Journal of the Bahamas Historical Society* 31 (October 2009): 43–56.

————. "The Conservation and Repair of Tabby Structures in Beaufort County, South Carolina." In *The Conservation and Preservation of Tabby: A Symposium on Historic Building Material in the Coastal Southeast,* ed. Jane Powers Welden, 61–74. Atlanta: Georgia Department of Natural Resources, 1998.

———. "A Forgotten Grandee, Lt. Gov. John Moultrie: His "People" and Bahamian Plantations, 1784–1800." *Journal of the Bahamas Historical Society* 32 (October 2010): 5–17.

———. "Haig Point House—the Architecture." in *Haig Point, a Nineteenth Century Plantation, Daufuskie Island, South Carolina, ed.* Michael Trinkley, 87–117. Chicora Research Series 15. Columbia, S.C.: Chicora Foundation, 1989.

———. "John Wood and the Late Eighteenth Century Development of Clifton Plantation, New Providence, the Bahamas." *Journal of the Bahamas Historical Society* 29 (October 2007): 19–26.

———. "The Major William Horton House, Jekyll Island, Georgia: Preservation, Interpretation and Display." Report presented to Jekyll Island Authority, Friends of Jekyll Island. Report on file, Jekyll Island Museum, Jekyll Island, Ga.

———. "Tabby Structures on Spring Island." In *The Second Phase of Archaeological Survey on Spring Island, Beaufort County SC,* Michael Trinkley. Chicora Research Series 20. Columbia, S.C.: Chicora Foundation, 1980.

Brooker, C., Bruce G. Harvey, David B. Schneider, and Sarah Fick. "Beaufort County Above-Ground Historic Resources Survey of Beaufort County, South Carolina." Brockington Associates Inc.; Brooker Architectural Design Consultants; Historic Beaufort Foundation; Preservation Consultants Inc. 1997. Report on file, S.C. Department of Archives and History, Columbia.

Brooker, C., and Eric Poplin. "The Historical Development of Dataw Island, Architectural and Archaeological Investigations at the Sams Plantation Complex." Brockington and Associates Inc. (Atlanta, Charleston); Brooker Architectural Design (Beaufort, S.C.), 1994. Report on file, S.C. Department of Archives and History, Columbia.

Brooker, C., and James Miller. "Evaluation of Heritage Resources at Old Freetown, Grand Bahama, the Bahamas." Report prepared for National Museum of the Bahamas; Freeport Authority, 2008. Report on file, National Museum of the Bahamas, Nassau.

———. "Survey of Heritage Resources at High Bank and Lantern Head, South Abaco, the Bahamas." Report prepared for Valencia Capital; Antiquities, Monuments and Museums Corp. (AMMC), Nassau, 2008. Report on file, National Museum of the Bahamas, Nassau.

Bush, Olga. "The writing on the wall: reading the decoration of the Alhambra." *Muqarnas,* 26 (2009): 119-147.

Cellauro, Louis, and Gilbert Richaud. "Thomas Jefferson and François Cointeraux, Professor of Rural Architecture in Revolutionary Paris." *Architectural History* 48 (2005): 173–206.

Crook, Morgan R., Jr., and, Patricia D. O'Grady. "Spalding's Sugar Works Site, Sapelo Island, Georgia." *West Georgia College Studies in the Social Sciences, Sapelo Papers: Studies in the History and Prehistory of Sapelo Island Georgia* 19 (June 1980): 9–35.

Crook, Ray. "Gullah-Geechee Archaeology: The Living Space of Enslaved Geechee on Sapelo Island." *African Diaspora Archaeology Network, Newsletter,* March 2008.

Curtis, William J. R. "Type and Variation: Berber Collective Dwellings of the Northern Sahara." *Murqanas* 1 (1983): 181–209.

Dickie, James. "The Hispano- Arab Garden: its philosophy and function." *Bulletin of the School of Oriental and African Studies University of London.* Vol. 31 No. 2 (1968): 237–248.

Elliot, Daniel T. *Archaeological Reconnaissance at the Drudi Tract, Tybee Island, Chatham County, Georgia.* LAMAR Institute Publication Series Report No. 127. Savannah, Ga., 2008.

Elliott, William. "On the Cultivation and High Prices of Sea-Island Cotton." *Southern Agriculturalist,* April 1828, 151–63.

Fischetti, David C. "Tabby: Engineering Characteristic of a Vernacular Construction Material." In *The Conservation and Preservation of Tabby: A Symposium on Historic Building Material in the Coastal Southeast*, ed. Jane Powers Welden, 55–59. Atlanta: Georgia Department of Natural Resources, 1998.

Floyd, Marmeduke. "Certain Tabby Ruins on the Georgia Coast." In *Georgia's Disputed Ruins*, ed. E. Merton Coulter, 4–189. Chapel Hill: University of North Carolina Press, 1937.

Graciani, Amparo. "Notes on 'Tapia' Walls in Seville (Spain) during the 16th Century in the Modern Age." *Second International Congress on Construction History, Proceedings* 2:1375–85. Queen's College, March 29–April 2, 2006.

Graciani García, A., and Tabales Rodriguez. "Typological Observations on Tapia Walls in the Area of Seville." *First International Congress on Construction History, Proceedings* 1:1193–1206. Madrid, January 20–24, 2003.

Harvey, Bruce G., Colin Brooker, David B. Schneider, and Sarah Fick. *Beaufort County Above-Ground Historic Resources Survey of Beaufort County, South Carolina*. Brockington Associates Inc.; Brooker Architectural Design Consultants; Historic Beaufort Foundation; Preservation Consultants Inc., 1998. Report on file, S.C. Department of Archives and History, Columbia.

Holland, Henry. "On Cottages." *Board of Agriculture on Subjects Relative to Husbandry and Internal Improvement of the Country* 1, no. 2 (London, 1797): 97–102.

———. "Pisé, or The Art of Building Strong and Durable Walls, to the Height of Several Stories, with Nothing but Earth, or the Most Common Materials." *Board of Agriculture Communications on Subjects Relative to Husbandry and Internal Improvement of the Country* 1, no. 3–4 (London, 1797): 387–403.

Hubka, Thomas. "Just Folks Designing: Vernacular Designers and the Generation of Form." In *Common Places, Readings in American Vernacular Architecture*, ed. D. Upton and M. Vlach, 426–32. Athens: University of Georgia Press, 1986.

Hughes, David, Simon Swann, and Alan Gardner. "Roman Cement, Part One: Its Origins and Properties." *Journal of Architectural Conservation* 13, no. 1 (March 2007): 21–36.

Lopez, Matilde Casares. "Documentos sobre le Torre de Comares." *Cuadernos de la Alhambra* 9 (1973): 37–52.

Macomb, Alexander. "Observations on the Art of Making the Composition Called Tapia." *American Medical and Philosophical Register or Annals of Medicine* 2 (July 1811): 128–32.

———. "On the art of constructing buildings with *tapia*." American Farmer, Vol. VIII (January 26, 1827): 353–354.

Manuel, Dale A. Forts Hampton, Winyaw, and Marion Second-System Coastal Forts of the Carolinas. *The Coast Defense Journal* (November 2003): 45.

Marder, Walter S. "Tabby Resources in Florida." In *The Conservation and Preservation of Tabby: A Symposium on Historic Building Material in the Coastal Southeast*, ed. Jane Powers Welden, 27–30. Atlanta: Georgia Department of Natural Resources, 1998.

McKinnen, Daniel. "General Description of the Bahama Islands." In *The History Civil and Commercial of the British Colonies in the West Indies*, by Bryan Edwards, vol. 4. Philadelphia: James Humphreys, 1806: 329–403.

Meraz Quintana, Leonardo & Guerrero Baca, Luis. "Calpan (Mexico) historia, urbanismo y tapial." In Félix Jové Sandaval & José Luis Sáiz Guerra (Editors) *Construción con tierra: technología y Arquitectura. Congreso de Arquitectura de tierra en Cuenca de Campos*. University of Valladolid, 2011: 33–46.

Meraz Quintana, Leonardo [and] Guerrero Baca, Luis. "Calpan (Mexico) history urban planning and tapial." In: Construction with Earth. Technology and Architecture Congresses in Cuenca de Campos, 2010. Valladolid, University of Valladolid, 2011, pp. 40–46, direct quote p. 46.

Moussatos, Martha Ann. "The Hants of Retreat Plantation." *Sandlapper Magazine,* October, 1980, 33–37.

O'Grady, Patricia D. "The Occupation of Sapelo Island since 1733." *West Georgia College Studies in the Social Sciences, Sapelo Papers: Studies in the History and Prehistory of Sapelo Island Georgia* 19 (June 1980): 1–8.

Ontiveros, Antonio Lopez. "Evolucion Urbana de Cordoba y de los Pueblos Campiñeses." *Colleccion de Estudios Cordobeses* 11 (Diputacion Provincial de Cordoba, 1981): 210–16.

Palazón, Julio Navarro et al. 'Aqua, aquitectura y poder en una capital del Islam, la finca real del Agdal de Marrakech (ss XII–XX). *Arquelogia de la Arquitectura,* 10 (2013).

Parker, James. Specifications for Cement for Building Purposes. Patent #2120, Granted June 28, 1796. English Patents 1612–1852, British Library, London.

Pavón Maldonado, Basilio. "Arte hispanomusulmán en Ceuta y Tetuán." *Cuadernos de la Alhambra* 6 (1970): 69–108.

———. "Contribución al estudio del arabismo de los castillos de la Penínsular Ibérica (Región levantina)." *Al-Andalus* 43, fasc. 1 (1977): 207–26.

Porcher, Richard D. "Rice Culture in South Carolina: A Brief History, the Role of the Huguenots, and Preservation of its Legacy." *Transactions of the Huguenot Society of South Carolina* 92 (1987): 1–22

Preservation Consultants, Inc. (Sarah Fick, Manager): 'Historical and Architectural Survey, Charleston County, South Carolina.' Charleston, S.C., August 1992.

Quesada-Garcia, S and Garcia-Pulido, L.J. "La torres medievales del Valle de Segura de la sierra o la constución del paisaje. Analisis de la morfologiía y fábricas de las tapias de tierra y cal empleadas de las torres norte y sur de Santa Catalina." In Félix Jové Sandaval & José Luis Sáiz Guerra (Editors) *Constructión con tierra: pasade, presente y futoro. Congresso de Arquitectura de tierra en Cuenca de Campos,* 2012. Valladolid, 2013, 109–122.

Ramagosa, Carol Walter. "Eliza Lucas Pinckney's Family in Antigua, 1668–1747." *South Carolina Historical Magazine* 99, no. 3 (July 1998): 238–58.

Salmon, R. "Improved Moulds and Description of Making Earth Walls." *Philosophical Magazine* 35 (January–June 1810): 265.

Schneider, David B., et al. "St. Helena Island Cemetery Survey (1999)." Survey report on file, Historic Beaufort Foundation, Beaufort, S.C.

Smith, Henry A. M. "Beaufort, the Original Plan and the Earliest Settlers." *South Carolina Historical and Genealogical Magazine* 9 (1908): 142–60.

Smith, G. B. "The Beaufort Volunteer Artillery." Address presented to the 17th Anniversary Meeting of the Hilton Head Historical Society, September 30, 1978. Copy, Historic Beaufort Foundation, Beaufort, S.C.

Spalding, Thomas. "Observations on the Method of Planting and Cultivating the Sugar-Cane in Georgia and South Carolina." Published by order of the Agricultural Society of South Carolina, Charleston, 1816. Rpt. in *Georgia's Disputed Ruins,* ed. E. Merton Coulter, 229–63. Chapel Hill: University of North Carolina Press, 1937.

———. "On the Mode of Constructing Tabby Buildings and the Propriety of Improving Our Plantations in a Permanent Manner." *Southern Agriculturalist,* December 1830, 617–23.

Starr, Rebecca K. "A Place Called Daufuskie, island Bridge to Georgia." M.A. Thesis, Department of History, University of South Carolina, Columia, S.C., 1984.

Stehelin, Col. C. R. E. "Gibraltar: Memorandum Relative to Preparation of, and Mode of Building with Tapia, as Executed under the Immediate Direction and Supervision of Moorish Masons, in Erection of a Traverse of the Grand Battery at Gibraltar Compiled by Mr. Henry C. Jago. Clerk of Works." *Papers on Subjects Connected with the Duties of the Corps of Royal Engineers,* n.s., 10 (ed. W. P. Jackson Woolwich, 1861): 25–27.

Stephenson, Keith, Bruce G. Harvey, and Eric C. Poplin. *Archaeological Survey of the Burlington Plantation Tract, Beaufort County, South Carolina.* Brockington and Associates, Atlanta and Charleston, S.C., 1998. Report on file, S.C. Department of Archives and History, Columbia.

Torres Balbas, Leopoldo. "Gibraltar, Llave y Guarda de Espana." *Obra Dispersal, Al-Andulus, Cronica de la Espana Musulmana* 2 (Instituto de Espana, 1981): 191–201.

Tuten, James H. "Live and die on Hobonny: The Rise, Decline and Legacy of Rice Culture on Hobonny Plantation, 1733-1980." M.A. Thesis, Department of History, Wake Forest University, Winston-Salem, NC, 1992.

UNESCO "Rabat capitale modern et ville historique Un patrimoine partage proposition d'inscription su a liste Mondial soumise por le Royaume du Maroc." June 20, 2011.

Villegas Cerado, Robert. "Análisis structural de Patrimonis Histórico Torre del Homenaje de la Alhambra." Final, Máster de Estructuras, University of Granada, 2011–2012.

Villegas, D., Cámara, M., Compán, V. Análisis structural de la torre de homenaje de la Alhambra de Granada. Informes de la Construcctión, Vol. 66, Extra 1, 2014.

Wade, Arthur P. "The Seacoast Fortifications of the Southern Department, 1708–1815." Compiled for the 16th Annual Military History Conference of the Council of America's Military Past, Charleston, S.C., 1982. Copy of unpaginated typescript on file, U.S. Marine Corps Museum, Paris Island, Beaufort, S.C.

Wells, David A. (Editor) "Cleaning of Sea Island Cotton." *Year Book of Agriculture or the Annual of Agricultural Progress and Discoveries, 1855–1856.* Childs and Peterson, Philadelphia, 1856: 23–24.

Werner, Louis. "Zillij in Fez." *Saudi Aramco World,* Vol. 52, No. 3 (May/June 2001): 18–31.

Williams, Harriet Simons. "Eliza Lucas and Her Family before the Letterbook. *South Carolina Historical Magazine* 99, no. 3 (July 1998): 259–79.

Wilson, Samuel. "Gulf Coast Architecture," Ernest F. Dibble and Earle W. Newton (Editors) *Spain and her Rivals on the Gulf Coast,* 78–126. Pensacola, Florida Department of State, Historic Pensacola Preservation Board, 1971.

Books

Adams, Natalie, and M. Trinkley. *Archaeological Testing of the Stoney/Baynard Plantation, Hilton Head Island, Beaufort County, South Carolina.* Research Series 28. Columbia, S.C.: Chicora Foundation, 1991.

Adderley, Rosanne Marion. *New Negroes from Africa: Slave Trade Abolition and Free African Settlement in the Nineteenth-Century Caribbean.* Bloomington: Indiana University Press, 2006.

Arana, Luis Rafael, and Albert Manucy. *The Building of Castillo de San Marcos.* Eastern National Park and Monument Association, for Castillo de San Marcos National Monument—1977 [St. Augustine, FL].

Archer, John. *The Literature of British Domestic Architecture 1715–1842.* Cambridge, Mass.: MIT Press, 1985.

Barnwell, Stephen B. *The History of an American Family.* Marquette: Privately printed, 1969.

Barrucand, Marianne, and Achim Bednorz. *Moorish Architecture in Andalusia.* Cologne: Taschen, 1992.

Beckford, William. *A Descriptive Account of the Island of Jamaica with Remarks of the Sugar Cane throughout the Different Seasons of the Year, and Chiefly Considered in a Picturesque Point of View.* 2 vols. London: T. and J. Egerton 1790.

Behan, William A. *Exploring the Sullivan Tabby Point Ruin, Callawassie Island, South Carolina.* South Carolina Institute of Archaelogoy and Anthropology, Research Manuscript Series 234. Columbia, S.C. College of Arts and Sciences, University of South Carolina, 2007.

Bell, Malcolm, Jr. *Major Butler's Legacy: Five Generation of a Slaveholding Family.* Athens: University of Georgia Press, 1987.

Berthier, Paul. *Un episode de l'historie de la canne a sucre. Les anciennes sucreries du maroc et leaurs reseaux hydrauliques.* 2 vols, Rabat, Imprimeries Francaises et Marocaines, 1966.

Bishir, Catherine W. *North Carolina Architecture.* Chapel Hill: University of North Carolina Press, 1990.

Braudel, Fernand. *The Mediterranean and the Mediterranean World in the Age of Phillip II.* 2 vols. Translated by Sian Reynolds. New York: Harper and Row, 1972.

Bridges, Anne Baker Leland and Williams, Roy III. *St. James Santee, Plantation Parish. History and Records, 1685-1925.* The Reprint Company, Spartanburg, S.C., 1997.

Brooker, Colin, Maygarden, Yakubik, Franks, Tavaszi, Spencer, et al. *Archaeological Data Recovery at Ashland-Belle Helene Plantation (16AN26), Ascension Parish, Louisiana, Investigations at the Sugar House.* New Orleans: Earth Search, 1994.

Brown, R.J. *The English Country Cottage.* London, Robert Hale, 1979.

Bruce, Peter Henry. *Memoirs of Peter Henry Bruce Esq. . . . Containing an Account of His Travels in Germany, Russia, Tartary, Turkey, the West Indies &c.* London: T. Payne, 1782.

Brunskill, R. W. *Illustrated Handbook of Vernacular Architecture.* London: Faber and Faber, 1971.

Buisseret, David, *Historic Architecture of the Caribbean.* Heinemann: London, 1980.Burn, Billie. *An Island Named Daufuskie.* Spartanburg, S.C.: Reprint Co., 1991.

Burn, Billie. *An Island Named Dafuskie.* The Reprint Company, Spartanburg, S.C., 1991.

Cointeraux, François. *Ecole d'Architecture rurale.* Paris: Chez l'Autéur, 1790; 1791.

[Cole, Cynthia]. *Historic Resources of the Low Country.* Yemassee, S.C.: Low Council of Governments, 1979.

Coulter, E. Merton, ed. *Georgia's Disputed Ruins.* Chapel Hill: University of North Carolina Press, 1937.

Coulter, E. Merton and Saye, Albert B. *A List of the Early Settlers of Georgia.* University of Georgia Press, Athens, Georgia, 1949.

Crane, Verner W. *The Southern Frontier, 1670-1732.* Durham, N.C.: Duke University Press, 1928. Rpt. Ann Arbor: Ann Arbor Books/University of Michigan Press, 1956.

Craton, Michael. *A History of the Bahamas.* London: Collins, 1962.

Craton, Michael, and Gail Saunders. *Islanders in the Stream: A History of the Bahamian People.* 2 vols. Athens: University of Georgia Press, 1992.

Cruickshank, Dan, and Peter Wylde. *London: The Art of Georgian Building.* London: Architectural Press, 1975.

Cummings, Abbott Lowell. *The Framed Houses of Massachusetts Bay, 1625-1725.* Cambridge, Mass.: Belknap Press of Harvard University Press, 1979.

Dabbs, Edith M. *Sea Island Diary: A History of St. Helena Island.* Spartanburg, S.C.: Reprint Co., 1983.

Daise, Ronald. *Reminiscences of Sea Island Heritage.* Orangeburg, S.C.: Sandlapper, 1986.

Deagan, Kathleen, and José María Cruxent. *Archaeology at La Isabela America's First European Town.* New Haven, Conn.: Yale University Press, 2002.

———. *Columbus's Outpost among the Taínos: Spain and America at La Isabela, 1493–1498.* New Haven, Conn.: Yale University Press, 2002.

de Molina, Goncalo Argote. *Nobleza del Andaluzia.* Seville: Fernando Diaz, 1588.

Doar, David. *Rice and Rice Planting in the South Carolina Low County.* Contributions from the Charleston Museum 8. Charleston, S.C., 1936.

Drinkwater. *A History of the Siege of Gibraltar, 1779–1783.* New Edition. London, John Murray, 1905.

Du Prey, Pierre de La Ruffiniere. *John Soane, the Making of an Architect.* Chicago: University of Chicago Press, 1982.

Du Tertre, Jean-Baptiste. *Histoire Générale des Antilles Francois.* Vol. 2. Paris, 1667.

Ebl, Martin Malcom. *Portuguese Tangier (1471–1662) Colonial Urban Fabric as Cross Cultural Skeleton.* Toronto and Peterborough, Baywolf Press, 2013.

Elliott, William. *Carolina Sports by Land and Water.* Burges and James, Charleston, 1846.

Encyclopedia of Islam. New Edition (B. Lewis, Ch. Pellat and J. Sacht, Editors). Leiden, E. J. Brill; London, Luzac and Co. 1965.

Fa, Darren, and Clive Finlayson. *The Fortifications of Gibraltar 1068–1945.* Oxford: Osprey, 2006.

Farnsworth, Paul, ed. *Island Lives: Historical Archaeologies of the Caribbean.* Tuscaloosa: University of Alabama Press, 2001.

Fernandez-Puertas, Antonio. *The Alhambra.* Vol. 1. London: Saqi Books, 1997.

Fischer, David Hackett. *Albion's Seed, Four British Folkways in America.* New York: Oxford University Press, 1989.

Fleetwood, Jr. W.C. *Tidecraft. The Boats of South Carolina, Georgia and Northwestern Florida, 1550-1950.* Tybee Island, Georgia, WBG Marine Press, 1995.

Fowler, Orson S. *A Home for All, or The Gravel Wall and Octagon Mode of Building.* New York, Fowler and Wells, 1853.

Fox-Genovese, Elizabeth. *Within the Plantation Household: Black and White Women of the Old South.* Chapel Hill: University of North Carolina Press, 1988.

Fox-Genovese, Eugene. *Roll, Jordan, Roll: The World the Slaves Made.* New York: Random House, 1988.Gasparini, Graziano. *Las Fortificaciones del Periodo Hispanico en Venezuela.* Caracas: Armitano, 1985.

Garlake, Peter S. *Early Islamic Architecture of the East African Coast.* Memoir No. 1 of the British Institute of History and Archaeology in East Africa. Oxford University Press, Nairobi and London, 1966.

Gasparini, Graziano. *Las Fortificaciones del Periodo Hispanico.* Caracas, Armitano, 1985.

Gilman, E. *The Economical Builder: A Treatise on Tapia and Pisé Walls.* Washington, D.C.: Jacob Gideon, 1839.

Glick, Thomas F. *Islamic and Christian Spain in the Early Middle Ages.* Princeton, N.J.: Princeton University Press, 1979.

Gordon, Elisabeth K. *Florida's Colonial Architectural Heritage.* Gainesville: University Press of Florida, 2002.

Goury, Jules and Jones, Owen. *Plans, Elevations, Sections and Details of the Alhambra*, 2 vols. Owen Jones, London, 1842–45.

Grabar, Oleg. *The Alhambra*. Cambridge, Mass.: Harvard University Press, 1978.

Granger, Mary, ed. *Savannah River Plantations*. Savannah: Georgia Historical Society, 1947.

Gray, Angelica. *Gardens of Marrakesh*. London, Frances Lincoln Limited, 2013.

Gutiérrez, Ramón. *Fortificiones en Ibero América*. Madrid: Fundacion Iberdeola/Ediciones el Viso, 2005.

Gwilt, Joseph. *Encyclopedia of Architecture, Historical, Theoretical and Practical*. Originally published by Longman Green, London, 1867, reprint Bonanza Books, New York, 1982.

Higman, B. W. *Jamaica Surveyed Plantation Maps and Plans of the Eighteenth and Nineteenth Centuries*. Kingston: Institute of Jamaica Publications, 1988.

Hoffman, Paul E. *The Spanish Crown and the Defense of the Caribbean, 1535–1585*. Baton Rouge: Louisiana State University Press, 1980.

Hunt, John Dixon, and Peter Willis. *The Genius of the Place: The English Landscape Garden 1620–1820*. London: Paul Elek, 1975.

Hakewill, James. *A Picturesque Tour of the Island of Jamaica from Drawings Made in the Years 1820 and 1821*. London: Hurts and Robinson, 1825.

Iniguez, Diego Angulo. *Bautista Antonelli las Fortificaciones Americanas del Siglo XVI*. Madrid: Hauser y Menet, 1942.

Ivers, Larry E. *Colonial Forts of South Carolina, 1670–1775*. Tricentennial Booklet 3. Columbia: University of South Carolina Press, 1970.

Jackson, James Grey. *An Account of the Empire of Morocco and the District of Susse . . . to which is added an Accurate and Interesting Acocount of Timbuctoo the Great Emporium of Central Africa*. London, Longman, Hurst, Reeve, and Brown, 1820.

James, Thomas. *The History of the Herculean Straits now called the Straits of Gibraltar including those parts of Spain and Barbary that lie contiguous thereto*. London, Charles Rivington for the Author, 1771.

Johnston, Frances Benjamin. *The Early Architecture of North Carolina*. Chapel Hill: University of North Carolina Press, 1947.

Johnston, Guion Griffis. *Social History of the Sea Islands*. Chapel Hill: University of North Carolina Press, 1930.

Jones, Edgar. *Industrial Architecture in Britain, 1750–1939*. New York: Facts on File, Inc. 1985.

Jones, Katherine M. *Port Royal under Six Flags: The Story of the Sea Islands*. Indianapolis: Bobbs-Merrill, 1960.

Joyner, Charles. *Down by the Riverside: A South Carolina Slave Community*. Urbana: University of Illinois Press, 1984.

Kelso, William M. *Captain Jones's Wormslow*. Athens: University Press, 1979.

Kemble, Frances Anne. *Journal of a Residence on a Georgian Plantation in 1838–39*. 1863. Edited by John A. Scott. Athens: University of Georgia Press, 1984.

Kent, Nathaniel. *Hints to Gentlemen of Landed Property*. London: J. Dodsley, 1775.

Knight, David W. *Understanding Annaberg, A Brief History of Estate Annaburg on St. John, U.S. Virgin Islands*. United States Virgin Islands, Little Northside Press, 2002.

Kubler, George. *Arquitectura Mexicana del siglo XVI*. Mexico City: Fondo Cultura Economica, 1984.

Kubler, George, and Martin Soria. *Art and Architecture of Spain and Portugal and Their American Dominions, 1500–1800*. Pelican History of Art. Harmondsworth: Penguin Books, 1959: 66.

Lane, Mills. *Architecture of the Old South, South Carolina.* Savannah, Ga.: Beehive, 1984.

———. *Architecture of the Old South, North Carolina.* Savannah, Ga.: Beehive, 1985.

———. *Architecture of the Old South, Georgia.* New York: Abbeville, 1990.

———. *General Oglethorpe's Georgia.* Savannah, Ga.: Beehive, 1975.

La Rochefaucault-Liancourt, François Alexandre Frédéric duc de, *Voyage dans les États-Unis d'Amérique, fait en 1795, 1796, 1797.* Paris, Du Poont, Buisson, Charles Pougens, l'an VII [1799 or 1800].

La Rochefaucault-Liancourt, F. and Newman, H. *Travels through the United States of North America, the country of the Iroquois, and Upper Canada, in the years 1796 and 1797; with an authentic account of Lower Canada.* R. Phillips, London: 1799, 2 vols.

Lattimore, Robin Spencer. *Rural Splendor: Plantation Houses of the Carolinas.* Rutherfordton, N.C.: Hilltop, 2010.

Lawrence, John. *The Modern Land Steward, etc.* 2nd ed. London: C. Whitingham, 1806.

Lawson, John. *History of North Carolina.* London, 1714. Edited by Frances Latham Harris. Richmond: Garrett and Massie, 1937.

Lengnick, Lena Wood. *Beaufort Memoirs.* Privately Printed, 1936.

Lewcock, Ronald. *Early Nineteenth Century Architecture in South Africa.* Cape Town: A. A. Balkema, 1963.

Linder, Suzanne Cameron. *Anglican Churches in Colonial South Carolina.* Charleston, S.C. Wyrick and Company, 2000.

———. *Historical Atlas of the Rice Plantations of the ACE River Basin, 1860.* Columbia: S.C. South Carolina Department of Archives and History, 1995.

Linder, Suzanne Cameron, and Marta Leslie Thacker. *Historical Atlas of Rice Plantations of Georgetown County and the Santee River.* Columbia, S.C. South Carolina Department of Archives and History, n.d.

Linley, John. *The Georgia Catalogue: Historic American Building Survey.* Athens: University of Georgia Press, 1982.

———. *Architecture of Middle Georgia: The Oconee Area.* Athens: University of Georgia Press, 1972.

Loring, Jessica Stevens. *Auldbrass, the Plantation Complex Designed by Frank Lloyd Wright.* Greenville, South Carolina: Southern Historical Press, Inc. 1992.

Mack, Angela D., and Stephen G. Hoffius. *Landscape of Slavery: The Plantation in American Art.* Columbia: University of South Carolina Press, 2008.

Maddox, Annette Milliken. *A Lamp unto the Lowcountry: The Baptist Church of Beaufort 1804–2004, Beaufort, South Carolina.* Brentwood, Tenn.: Baptist History and Heritage Society, 2004.

Mandar, C.F. *De l'architecture des Fortresses,* Paris, Chez Magimel, An IX -1801.

Mansbridge, Michael. *John Nash A Complete Catalogue.* London, Phaidon, 1991, 2004.

Manucy, Albert. *Houses of St. Augustine 1565–1821.* 1962. Tallahassee, Fla.: St. Augustine Historical Society/Rose, 1978.

Marçais, William, Marçais, Georges. *Les monuments arabes de Tlemcen.* Paris, Thoran et Fiuls, 1903.

Markman, Sidney David. *Architecture and Urbanization in Colonial Chiapas, Mexico.* Philadelphia: American Philosophical Society, 1984.

———. *Colonial Architecture of Antigua Guatemala.* Philadelphia: American Philosophical Society, 1966.

Mathews, Edward Bennett and Grasty, John Sharshall. *Limestones of Maryland with Special*

Reference to their use in the Manufacture of Lime and Cement. Maryland Geological Survey, Baltimore, 1910.

McCann, John. *Clay and Cob Buildings.* Princess Risborough, Shire Publications, 2004.

McCash, William Barton, and June Hall McCash. *The Jekyll Island Club, Southern Haven for America's Millionaires.* Athens and London: University of Georgia Press, 1989.

McKee, Harley J. *Introduction to Early American Masonry, Stone, Brick, Mortar and Plaster.* National Trust, Columbia University Series on Technology of Early Amreican Building 1 Washington, D.C., National Trust for Historic Preservation, 1973.

McKinnen, Daniel. "General Description of the Bahama Islands" in Edwards, Bryan. *The History Civil and Commerical of the British Colonies in the West Indies.* Vol. 4. Philadelphia, James Humphreys, 1806.

Mercer, Eric. *English Vernacular Houses.* Royal Commission on Historical Monuments England. London: H.M. Stationery Office, 1975.

Miller, James J. *An Environmental History of Northeast Florida.* Gainesville: University Press of Florida, 1998.

Moore, Francis. *A Voyage to Georgia begun in the year 1735.* Reprinted from *Collections of Georgia Historical Society,* Vol. 1, Fort Frederica Association, 1983.

Moultrie, William. *Memoirs of the American Revolution.* Vol. 1. S. Longworth, New York 1802.

Newman, John, and Nikolaus Pevsner. *Buildings of England: Dorset.* Harmondsworth: Penguin, 1972.

Nichols, Frederick Doveton. *The Early Architecture of Georgia.* University of North Carolina Press, Chapel Hill, 1957.

Page, Marian. *Historic Houses Restored and Preserved.* New York: Whitney Library of Design/ Watson-Gumptill, 1976.

Palladio, Andrea. *The Four Books of Architecture.* 1570. Translated by Robert Tavernor and Richard Schofield. Cambridge, Mass.: MIT Press, 1997.

Pepper, Simon and Adams, Nicholas. *Firearms and Fortifications. Military Architecture and Siege Warfare in Sixteenth-Century Sienna.* Chicago and London, Chicago University Press, 1986.

Plaw, John. *Ferme ornée; or Rural Improvement. . . .* London: I. and J. Taylor, 1795.

Porcher, Richard Dwight, and Sarah Fick. *The Story of Sea Island Cotton.* Charleston, S.C.: Wyrick, 2005.

Powell, J. W., ed. *Ninth Annual Report of the United States Geological Survey to the Secretary of the Interior 1887–'88.* Washington, D.C.: Government Printing Office, 1889.

Quinn, Charlotte A. *Mandingo Kingdoms of Senegambia.* Evanston, Ill.: Northwestern University Press, 1972.

Raines, C. W., ed. *Six Decades in Texas or Memoirs of Francis Richard Lubbock.* Austin, Tex., B.C. Jones and Co. 1900.

Robinson, John Martin. *Georgian Model Farms: A Study of Decorative and Model Farm Building in the Age of Improvement, 1700–1846.* Oxford: Clarendon, 1983.

Rose, Willie Lee. *Rehearsal for Reconstruction: The Port Royal Experiment.* London: Oxford University Press, 1967.

Rosengarten, Theodore. *Tombee, Portrait of a Cotton Planter with the Journal of Thomas B. Chaplin (1822–1890).* New York: William Morrow, 1986.

Rosenthal, Earl E. *Palace of Charles V in Granada.* Princeton, N.J.: Princeton University Press, 1985.

Rowland, Lawrence S., Alexander Moore, and George C. Rogers Jr. *The History of Beaufort County, South Carolina.* Vol. 1, *1514–1861.* Columbia: University of South Carolina Press, 1996.

Salley, A. S., ed. *Minutes of the Vestry of St. Helena's Parish, South Carolina 1726–1812.* Columbia: State, 1919; rpt. South Carolina Archives Department, 1958.

Saunders, Gail. *Bahamian Loyalists and Their Slaves.* London, Macmillan, Caribbean, 1983.
———. *Slavery in the Bahamas, 1648–1838.* Nassau: Guardian, 1985.

Schafer, Daniel L. *Zephaniah Kingsley Jr. and the Atlantic World. Slave Trader, Plantation Owner, Emancipator.* Gainesville, Tallahassee, etc. University Press of Florida, 2013.

Shaw, Thomas. *Travels or Observations Relating to Several parts of Barbary and the Levant.* Oxford, at the Theatre, 1738.

Sickels-Taves, Lauren B., and Michael S. Sheehan. *The Lost Art of Tabby Redefined.* Southfield, Mich.: Architectural Conservation, 1999.

Siebert, Wilbur Henry. *Loyalists in East Florida 1774–1785.* 2 vols. Deland: Florida Historical Society, 1929.

Simons, Albert, and Samuel Lapham Jr. *Octagon Library of Early American Architecture.* Vol. 1, edited by Charles Harris Whitaker. New York: Press of American Institute of Architects, 1927.

Smith, Alice R. Huger, and D. E. Huger Smith. *Dwelling Houses of Charleston, South Carolina.* Philadelphia: J. B. Lippincott, 1917.

South, Stanley, Stoner Michael with contributions by William Behan, Colin Brooker, William Sullivan, *The Sullivan Tabby Point Ruin, Calawassie Island, South Carolina.* S.C. Institute of Archaeology and Anthropology, Research Manuscript Series 233. Columbia, S.C. College of Arts and Sciences, University of Sotuh Carolina, 2007.

Stephens, John Lloyd. *Incidents of Travel in Central America, Chiapas and Yucatan.* Two Volumes. New York Harper and Brother, 1841.

Stoney, Samuel Gaillard. *Plantations of the Carolina Low Country.* 1938. Charleston, S.C.: Carolina Art Association, 1964.

Summerson, John. *Georgian London.* New York: Praeger, 1970.

Swanson, Drew A. *Remaking Wormsloe: The Environmental History of a Lowcountry Landscape.* Athens: University of Georgia Press, 2011.

Taylor, C. R. *Biographical Directory of the South Carolina Senate, 1776–1985.* Columbia: University of South Carolina Press, 1986.

Thomas, David Hurst. *St. Catherines An Island in Time.* Atlanta, Georgia Endowment for the Humanities, 1977.

Thompson, Alicia 'Lish' Anderson. *Images of America, Rockville.* Arcadia, Charleston, S.C., etc. 2006.

Todd, John R., and Francis M. Hutson. *Prince William's Parish and Plantations.* Richmond, Va.: Garret and Massie, 1935.

Torres Balbas, Leopoldo. *Ciudades Hispano-Musulmanas.* 2nd ed. Madrid: Instituto Hispano Arabe de Cultura, 1985.

Toussaint, Manuel. *Colonial Art in Mexico,* translated by Elizabeth Wilder Weismann. Austin, University of Texas Press, 1964.

Trinkley, M. *Archaeological Excavation of 38BU96 a Portion of Cotton Hope Plantation, Hilton Head Island, Beaufort County, South Carolina.* Chicora Research Series 21. Columbia, S.C.: Chicora Foundation, 1990.

———. *An Archaeological Survey of Phase I of the Walling Grove Plantation Development, Lady's Island, Beaufort County, South Carolina.* Chicora Research Series 18. Columbia, S.C.: Chicora Foundation, 1989.

———. *Further Investigation of the Stoney/Baynard Main House, Hilton Head Island, Beaufort County, South Carolina.* Chicora Research Series 47. Columbia, S.C.: Chicora Foundation, 1996.

———, ed. *The Second Phase of Archaeological Survey on Spring Island, Beaufort County SC.* Chicora Research Series 20. Columbia, S.C.: Chicora Foundation, 1990.

Trinkley, Michael, Natalie Adams, and Deby Hacker. *The Property Nobody Wanted: Archaeological and Historical Investigations at Fort Johnson, S.C.* Chicora Research Series 48. Columbia, S.C.: Chicora Foundation, 1994.

Trinkley, Michael, and Debbie Hacker. *An Investigation of the St. Queuntens Plantation Main House, Beaufort County, South Carolina.* Chicora Research Series 51. Columbia, S.C.: Chicora Foundation, 1996.

———. *Preliminary Archaeological and Historical Investigations at Old House Plantation, Jasper County, South Carolina.* Chicora Research Series 49. Columbia, S.C.: Chicora Foundation, 1996.

Tucker, Spencer C. *The Jeffersonian Gunboat Navy.* Columbia: University of South Carolina Press, 1993.

Turner, J. A. *The Cotton Planters Manual, Being a Compilation of Facts from the Best Authorities on the Culture of Cotton, Its Natural History, Chemical Analysis, Trade and Consumption and Embracing a History of the Cotton Gin.* New York: C. M. Saxton, 1857.

Vanstory, Burnette. *Georgia's Land of the Golden Isles.* Athens: University of Georgia Press, 1956.

Vlach, John Michael. *Back of the Big House: The Architecture of Plantation Slavery.* Chapel Hill: University of North Carolina Press, 1993.

Wayne, Lucy B. *Sweet Cane: The Architecture of the Sugar Works of East Florida.* Tuscaloosa: University of Alabama Press, 2010.

Welden, Jane Powers, ed. *The Conservation and Preservation of Tabby: A Symposium on Historic Building Material in the Coastal Southeast.* Atlanta: Georgia Department of Natural Resources, 1998.

Westcoast, James L. Jr. and Wolschke-Bulmahn, Joachim, *Mughal Gardens, Sources, Places, Representations and Prospects,* Washington, DC. Dumbarton Oaks Research Library and Collection, 1996.

Westermark, Edward. *Ritual and Belief in Morocco.* London. Macmillan and Co. 1926.

Wharton, Edith. *In Morocco,* New York, Scribners's Sons, 1920.

White, Laura A. *Robert Barnwell Rhett: Father of Secession.* New York, London: the CenturyCo., 1931.

Willen, T. S. *Studies in Elizabethan Foreign Trade.* Manchester: Manchester University Press, 1968.

Wood, Virginia Steele, Bullard Mary R. (Editors) *Journal of a Visit to Georgia.* Mercer University Press, 1996 (no specific publication date given).

Wylly, William. *Regulations for the Governance of Slaves at Clifton and Tusculum in New Providence.* Nassau: Royal Gazette, 1815.

INDEX